Applied Mathematical Sciences | Volume 5

R. von Mises
K. O. Friedrichs

Fluid Dynamics

With 216 Illustrations

Springer-Verlag New York · Heidelberg · Berlin

1971

Richard von Mises†
Kurt O. Friedrichs
Courant Institute of Mathematical Sciences
New York University, New York

Library of Congress Catalog Card Number 73-175242.
Printed in the United States of America.

ISBN 0-387-90028-4 Springer-Verlag New York • Heidelberg • Berlin
ISBN 3-540-90028-4 Springer-Verlag Berlin • Heidelberg • New York

PREFACE

In the summer of 1941 Brown University undertook a Program of Advanced Instruction and Research in Mechanics. This in fact was the precursor to the present day Division of Applied Mathematics. Certainly an outstanding feature of this program must have been the lectures in Fluid Dynamics by Professor Friedrichs and the late Professor von Mises. Their notes were prepared in mimeograph form and given a wide distribution at that time. Since their appearance these lectures have had a strong influence on teaching and research in the subject.

As the reader soon learns the notes have lost none of their vitality over the years. Indeed in certain instances only in the last few years has the field caught up with the ideas developed in the course of these lectures. Many ideas of value are still to be found in these notes and the Editors are most happy to be able to include this volume in the series.

The corrections which have accumulated over the years have been incorporated, and in addition an index has been added. With these exceptions all desire to revise has been resisted. Also in this connection we are very grateful to Dr. T. H. Chong for carefully overseeing the preparation of the present manuscript.

The Editors
August, 1971

TABLE OF CONTENTS

PROBLEMS

INTRODUCTION

Fluid dynamics, including mechanics of liquids and of gases, is a part of physics. We can divide the processes of any physical theory into four steps:

1. Placing observed data into general physical laws, assuming the uniformity of nature.

2. Transferring the physical laws into mathematical form, thus obtaining a logically consistent system of axioms which usually consists of a system of differential equations.

3. Drawing conclusions from these differential equations. This is a purely mathematical step.

4. Check on the results obtained in (3) by experiments.

It is often assumed that the first two steps are completed for all branches of mechanics and that, therefore, mechanics at present offers only mathematical difficulties. This, however, is not so for certain facts in fluid dynamics (and elsewhere).

The first two steps above have been carried out, as far as fluids are concerned, in two forms. First a theory of "perfect fluids" has been developed, and then it was supplemented by a theory of "viscous fluids". However, these theories do not explain all the known facts in fluid dynamics. They cover the explanation of so-called laminar flow. Here the particles travel along smooth curves, for example the slow motion of a fluid through a pipe of small cross-section, or the jet emanating from the opening of a tank. On the other hand, motion in larger channels, like a river, may look smooth as a whole, but each particle by itself has a complicated irregular motion. It fluctuates violently and may exhibit chaotic behavior. Such a motion is called turbulent.

We may say that the first two steps of the physical investigation have been completed for laminar motion; the theory of perfect fluids was given by Euler about 1760, that for viscous fluids by Navier and by Stokes about 1850. In both cases, we have a definite system of partial differential equations. Now the third step consists of the solution of certain boundary value problems of these partial differential equations. This mathematical investigation is still going on. However, in the case of turbulent motion hardly the first step has been completed. Some general rules have been derived, but at present there is no complete and consistent set of axioms (principal equations) which would explain all particular problems of turbulent motion.

In these lectures, we first consider the theory of perfect fluids, as this is the simplest case and the basis for all further developments. A great deal of practical results can be obtained within this theory and its usefulness is broader than usually assumed. It is not correct to say that there is no resistance to a body moving through a fluid when considered as a perfect fluid. Thus, we can derive, in the case of an airwing, a lift force and a drag even if we assume a perfect fluid. Hence the results are better than would be expected and the simpler assumptions make it worthwhile studying.

CHAPTER I

GENERAL THEORY OF PERFECT FLUIDS

1. Equations of Motion.

We consider an element of volume anywhere in a fluid. Let dA be an element of area of the surface of this element of volume. Let dF be the force exerted upon dA by the surrounding particles. It is the essential hypothesis by which the perfect fluid is defined that this force is always acting normal to dA. Then it follows by mathematical reasoning, which we omit here, that the value of dF/dA is the same for all elements of surface passing through a given point. If we consider two elements dA and dA' belonging to the boundary surfaces of two different elements of volume, but both having a point P in common, then the differential quotients dF/dA and dF'/dA' have the same value. This quotient (of dimension FL^{-2}) is called the hydraulic pressure p. We shall have to consider p as a function of x, y, z, and eventually of the time, t, but p will not depend upon the orientation of the surface element dA.

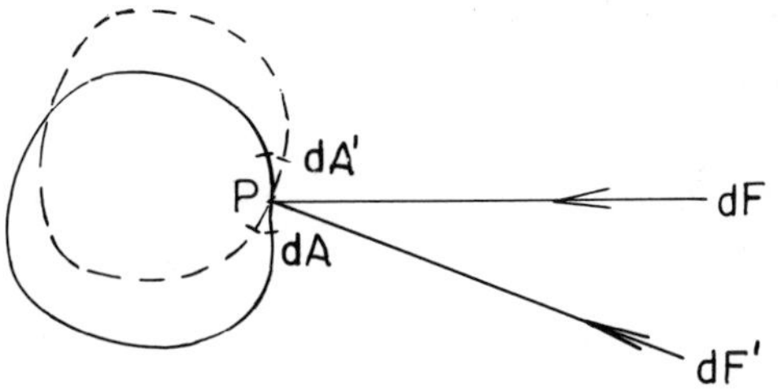

In the case of viscous fluids the force dF is no longer normal to dA. In this case, we admit a tangential component and then the relation between the amount of these normal and tangential components and the orientation of the element dA becomes complicated. However, in the state of equilibrium, (hydrostatics) we have in no case to consider shearing stresses. Hence the distinction between perfect and viscous fluids vanishes here. In problems of equilibrium the hydraulic pressure is a function of position alone.

In order to set up equations of motion for the perfect fluid we have to apply Newton's second law of motion and to combine it with the concept of hydraulic pressure. In the mechanics of small rigid bodies or particles the second law states that at any instant the acceleration of the body is given by

$$\text{acceleration} = \frac{\text{sum of forces}}{\text{mass}} = \frac{\sum_i F_i}{m}$$

where $F_1, F_2, \ldots$ are the external forces acting upon the body and m is a positive constant, the mass of the body.

Such a material body of small dimensions has just one velocity and acceleration. But in a fluid we may have at each point a different velocity and acceleration. Thus the second law needs a certain adaptation to this case. Forces usually acting upon a fluid element are 1) gravity and 2) the force resulting from the pressures which act upon the surface elements of the element of volume under consideration. (For viscous fluids we have also to include shearing forces, but at present we have only to do with the weight force and pressure forces.)

Denote the velocity vector of a point P in the fluid at time t by $\vec{q}$ whose components q_x, q_y, q_z we shall denote by u, v, w respectively. Then at a slightly later time $t + dt$ the point P will be at P' and will have the velocity $\vec{q} + d\vec{q}/dt\, dt$. The derivative $d\vec{q}/dt$ is called "material" derivative as it refers to a determined material point and includes the change in $\vec{q}$ due to change in position as well as to change in time. $d\vec{q}/dt$ is then the acceleration at P at time t.

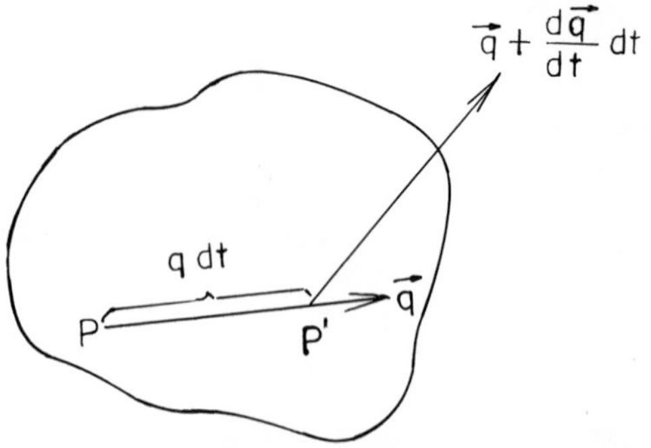

The form of Newton's second law as adapted to the case of fluid mechanics would be

$$\text{acceleration} = \frac{d\vec{q}}{dt} = \lim_{\text{vol.} \to 0} \frac{\text{sum of forces}}{\text{mass}} .$$

The limit is understood to mean that the dimensions of the element of volume, that includes the point P, are more and more reduced to zero. Assuming that the limit of the sum in our expression is the sum of the limits and the limit of the product in the following is the product of the limits, we write

$$\frac{d\vec{q}}{dt} = \lim \frac{\text{weight}}{\text{mass}} + \lim \frac{\text{pressure forces}}{\text{vol.}} \lim \frac{\text{vol.}}{\text{mass}} .$$

Now, for any body, we know that $\frac{\text{weight}}{\text{mass}} = \vec{g}$, the acceleration of gravity, which is a vector directed vertically downwards and of constant amount about 32 ft./sec.^{2}. The limit of the quotient $\frac{\text{mass}}{\text{vol.}}$ is called the <u>density</u> or specific mass of the fluid at the point P and will be denoted by ρ. In general, ρ is not a constant.

In computing $\lim_{\text{vol.} \to 0} \frac{\text{pressure forces}}{\text{vol.}}$ we may assume here that the limit exists, no matter how the small volume about P is reduced to P. (The existence of the limit, however, can be proved mathematically.) We take the small volume about P in the form of a right cylinder of base dA and length dx and let the generators of the cylinder be parallel to the x-axis. Consider the x component of the pressure forces. The pressures on the convex portion of the cylinder are normal to the x-axis and contribute nothing. The pressures on the two bases are shown in the figure, and give the total pressure force

$$p\,dA - \left(p + \frac{\partial p}{\partial x}\,dx\right)dA = -\frac{\partial p}{\partial x}\,dx\,dA .$$

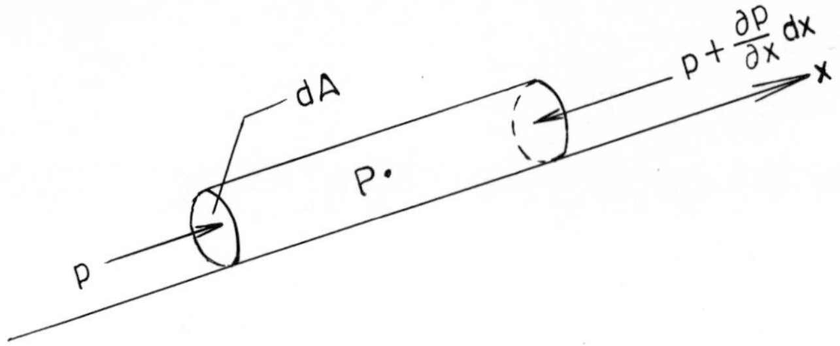

Since the volume of the cylinder is dx dA the limit of the x component of the ratio $\frac{\text{pressure forces}}{\text{vol.}}$ is $-\frac{\partial p}{\partial x}$. Similarly, by taking the generators of the cylinder parallel to the y- and z-axes, the y and z components are found to be $-\partial p/\partial y$ and $-\partial p/\partial z$ respectively. But $\partial p/\partial x$, $\partial p/\partial y$, $\partial p/\partial z$ are the components of a vector called the gradient of p and written as grad p.

Substituting these results back into the equation for $d\vec{q}/dt$ we have Euler's equation of motion

$$\frac{d\vec{q}}{dt} = \vec{g} - \frac{1}{\rho} \text{grad } p. \tag{1}$$

A few words have to be said about ρ and p.

We call incompressible fluid a fluid in which ρ can be considered as constant. In real liquids (water, oil, mercury, etc.) ρ is not exactly constant but almost so.

We call compressible fluid a fluid in which ρ cannot be treated as constant. In all mechanical problems we shall assume that ρ is a function of p. In general we should use the equation of state for the fluid which is an equation of the form $F(p,\rho,T) = 0$, where T is the temperature. (ρ is often replaced by the specific volume* $v = 1/\rho g$). As long as we remain within the domain of mechanics, we may either take T a constant, isothermal case, or else introduce another

*Note that this definition of specific volume differs from the customary definition of volume per unit mass.

relation between p, ρ and T which allows us to eliminate T, e.g., adiabatic case. In either case ρ becomes a function of p alone.

In general, liquids like water, etc. will be considered as incompressible and gases like air, etc. as compressible. But there are exceptions. For instance in the study of the vibrations of a column of water in a long pipe line the effect of change in ρ due to the compressibility is not negligible. On the other hand in the study of airfoils moving through air over a wide range of speeds the air can be assumed to have a constant ρ without any great error.

We are now going to study some transformations of Equation (1). Consider first the vector $\vec{g}$. We can write it as

$$\vec{g} = -\text{grad}\ (gh)$$

whose components are $-g\frac{\partial h}{\partial x}$, $-g\frac{\partial h}{\partial y}$, $-g\frac{\partial h}{\partial z}$ where h is the height of the point P above an arbitrary horizontal plane, measured positive upward. In fact, as shown in the figure we have $\partial h/\partial x = \cos(x,h)$ and therefore $-g\frac{\partial h}{\partial x}$ is the component of the acceleration of gravity along the x-axis If the x- and y-axes are horizontal, $\partial h/\partial x = \partial h/\partial y = 0$ and $\partial h/\partial z = 1$ as expected.

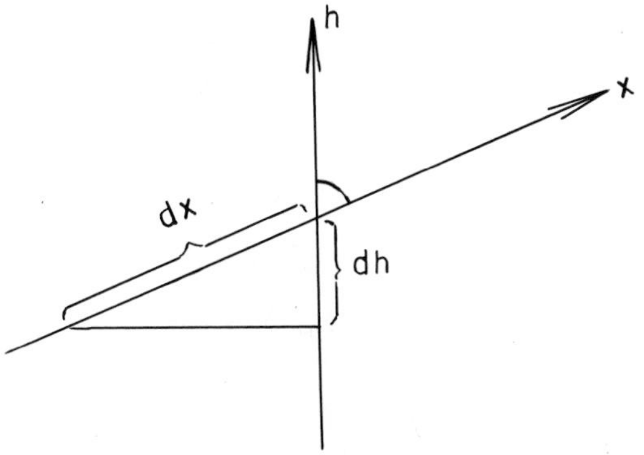

As ρ is assumed to be a function of p, we may consider the integral

$$\int^{p} \frac{dp}{\rho(p)} = \mathcal{P}(p) \tag{2}$$

where $\mathcal{P}(p)$ is defined except for an additive constant. For the imcompressible case, ρ = const. and $\mathcal{P}(p) = p/\rho$ while in the isothermal case for a so-called perfect gas with $pv = RT$ as equation of state, we draw the conclusion that

$$\rho = \frac{1}{gv} = \frac{p}{gRT}$$

where the denominator is a constant. Therefore, $\rho = \frac{\rho_0}{p_0} p$ and

$$\mathcal{P}(p) = \frac{p_0}{\rho_0} \log p.$$

Now it follows from (2) that

$$\text{grad } \mathcal{P}(p) = \frac{1}{\rho} \text{ grad } p. \tag{2a}$$

To show this, we differentiate (2) with respect to x and find

$$\frac{\partial \mathcal{P}}{\partial x} = \frac{\partial}{\partial x} \int^{p} \frac{dp}{\rho(p)} = \frac{d}{dp} \int^{p} \frac{dp}{\rho(p)} \cdot \frac{\partial p}{\partial x} = \frac{1}{\rho} \frac{\partial p}{\partial x} .$$

The same is true for the derivatives with respect to y and z.

Using the above given expression for $\vec{g}$ and the now found expression for $1/\rho$ grad p, we can write Euler's equation (1) in the form

$$\frac{d\vec{q}}{dt} = -\text{grad}(gh) - \text{grad } \mathcal{P} = -\text{grad}(gh+\mathcal{P}) = -g \text{ grad}(h + \frac{\mathcal{P}}{g}). \tag{3}$$

$\frac{\mathcal{P}}{g}$ is called the <u>pressure head</u>. It is equal to $\frac{p}{\rho g}$ for incompressible fluids and equal to a constant times $\log p$ for the isothermal case of a compressible fluid. Equation (3) thus gives the acceleration at P equal to the negative of the acceleration of gravity multiplied by the dimensionless quantity, gradient of height plus pressure head.

According to Equation (3) the vector field of acceleration is the gradient of a potential field or

$$gh + \mathscr{P} = \text{-potential of } \frac{d\vec{q}}{dt} .$$

In the case of equilibrium where $\vec{q} = 0$ or at most $\vec{q}$ = constant, we have $gh + \mathscr{P}$ = constant.

It is to be noticed that Euler's equations are three in number with four dependent variables u, v, w and p (as ρ is a given function of p). Thus we need a fourth equation in order that these four quantities can be determined. We proceed to do this by using the fact that in Newton's second law the mass remains a constant for any definite body. Thus, if we consider a portion of the fluid arbitrarily cut out from the whole, the mass of which is $\int \rho \, dV$, we have to state that this value does not change throughout the motion.

Let dA be an element of the boundary surface of the portion under consideration. Let $\vec{q}$ be the velocity vector at a point of dA and $\vec{n}$ the normal to dA. After time dt this element will be displaced to a new position and the volume swept out by dA is given by $dA \, q \cos(n,q)dt$. Since the mass is given by $\int \rho \, dV$, our statement reads

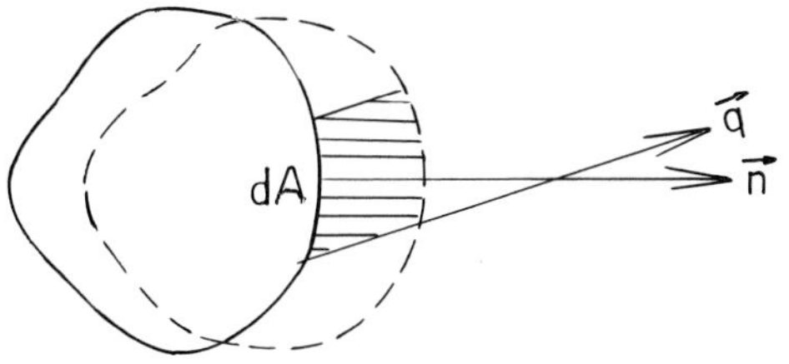

$$\frac{d}{dt} \int \rho \, dV = 0.$$

Carrying out the differentiation we have to consider the change of ρ in time and the above mentioned shifting of the boundary. Thus we get

$$\int \frac{\partial \rho}{\partial t}\, dV + \int \rho q \cos(n,q) dA = 0 \tag{4}$$

where the first integral is a volume integral accounting for the change in ρ at each point, and the second is a surface integral taking into account the change of the boundary.

Now using Gauss' formula relating volume and surface integrals, $\int_{(V)} \frac{\partial f}{\partial x} dV = \int_{(A)} f \cos(n,x)dA$, and considering that $q \cos(n,q) = u \cos(n,x) + v \cos(n,y) + w \cos(n,z)$, we get

$$\int_{(A)} \rho q \cos(n,q)dA = \int_{(V)} \left[\frac{\partial(\rho u)}{\partial x} + \frac{\partial(\rho v)}{\partial y} + \frac{\partial(\rho w)}{\partial z}\right] dV.$$

The expression $\frac{\partial(\rho u)}{\partial x} + \frac{\partial(\rho v)}{\partial y} + \frac{\partial(\rho w)}{\partial z}$ is known as the divergence of $\rho\vec{q}$ and written $\text{div}(\rho\vec{q})$. Thus (4) takes the form $\int [\frac{\partial \rho}{\partial t} + \text{div}\, \rho\vec{q}]dV = 0$, and this must hold for any volume. In particular, for volumes closing down on a point P we see that finally the expression under the sign of integration must vanish. Hence at every point P we have

$$\text{div}(\rho\vec{q}) = -\frac{\partial \rho}{\partial t}. \tag{5}$$

This is usually called the <u>continuity equation</u>. By calculating $\text{div}(\rho\vec{q})$ we get another form of this equation:

$$\rho \,\text{div}\, \vec{q} + u \frac{\partial \rho}{\partial x} + v \frac{\partial \rho}{\partial y} + w \frac{\partial \rho}{\partial z} + \frac{\partial \rho}{\partial t} = 0.$$

The last four terms of this equation form the "material derivative" of ρ. For if f is any function of the particle, then by the so-called <u>Euler rule for differentiation</u> we get the material derivative

$$\frac{df}{dt} = \frac{\partial f}{\partial t} + q\frac{\partial f}{\partial s} = \frac{\partial f}{\partial t} + u\frac{\partial f}{\partial x} + v\frac{\partial f}{\partial y} + w\frac{\partial f}{\partial z} \quad \text{where} \quad q = \frac{ds}{dt}.$$

Applying this to the preceding we get the continuity equation in the form

$$\rho \operatorname{div} \vec{q} + \frac{d\rho}{dt} = 0, \quad \text{or} \quad \operatorname{div} \vec{q} = -\frac{d}{dt} \log \rho. \tag{5a}$$

We note that if the fluid is incompressible then ρ = constant, and the equation becomes $\operatorname{div} \vec{q} = 0$.

2. Bernoulli Equation.

Let P be a point of the fluid and $\vec{q}$ the velocity vector. The direction of this vector is independent of the coordinate system; we will call ds the element of length in this direction. The x-component of Euler's equation is

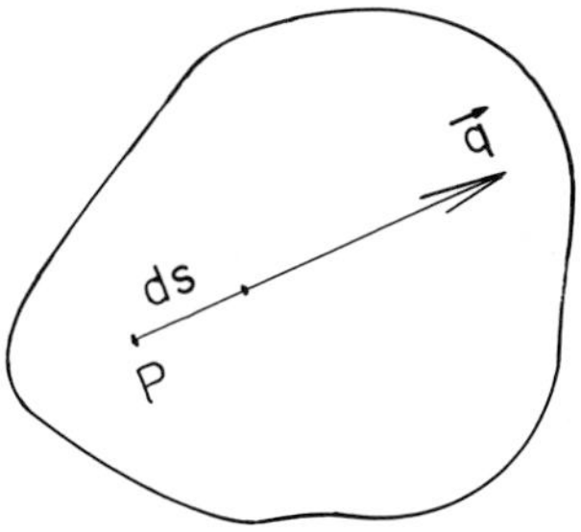

$$\frac{du}{dt} = -g\frac{\partial}{\partial x}\left(h + \frac{\mathscr{P}}{g}\right),$$

or using Euler's rule for differentiation

$$\frac{\partial u}{\partial t} + q\frac{\partial u}{\partial s} = -g\frac{\partial}{\partial x}\left(h + \frac{\mathscr{P}}{g}\right).$$

Now we set up the equation for the s-direction instead of the x-direction. Then we have, since q is the component of $\vec{q}$ in the s-direction

$$\frac{\partial q}{\partial t} + q\frac{\partial q}{\partial s} = -g\frac{\partial}{\partial s}\left(h + \frac{\mathscr{P}}{g}\right),$$

which becomes

$$\frac{\partial q}{\partial t} + g \frac{\partial}{\partial s}\left[\frac{q^2}{2g} + h + \frac{\mathcal{P}}{g}\right] = 0. \tag{6}$$

This is known as the Bernoulli equation. It is used mostly in so-called "steady" or "permanent" motion. We say that a motion is steady (permanent, stationary) if all the partial derivatives with respect to time vanish. Hence for steady motion $\frac{\partial q}{\partial t} = 0$ and (6) becomes

$$\frac{\partial}{\partial s}\left[\frac{q^2}{2g} + h + \frac{\mathcal{P}}{g}\right] = 0. \tag{6a}$$

For steady motion we have definite streamlines, i.e., curves which have two properties: they are the pathways of the particles and they are curves such that the tangent at any point is in the same direction as the velocity vector at that point. In the case of unsteady motion, however, the curves which are the paths of the particles do not coincide with the curves which have at any moment the property that the tangent at any point is in the same direction as the velocity vector at that point.

It follows from (6a) that in steady motion we have

$$\frac{q^2}{2g} + h + \frac{\mathcal{P}}{g} = H = \text{const.} \tag{7}$$

along a streamline. We call $\frac{q^2}{2g}$ the velocity head. Then this last equation states that for steady motion: the sum of the velocity head, the geometrical height, and pressure head is a constant along a streamline. This equation must not be confused with the energy equation. While it is true that $\frac{q^2}{2g}$ is a kinetic energy and h is a measure of potential energy, there exists no third kind of "pressure energy".

3. Circulation.

Let AB be a curve in the fluid, P a point on the curve, $\vec{q}$ the velocity vector at P, and $d\ell$ be an element of arc along the curve. Then we can take the line integral of the velocity vector along AB

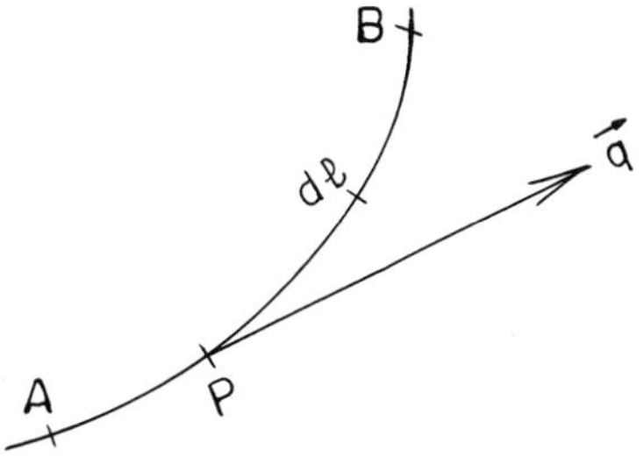

$$\int_{(A)}^{(B)} q \cos(q,\ell)\,d\ell = \int_{(A)}^{(B)} \vec{q} \cdot d\vec{\ell}.$$

In particular, if we have a closed curve or a circuit we may consider the integral from A to A, i.e., over the whole circuit. Then we write it in the form

$$\oint \vec{q} \cdot d\vec{\ell}$$

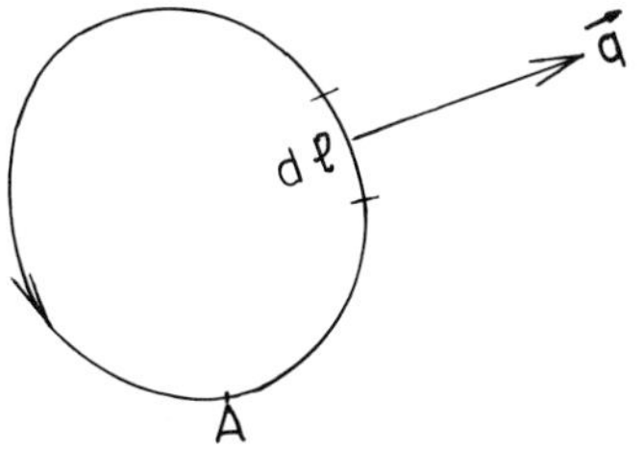

We are going to prove the following:

Theorem. The value of the line integral of q for a closed curve does not change during the motion if the curve is always composed of the same particles:

$$\frac{d}{dt} \oint \vec{q} \cdot d\vec{\ell} = 0 \tag{8}$$

where as before $\frac{d}{dt}$ is the material derivative. This statement is equivalent to the Euler equation.

Proof: The change in $\oint \vec{q} \cdot d\vec{\ell}$ is due first to the change in $\vec{q}$ for each point and then to the change of the curve involved. We have

$$\frac{d}{dt}\oint \vec{q} \cdot d\vec{\ell} =$$

$$\oint \frac{d\vec{q}}{dt} \cdot d\vec{\ell} + \oint \vec{q} \cdot \frac{d\vec{\ell}' - d\vec{\ell}}{dt}$$

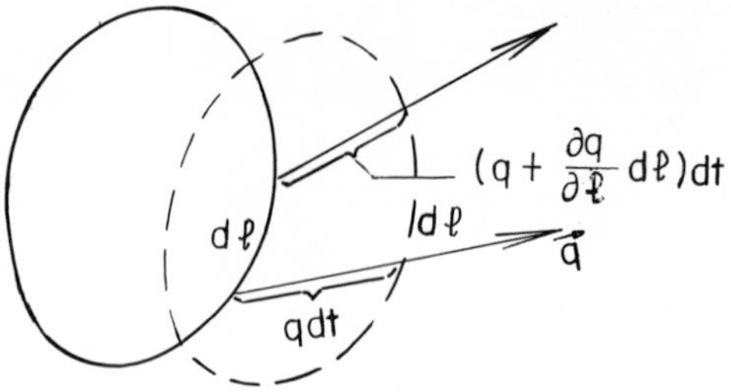

Here $d\vec{\ell}'$ is the new element of arc into which $d\vec{\ell}$ is transformed. From the figure it is apparent that

$$\vec{q}\,dt + d\vec{\ell}' = d\vec{\ell} + (\vec{q} + \frac{\partial \vec{q}}{\partial \ell}\,d\ell)dt.$$

Hence $d\vec{\ell}' - d\vec{\ell} = \frac{\partial \vec{q}}{\partial \ell}\,d\ell\,dt$. Substituting above: $\frac{d}{dt}\oint \vec{q} \cdot d\vec{\ell} =$ $\oint [\frac{d\vec{q}}{dt} \cdot d\vec{\ell} + \vec{q} \cdot \frac{\partial \vec{q}}{\partial \ell}\,d\ell]$, and applying Euler's equation,

$$\frac{d}{dt}\oint \vec{q} \cdot d\vec{\ell} = \oint [-\text{grad}(gh+\mathscr{P}) \cdot d\vec{\ell} + \frac{\partial}{\partial \ell}(\frac{1}{2}q^2)d\ell].$$

The scalar product in the first term on the right hand side above equals the product of $d\ell$ by the ℓ-component of the gradient. Thus we find

$$\frac{d}{dt}\oint \vec{q} \cdot d\vec{\ell} = \oint [-\frac{\partial}{\partial \ell}(gh+\mathscr{P}) + \frac{\partial}{\partial \ell}(\frac{1}{2}q^2)]d\ell = g\oint \frac{\partial}{\partial \ell}(\frac{q^2}{2g} - h - \frac{\mathscr{P}}{g})d\ell.$$

This integral, if it were taken from A to B would be $\left.\frac{q^2}{2g} - h - \frac{\mathscr{P}}{g}\right|_{(A)}^{(B)}$. As A and B coincide in the case of a closed circuit, and since $\frac{q^2}{2g} - h - \frac{\mathscr{P}}{g}$ is clearly single-valued, the integral around a closed curve is zero. Q.E.D.

It is understood that this theorem holds true within a region where $\vec{q}$ is a continuous function of x, y, z. In the case of so-called discontinuous surfaces we have to examine each particular situation.

This theorem of Lord Kelvin has a converse which states that if the integral

$$\Gamma = \oint \vec{q} \cdot d\vec{\ell}$$

which is called the circulation, has a constant value for each curve during the motion, then we have what is essentially Euler's equation. To show this, suppose we have a given vector field $\vec{q}(x,y,z,t)$ and assume that $\frac{d\Gamma}{dt} = 0$ for each closed curve in the field. Now we found that

$$\frac{d\Gamma}{dt} = \oint \left[\frac{d\vec{q}}{dt} \cdot d\vec{\ell} + \vec{q} \cdot \frac{\partial \vec{q}}{\partial \ell}\, d\ell\right]$$

where the second term under the integral sign, $\vec{q} \cdot \frac{\partial \vec{q}}{\partial \ell}\, d\ell = \frac{\partial}{\partial \ell}\left(\frac{1}{2} q^2\right) d\ell$, is a total differential in ℓ. Hence the integral of the second term is zero. For the integral of the first term to vanish also, the component of $\frac{d\vec{q}}{dt}$ along the direction of $d\vec{\ell}$ or $\left(\frac{d\vec{q}}{dt}\right)_\ell$ multiplied by $d\ell$ must be a total differential, since $\frac{d\vec{q}}{dt} \cdot d\vec{\ell} = \left(\frac{d\vec{q}}{dt}\right)_\ell d\ell$. But the direction of $d\vec{\ell}$ is arbitrary, so the component of the acceleration in any direction $\vec{\ell}$ must be a differential quotient with respect to ℓ. This is true only if the acceleration is the gradient of a function V or $\frac{d\vec{q}}{dt} = \operatorname{grad} V$. This is essentially Euler's equation for all we have to do is to identify $-V$ with $gh + \mathcal{P}$.

For an incompressible fluid the continuity equation is simply $\operatorname{div} \vec{q} = 0$. This equation does not depend on p. Thus a velocity field satisfying the two conditions

$$\frac{d\Gamma}{dt} = 0 \quad \text{for each curve,} \quad \operatorname{div} \vec{q} = 0 \tag{9}$$

gives in any case a physically possible perfect fluid motion. The corresponding pressure distribution can be deduced from the Bernoulli equation (6). We can say, that in the case of incompressibility, (9) follows from (1) and (5) by eliminating p.

In the case of a compressible fluid we have the function ρ of p in the continuity equation. The circulation theorem states the existence of a function $-V$, which we have to identify with $gh + \mathscr{P}$. In this way we find the value of p and, therefore, of ρ at any point and we have to check whether the velocity function $\vec{q}$ then satisfies the continuity equation.

4. Helmholtz' Vortex Theory.

Given any vector field $\vec{v}$ we can derive a related vector field by taking as components of the new vector

$$\frac{\partial v_z}{\partial y} - \frac{\partial v_y}{\partial z}, \quad \frac{\partial v_x}{\partial z} - \frac{\partial v_z}{\partial x}, \quad \frac{\partial v_y}{\partial x} - \frac{\partial v_x}{\partial y} .$$

Such a vector is called the curl of $\vec{v}$. We omit the proof that this is, in fact, a vector, i.e., that after a transformation of axes the new components are given in the same form in terms of the new components of $\vec{v}$.

For the vector field $\vec{q}$ we set $\vec{\omega} = \text{curl}\,\vec{q}$ with components

$$\omega_x = (\text{curl}\,\vec{q})_x = \frac{\partial w}{\partial y} - \frac{\partial v}{\partial z}$$
$$\omega_y = (\text{curl}\,\vec{q})_y = \frac{\partial u}{\partial z} - \frac{\partial w}{\partial x}$$
$$\omega_z = (\text{curl}\,\vec{q})_z = \frac{\partial v}{\partial x} - \frac{\partial u}{\partial y} .$$

The vector $\vec{\omega}$ is called the vortex vector. The reason for thus calling $\vec{\omega}$ a vortex vector may be seen by considering the value of ω_z for flow in the xy-plane. The figure shows a small rectangle with one corner at the point A: (x,y), with sides dx and dy. As we move from A in the x direction along AB, $\frac{\partial v}{\partial x}$ gives the rate of increase in the y component, v, of the velocity. As we move from A in the y direction to D, $\frac{\partial u}{\partial y}$ gives the rate of increase in the x component, u, of the velocity. The first increase means a rotation of AB in the positive sense while the second increase means a rotation of AD in the negative sense. Then $\omega_z = \frac{\partial v}{\partial x} - \frac{\partial u}{\partial y}$ is the sum of the two rotations and $\frac{1}{2}\omega_z$ will be a mean rotation or the rotation of the bisector AP of the angle between AB and AD into the bisector AP' of the new angle. Thus $\frac{1}{2}\vec{\omega}$ can be considered as a kind of average rotation of the element of volume.

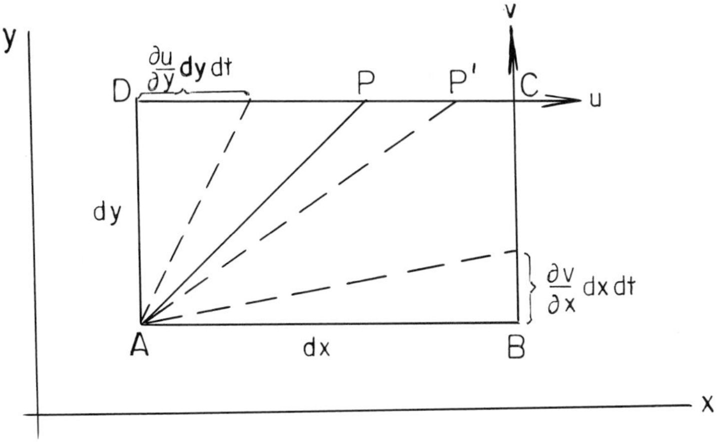

However, a fluid motion with the value of $\vec{\omega}$ being nonzero may look quite different than what we think of ordinarily in speaking of a "vortex". Consider for example the flow in streamlines parallel to the x-axis such that $u \neq 0$, $\frac{\partial u}{\partial y} > 0$ while $v = 0$, $\frac{\partial v}{\partial x} = 0$. Then $\omega_z = -\frac{\partial u}{\partial y} < 0$ gives a mean rotation of the particles in the negative sense as shown by the rotation of the bisector at P_1 into the bisector at P_2. Thus this straight line flow is a vortex motion

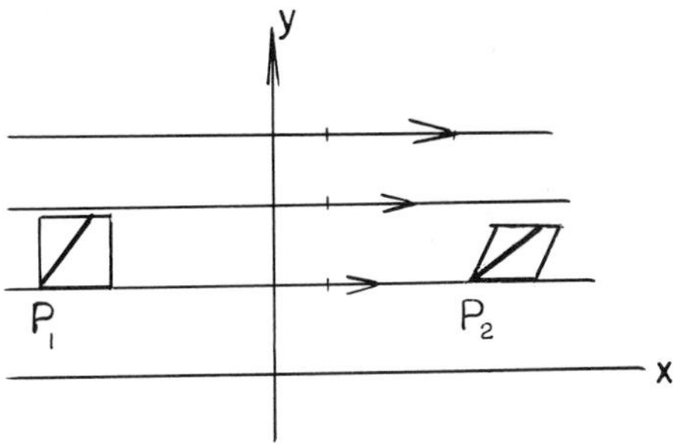

according to our definition.

The circulation Γ for any closed curve C equals the sum of the circulations for any network of meshes into which C is divided up. For, if we draw an arc joining two distinct points B and E of C as shown in the figure, we have Γ_1 as the circulation for the closed curve ABEA and Γ_2 as the circulation for the closed curve DEBD, and we find that $\Gamma = \Gamma_1 + \Gamma_2$ since

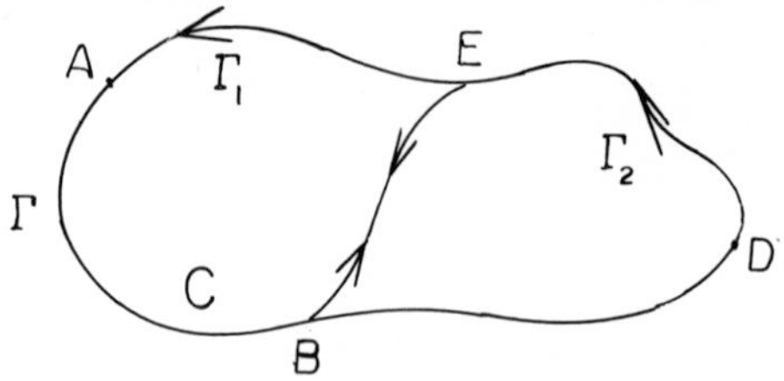

$$\Gamma_1 + \Gamma_2 = \left(\int_{(EAB)} \vec{q}\cdot d\vec{\ell} + \int_{(BE)} \vec{q}\cdot d\vec{\ell} \right) + \left(\int_{(BDE)} \vec{q}\cdot d\vec{\ell} + \int_{(EB)} \vec{q}\cdot d\vec{\ell} \right)$$

$$= \int_{(EAB)} \vec{q}\cdot d\vec{\ell} + \int_{(BDE)} \vec{q}\cdot d\vec{\ell} = \Gamma,$$

while $\int_{(BE)} \vec{q}\cdot d\vec{\ell} = -\int_{(EB)} \vec{q}\cdot d\vec{\ell}.$

In the same way if we break up any surface A of which C is the boundary into a large number of small elements we can write Γ as the sum of the circulations for each element of the surface. Finally, passing to the limit as the number of such elements becomes infinite and the dimensions of each element becomes zero we have

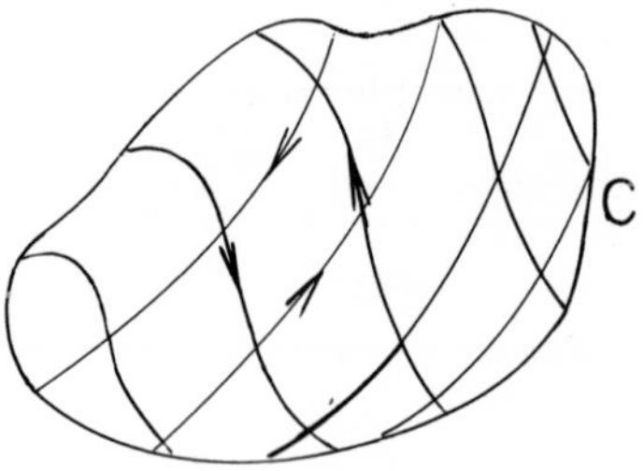

$$\Gamma = \int_{(A)} d\Gamma .$$

To find $d\Gamma$ we consider an element of area in the shape of a rectangle and take its sides parallel to the x and y-axes. The sides then are dx and dy as shown in the figure. For this rectangle

$$d\Gamma = \oint_{(dA)} \vec{q} \cdot d\vec{\ell}$$

$$= u\, dx + (v + \frac{\partial v}{\partial x} dx)dy - (u + \frac{\partial u}{\partial y} dy)dx - v\, dy$$

$$= (\frac{\partial v}{\partial x} - \frac{\partial u}{\partial y})dx\, dy = \omega_z dA.$$

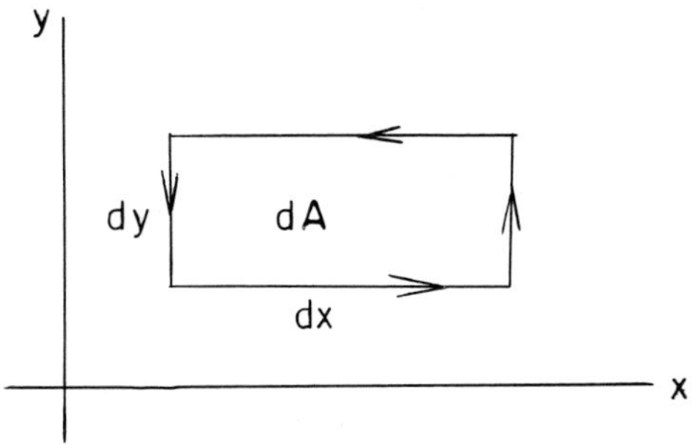

Here ω_z is the component of $\vec{\omega}$ normal to the surface element dA. For an arbitrary surface A stretched across the curve C this expression for $d\Gamma$ will then read $d\Gamma = \omega_n dA$ where ω_n is the component of $\vec{\omega}$ normal to dA. If we let $\vec{n}$ be a unit vector normal to A and pointing in the direction in which a right handed screw would move when turning in the direction in which we take the integral Γ along C, we have $\omega_n = \vec{n} \cdot \omega$, and using $d\vec{A} = \vec{n}dA$ we can write

$$\Gamma = \int_{(A)} d\Gamma = \int_{(A)} \omega_n dA = \int_{(A)} \vec{\omega} \cdot d\vec{A}. \qquad (10)$$

This equation is known as the theorem of Stokes. It gives a formal relation between the values $\vec{q}$ of a vector field and the deduced field of curl $\vec{q} = \vec{\omega}$.

At each point of the fluid we have the vector $\vec{\omega}$ as well as the vector $\vec{q}$. A line such that at each point in the fluid its tangent line coincides with the $\vec{\omega}$ vector at that point is called a vortex line. All vortex lines through the points of a closed curve C_1 give us a vortex tube. The vortex lines form the wall of the tube and, if C_2 is another curve through all points of which the same lines pass, any two surfaces A_1 and A_2 of which C_1 and C_2 respectively are the boundaries may be considered as cross sections of the tube.

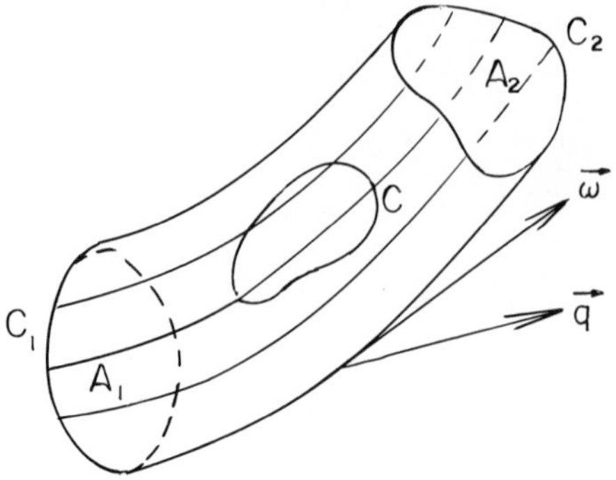

Take a curve C which lies completely on the wall of the vortex tube. For this curve we have $\Gamma = 0$ according to (10) as $\omega_n = 0$ at each point of the wall. At some later instant our vortex tube will be at a new position. The particles which formed the curve C will now form a new curve lying on the new position of the wall. By Kelvin's theorem (8) the circulation has not changed, thus we have again $\Gamma = 0$ for the curve into which C is transformed. Since C was any curve on the wall of the original tube, we have $\Gamma = 0$ for any curve on the new surface into which the wall has changed. This must hold for each infinitely small circuit too, therefore, we have $d\Gamma = 0$ or $\omega_n = 0$ at any point of the new surface. It follows that the new surface is again the wall of a vortex tube. In other words, fluid particles which once form a vortex tube do so permanently. Any vortex line can be considered as the limit of a vortex tube of small cross section. After any interval of time this vortex line will thus still be a vortex line. So we have the first theorem of Helmholtz which states: Vortex lines are material lines; they consist permanently of the same particles. In other words, a vortex tube remains always a vortex tube.

Next, consider the circulation for any cross sectional curve of the tube, such as C_1 or C_2 in the last figure. Since $\omega_n = 0$ over the wall of the tube connecting C_1 and C_2 we have by Stoke's theorem that the circulation Γ_1 for C_1 equals the circulation Γ_2 for C_2. In fact, we can choose for computing Γ_1, according to (10), as our surface A either the cross section A_1 or the wall of the tube together with the cross section A_2. In the second case the integral reduces to the integral over A_2 since $\omega_n = 0$ on the wall. As, therefore, $\int \omega_n dA$ has the same value for each cross section of the tube we may call this value the vorticity of the vortex tube under consideration. By the way, this expresses the fact that $\operatorname{div} \vec{\omega} = \operatorname{div} \operatorname{curl} \vec{q} = 0$ in all cases.

Now we apply again Kelvin's theorem (8) to the Γ value of a cross sectional curve like C_1 or C_2. Each such curve becomes a cross sectional curve of the new vortex tube. The theorem states that the Γ value remains unchanged and thus we have the second theorem of Helmholtz which states: The vorticity of any vortex tube is unchaged during the motion.

Both of the theorems of Helmholtz which have been proven here hold for steady or unsteady motion and for compressible or incompressible fluids. The only condition which was used in the proof as far as compressible fluids are concerned is that ρ and p are connected by a single relation. In certain meteorological problems where this assumption does not hold the Helmholtz theory has to be replaced by a more general one.

Conversely, from the two theorems of Helmholtz we can deduce the theorem of Kelvin which states that the circulation is constant during the motion. Let C be a closed curve. Through each point of C will pass a vortex line. Thus this curve C can be considered in general as a cross sectional curve of a vortex tube. By Helmholtz's first theorem, at a later instant this vortex tube will be transformed into a vortex tube.

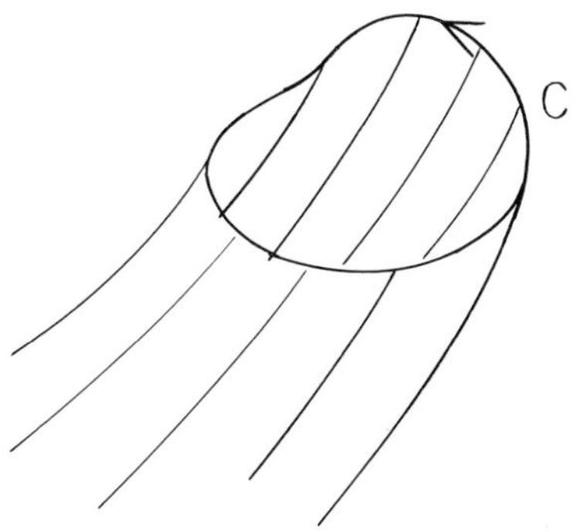

By Helmholtz's second theorem, the vorticity, $\int \omega_n \cdot dA$ (= Γ by Stokes theorem) is the same for the new tube as for the original one. Hence the circulation for C is not changed. The only case we still have to examine is the one in which all vortex lines through one point of C pass through another point of C. In this case the surface bounded by C will be part of a wall of a vortex tube and the circulation here will be zero. The surface will be transformed into a new surface which is, by virtue of the first Helmholtz theorem, part of the wall of a new vortex tube, and, by virtue of the second Helmholtz theorem, the circulation will be zero here again. Thus in this case too Kelvin's theorem follows.

As we showed that Kelvin's theorem with the continuity equation is an equivalent of the Euler theory of perfect fluids, we see that Helmholtz's two theorems together with the continuity equation also form a complete basis for the investigation of perfect fluid dynamics. If, in particular, we have an incompressible fluid, then the continuity equation becomes $\text{div}\, \vec{q} = 0$, and a velocity field satisfying this condition and the two theorems of Helmholtz is physically possible. A particular case is the one where the velocity field is such that $\omega = 0$ for all x, y, z and t. In this case, for any closed curve which can be made to be the boundary of a surface of fluid the circulation is zero. However, if we have a solid cylinder of infinite length immersed in the fluid, (so as to leave a doubly connected space for the fluid), and a closed curve C around this cylinder then the circulation must not vanish for C. If we consider the boundary of a fluid surface to consist of two closed curves C and C' around the cylinder, then it follows from $\omega = 0$ that the circulation around the complete boundary is zero (where the integration is understood to be taken in the sense shown in the figure).

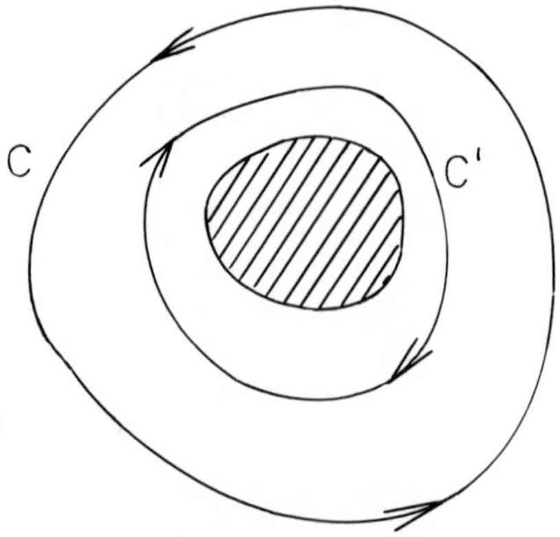

It means that in this case all circuits surrounding the immersed cylinder, and taken in the same sense, have the same circulation value. From the preceeding we draw the conclusion that a velocity distribution with div $\vec{q} = 0$ and circulation values equal for all circuits surrounding the same body is a physically possible motion of an incompressible fluid.

5. Relation Between Vorticity and Bernoulli Function.

Consider the Bernoulli function

$$\frac{q^2}{2g} + h + \frac{\mathscr{P}}{g} = H. \tag{11}$$

We showed that for steady motion $H = \text{const.}$ along a streamline. Now we wish to examine the general behavior of H. For this purpose we take the gradient of H

$$g \text{ grad } H = \text{grad}(\frac{q^2}{2}) + g \text{ grad}(h + \frac{\mathscr{P}}{g}). \tag{12}$$

Applying the equation of motion (1),

$$\begin{aligned} g \text{ grad } H &= \text{grad}(\frac{q^2}{2}) - \frac{d\vec{q}}{dt} \\ &= \text{grad}(\frac{u^2 + v^2 + w^2}{2}) - \frac{\partial \vec{q}}{\partial t} - q \frac{\partial \vec{q}}{\partial s} \\ &= \text{grad}(\frac{u^2 + v^2 + w^2}{2}) - \frac{\partial \vec{q}}{\partial t} - u \frac{\partial \vec{q}}{\partial x} - v \frac{\partial \vec{q}}{\partial y} - w \frac{\partial \vec{q}}{\partial z} . \end{aligned}$$

Now let us compute the x-component of this last expression

$$\begin{aligned} g \frac{\partial H}{\partial x} &= u \frac{\partial u}{\partial x} + v \frac{\partial v}{\partial x} + w \frac{\partial w}{\partial x} - \frac{\partial u}{\partial t} - u \frac{\partial u}{\partial x} - v \frac{\partial u}{\partial y} - w \frac{\partial u}{\partial z} \\ &= - \frac{\partial u}{\partial t} + v(\frac{\partial v}{\partial x} - \frac{\partial u}{\partial y}) + w(\frac{\partial w}{\partial x} - \frac{\partial u}{\partial z}) \\ &= - \frac{\partial u}{\partial t} + v\omega_z - w\omega_y . \end{aligned}$$

The expression $v\,\omega_z - w\,\omega_y$ is well known as the x-component of the vector which is the vector product of $\vec{q}$ and $\vec{\omega}$. For vector (or cross) product we will write $(\vec{q} \times \vec{\omega})$. Now repeating the same process for the y-components and z-components we get

$$g \text{ grad } H = -\frac{\partial \vec{q}}{\partial t} + (\vec{q} \times \vec{\omega}). \tag{13}$$

In the case of steady motion the partial derivatives with respect to time are zero and (13) becomes

$$g \text{ grad } H = \vec{q} \times \vec{\omega}. \tag{14}$$

Now it is well known that the vector obtained as the vector product of two vectors is normal to the plane formed by these two vectors, and hence (14) states that grad H at each point is normal to the plane formed by $\vec{q}$ and $\vec{\omega}$ at that point. Since in general the gradient of a function is directed so as to be normal to the surface where this function is constant, we have that H is constant along the vortex lines as well as along the streamlines. In fact, for steady motion we can find in this way a new proof of Helmholtz's first theorem. If we consider the streamlines through a vortex line, grad H is normal to the surface formed by these streamlines, and H is constant on this surface. Now if we consider the vortex lines through the original streamline we see that they must lie on the same surface. Hence the original vortex line moves along the streamlines on this surface.

In steady motion, if $\vec{\omega} = 0$ everywhere, then by (14), grad H = 0. This means H = const. everywhere throughout the fluid. Thus in steady motion with $\vec{\omega} = 0$, H has the same value along all streamlines. Now conversely, if H has the same value along all streamlines, then we cannot say definitely that $\vec{\omega} = 0$ everywhere. However, we can say that either $\vec{\omega} = 0$ or $\vec{\omega}$ is parallel to $\vec{q}$ at each point of the fluid. This last case certainly cannot happen in two dimensional flow, since here $\vec{\omega}$ is always normal to $\vec{q}$ by definition.

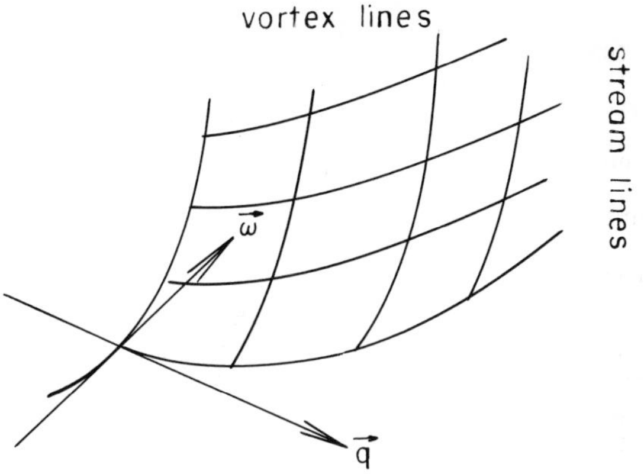

6. <u>Equation for Vortex-Free Motion.</u>

A motion is called <u>vortex-free</u> (irrotational) if $\vec{\omega} = 0$ at all points of the fluid. In this case $\omega_x = \omega_y = \omega_z = 0$ and we have

$$\omega_z = \frac{\partial v}{\partial x} - \frac{\partial u}{\partial y} = 0. \tag{15}$$

Now if there exists a function Φ such that

$$u = \frac{\partial \Phi}{\partial x}, \quad v = \frac{\partial \Phi}{\partial y} \tag{16a}$$

then (15) will automatically be satisfied. Also by considering $\omega_x = \frac{\partial w}{\partial y} - \frac{\partial v}{\partial z} = 0$, we get

$$w = \frac{\partial \Phi}{\partial z} . \tag{16b}$$

Here (16a) and (16b) state that $\vec{q} = \text{grad } \Phi$, which may be changed to the statement $\Phi = \text{pot } \vec{q}$. Thus we see that the statement that a motion is vortex-free is equivalent to each of the following equations

$$\text{curl } \vec{q} = 0; \quad \vec{q} = \text{grad } \Phi; \quad \Phi = \text{pot } \vec{q}. \tag{16}$$

If we assume a function Φ of x, y, z, t, and take $\vec{q} = \operatorname{grad} \Phi$, then we have a vortex-free velocity field. In order to have a vortex-free velocity field satisfying all physical conditions we only have to insure, in addition, that $\vec{q} = \operatorname{grad} \Phi$ fulfills the continuity equation. Here we have to deal separately with compressible and incompressible fluids.

(1) Incompressible Fluid. In this case the continuity equation becomes $\operatorname{div} \vec{q} = 0$. Hence $\frac{\partial u}{\partial x} + \frac{\partial v}{\partial y} + \frac{\partial w}{\partial z} = 0$, and since $\vec{q} = \operatorname{grad} \Phi$

$$\frac{\partial^2 \Phi}{\partial x^2} + \frac{\partial^2 \Phi}{\partial y^2} + \frac{\partial^2 \Phi}{\partial z^2} = 0 \quad \text{or} \quad \Delta\Phi = 0. \tag{17}$$

Equation (17) is known as the Laplace equation. If we take the integral of the Laplace equation which satisfies the particular boundary conditions of a problem, we have the problem solved. Equation (17) may also be written in the form

$$\operatorname{div} \operatorname{grad} \Phi = 0.$$

We note that (17) does not involve the time t; the dependency upon t is only brought in by the boundary conditions.

(2) Compressible Fluid. This case is more complicated and for the present we restrict ourselves to steady motion. Also we will neglect the influence of gravity, i.e., in the Bernoulli equation we will omit the $\operatorname{grad} h$ (take $h = 0$ or constant). Our justification for this is that in aero-dynamical problems the weight of the fluid particles in most cases is of no importance. Now, since we are considering vortex-free motion we have $\vec{q} = \operatorname{grad} \Phi$. Or, what is equivalent $\vec{\omega} = 0$, and this implies $H = \text{const}$. Hence

$$\frac{q^2}{2} + \mathscr{P} = gH = \text{const}. \tag{18}$$

Now, since $\mathcal{P} = \int \frac{dp}{\rho}$, and ρ is a function of p, (18) gives us a relation between p and q. Taking the derivative in (18), we have

$$qdq + d\mathcal{P} = 0, \quad qdq + \frac{dp}{\rho} = 0, \quad \frac{dp}{dq} = -\rho q. \tag{19}$$

Since our function Φ must also satisfy the continuity equation, we consider the latter, which for compressible fluids and steady motion states

$$\operatorname{div}(\rho\vec{q}) = 0.$$

By appealing to the definition of divergence it can easily be shown that this last equation can be transformed to

$$\rho \operatorname{div} \vec{q} + q \frac{\partial \rho}{\partial s} = 0.$$

Now for vortex-free motion $\operatorname{div} \vec{q} = \Delta\Phi$, whence

$$\Delta\Phi = -\frac{q}{\rho}\frac{\partial \rho}{\partial s}. \tag{20}$$

Also since ρ is a function of p, and p a function of q as shown above, we have, by the rule for implicit differentiation

$$\frac{\partial \rho}{\partial s} = \frac{d\rho}{dp} \cdot \frac{dp}{dq} \cdot \frac{\partial q}{\partial s}.$$

If we introduce for brevity

$$\frac{dp}{d\rho} = c^2 \quad (\text{of dimension } \frac{L^2}{T^2}) \tag{21}$$

and substitute this and (19) in the last equation we get

$$\frac{\partial \rho}{\partial s} = \frac{1}{c^2} (-\rho q) \cdot \frac{\partial q}{\partial s} .$$

The expression (20) for $\Delta\Phi$ now becomes

$$\Delta\Phi = \frac{q^2}{c^2} \frac{\partial q}{\partial s} . \tag{22}$$

The quantity c defined by (21) is often called the velocity of sound, since in a tube where $\frac{dp}{d\rho}$ has the constant value c^2 the propagation of sound waves has the velocity c. We remark that for the incompressible case we have $d\rho = 0$ and hence $c = \infty$. Thus equation (22) gives $\Delta\Phi = 0$ which was previously obtained. If we now consider (a) the isothermal case, then $p = \frac{p_0}{\rho_0} \rho$ and $c^2 = \frac{p_0}{\rho_0} = \text{const.}$ In this case (22) becomes

$$\Delta\Phi = \frac{\rho_0}{p_0} q \frac{\partial}{\partial s} (\frac{q^2}{2}). \tag{23}$$

Thus since $\vec{q} = \text{grad } \Phi$ we have here a differential equation of second order and third degree in Φ. In the isothermal case this equation replaces the simple $\Delta\Phi = 0$ which we had for the incompressible case.

(b) The adiabatic case is a particular case of the so-called polytropic case for which

$$p = C\rho^{\kappa} \qquad \kappa > 1.$$

For adiabatic processes, $\kappa = 1.4$. Here

$$c^2 = \frac{dp}{d\rho} = \kappa C \rho^{\kappa - 1}$$

so

$$\mathcal{P} = \int \frac{dp}{\rho} = \kappa C \int \rho^{\kappa-2} d\rho = \frac{\kappa C}{\kappa-1} \rho^{\kappa-1} = \frac{c^2}{\kappa-1} .$$

Substituting this expression for $\mathcal{P}$ into

$$\frac{q^2}{2} + \mathcal{P} = gH = \text{const.}$$

we have

$$\frac{c^2}{\kappa-1} = gH - \frac{q^2}{2} .$$

If we let c_0 be the value of c at a point where the value of q is q_0 this equation becomes

$$\frac{c_0^2}{\kappa-1} = gH - \frac{q_0^2}{2} .$$

Subtracting the last two equations eliminates gH giving

$$c^2 = c_0^2 - \frac{\kappa-1}{2} (q^2 - q_0^2)$$

which substituted back into (22) yields the final form of the differential equation for Φ in the polytropic case:

$$\Delta\Phi = \frac{1}{c_0^2 - \frac{1}{2} (\kappa-1)(q^2 - q_0^2)} q \frac{\partial}{\partial s} (\frac{1}{2} q^2)$$

$$\vec{q} = \text{grad } \Phi. \tag{24}$$

When the right hand side of (24) is small in magnitude we can find approximate solutions. First take the right hand side zero and solve the resulting equation, $\Delta\Phi = 0$. Call the solution, which satisfies the boundary conditions, q_1 and introduce this q_1 into the right hand side of (24). Then we have to solve an equation of the form $\Delta\Phi = f(x,y,z)$ where f is a known function. So we continue step by step. Such a method of solution offers difficulty, however, when q approaches c in magnitude.

7. Momentum Theorems for Perfect Fluid Motion.

We now proceed to derive two integrals of the Euler equation (1). Multiplying Equation (1) by ρ it becomes

$$\rho \frac{d\vec{q}}{dt} = -\rho g \text{ grad } (h + \frac{\mathscr{P}}{g}).$$

Now $\vec{g} = -g$ grad h is a vector pointing vertically downwards. Denoting by γ the specific weight ρg, we may write $\vec{\gamma}$ for $-\rho g$ grad h. Then the above equation can be written

$$\rho \frac{d\vec{q}}{dt} = \vec{\gamma} - \text{grad } p \tag{25}$$

since grad $\mathscr{P} = \frac{1}{\rho}$ grad p as was shown in Section 1.

If we integrate this equation over a volume V of the fluid and write $\vec{W}$ for the vector $\int_{(V)} \vec{\gamma}\, dV$, the weight of V, we have the first Momentum equation

$$\int_{(V)} \rho \frac{d\vec{q}}{dt} dV = \vec{W} - \int_{(V)} \text{grad } p \, dV.$$

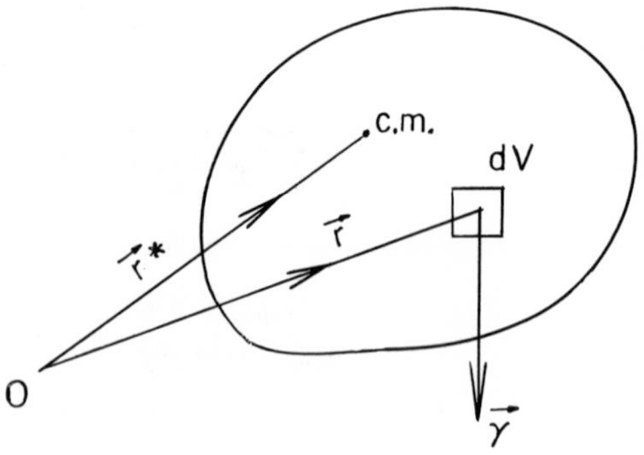

Let $\vec{r}$ be the position vector of the element of volume dV as shown in the figure, and let $\vec{r}*$ be the position vector of the center of mass c.m. of the whole volume V. Take the cross product of $\vec{r}$ with each term of Equation (25) and integrate the result over V getting

$$\int_{(V)} \rho(\vec{r} \times \frac{d\vec{q}}{dt})dV = \vec{r}* \times \vec{W} - \int_{(V)} (\vec{r} \times \text{grad}\ p)dV \tag{27}$$

where $\vec{r}* \times \vec{W} = \int_{(V)} \vec{r} \times \vec{\gamma}\, dV$. This is the second momentum equation or the equation of moment of momentum.

We want to transform the momentum equations (26) and (27) over into equations containing surface integrals. Take (26) first. In this equation

$$\rho \frac{d\vec{q}}{dt} = \rho \frac{\partial \vec{q}}{\partial t} + \rho q \frac{\partial \vec{q}}{\partial s}$$

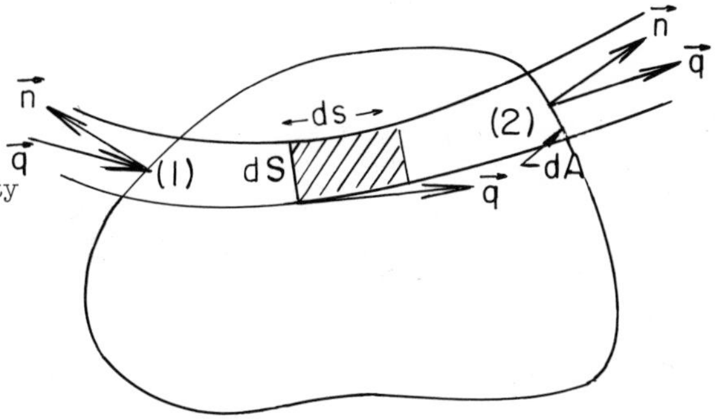

and if $dV = ds\, dS$ as shown in the figure, where ds is the element of length along a velocity line and dS is the area of the cross section of the bundle of such lines shown, we have

$$\int_{(V)} \rho \frac{d\vec{q}}{dt} dV = \int_{(V)} \rho \frac{\partial \vec{q}}{\partial t} dV$$

$$+ \int_{(V)} \rho q \frac{\partial \vec{q}}{\partial s} ds\, dS.$$

Now $\rho q\, dS$ is the mass crossing the face dS of the element of volume in unit time. Since the velocity line is tangent to $\vec{q}$ at each point, no fluid crosses the walls of the filament or bundle. Then by the continuity equation the same amount $\rho q\, dS$ must leave the opposite face in unit time. Hence $\rho q\, dS$ is constant along a filament. At the point where the filament leaves the volume V we see from the figure, if $\vec{n}$ is a unit vector parallel to the outward normal of the element of surface dA, that $\rho q\, dS = \rho q_n dA$. The component of $\vec{q}$ in the direction of $\vec{n}$ is written q_n. The second integral on the right hand side of the foregoing equation can thus be written

$$\int \rho q \, dS \int \frac{\partial \vec{q}}{\partial s} ds = \int \rho q \, dS[\vec{q} \text{ at point } (2) - \vec{q} \text{ at point } (1)].$$

The positive value $\rho q \, dS$, as was shown above, equals the product $\rho q_n dA$ at the point (2). On the other hand, it equals the product $-\rho q_n dA$ at the point (1), since there the component q_n is negative. If we introduce this into the last expression we see that

$$\int_{(V)} \rho q \frac{\partial \vec{q}}{\partial s} ds \, dS = \int_{(A)} \rho \vec{q} \, q_n dA.$$

On the other hand, it follows from the Gauss transformation

$$\int_{(V)} \text{grad } p \, dV = \int_{(A)} p\vec{n} \, dA.$$

Hence (26) becomes

$$\int_{(V)} \rho \frac{\partial \vec{q}}{\partial t} dV + \int_{(A)} \rho \vec{q} \, q_n dA = \vec{W} - \int_{(A)} p\vec{n} \, dA. \tag{28}$$

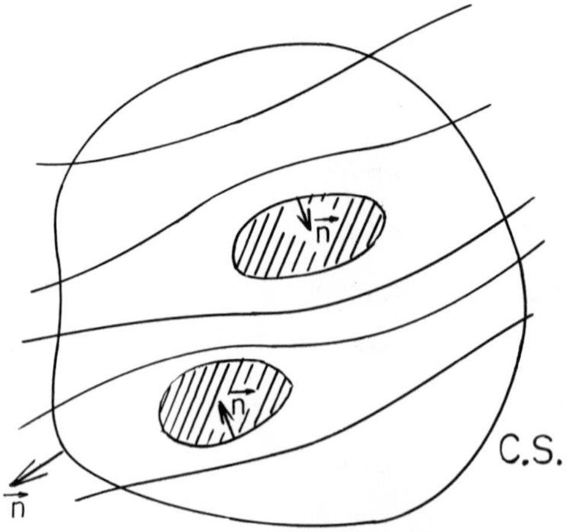

Suppose we have solid bodies immersed in the fluid as shown in the figure. We enclose the bodies inside a control surface which we denote by C.S. Then in equation (28) the surface A consists of the C.S. plus the surfaces of the bodies with the unit normal vectors $\vec{n}$ in the directions shown. Let $\vec{F}$ be force exerted by the fluid on the solid bodies, which is obviously $\int_{(\text{body surf.})} p\vec{n}\, dA$.

Then

$$\int_{(A)} p\vec{n}\, dA = \vec{F} + \int_{C.S.} p\vec{n}\, dA.$$

Also on a solid body $q_n = 0$ so

$$\int_{(A)} \rho\vec{q}\, q_n dA = \int_{C.S.} \rho\vec{q}\, q_n dA.$$

Substituting these two expressions into (28) and solving for $\vec{F}$ we have the first Momentum equation in the form which gives us the force exerted by the fluid upon any solid bodies immersed in it

$$\vec{F} = \vec{W} - \int_{(V)} \rho \frac{\partial\vec{q}}{\partial t}\, dV - \int_{C.S.} \rho\vec{q}\, q_n dA - \int_{C.S.} p\vec{n}\, dA. \tag{29}$$

Next, we consider Equation (27). We have to transform first, the volume integral of $\vec{r} \times \rho q \frac{\partial\vec{q}}{\partial s}$ and second, the volume integral of $\vec{r} \times \text{grad } p$. The z component of $\vec{r} \times \rho q \frac{\partial\vec{q}}{\partial s}$ is $x\rho q \frac{\partial v}{\partial s} - y\rho q \frac{\partial u}{\partial s}$ which equals

$$\rho q[\frac{\partial}{\partial s} (xv-yu) - v \frac{\partial x}{\partial s} + u \frac{\partial y}{\partial s}].$$

But $u = q \cos(x,q)$, $v = q \cos(y,q)$ while

$$\frac{\partial x}{\partial s} = \cos(x,q), \quad \frac{\partial y}{\partial s} = \cos(y,q)$$

so

$$-v \frac{\partial x}{\partial s} + u \frac{\partial y}{\partial s} \equiv 0.$$

Hence

$$x\,\rho q\,\frac{\partial v}{\partial s} - y\,\rho q\,\frac{\partial u}{\partial s} = \rho q\,\frac{\partial}{\partial s}\,(xv-yu)$$

and

$$\vec{r} \times \rho q\,\frac{\partial \vec{q}}{\partial s} = \rho q\,\frac{\partial}{\partial s}\,(\vec{r} \times \vec{q}).$$

Then by the same method used to integrate $\rho q\,\frac{\partial \vec{q}}{\partial s}$

$$\int_{(V)} \rho(\vec{r} \times \frac{d\vec{q}}{dt})dV = \int_{(V)} \rho(\vec{r} \times \frac{\partial \vec{q}}{\partial t})dV + \int_{(A)} \rho(\vec{r} \times \vec{q})q_n dA$$

while again by the Gauss transformation:

$$\int_{(V)} (\vec{r} \times \text{grad } p)dV = \int_{(A)} p(\vec{r} \times \vec{n})dA.$$

Hence (27) becomes

$$\int_{(V)} \rho(\vec{r} \times \frac{\partial \vec{q}}{\partial t})dV + \int_{(A)} \rho(\vec{r} \times \vec{q})q_n dA = \vec{r}* \times \vec{W} - \int_{(A)} p(\vec{r} \times \vec{n})dA.$$

For bodies immersed in the fluid the moment of the forces is

$$\vec{M} = \int_{(\text{body surfs.})} p(\vec{r} \times \vec{n})dA\ .$$

Then in the last equation

$$\int_{(A)} p(\vec{r} \times \vec{n})dA = \vec{M} + \int_{\text{C.S.}} p(\vec{r} \times \vec{n})dA,$$

which gives us the second momentum equation in the final general form

$$\vec{M} = \vec{r}* \times \vec{W} - \int_{(V)} \rho(\vec{r} \times \frac{\partial \vec{q}}{\partial t})dV - \int_{C.S.} \rho(\vec{r} \times \vec{q})q_n dA - \int_{C.S.} p(\vec{r} \times \vec{n})dA. \qquad (30)$$

This equation gives the moment of the forces which the fluid exerts upon the solid bodies inside the controlling surface.

These equations for $\vec{F}$ and $\vec{M}$ will be used primarily for cases of steady motion when $\frac{\partial \vec{q}}{\partial t} = 0$ and under conditions when $\vec{W}$ can be neglected. Then

$$\vec{F} = -\int_{C.S.} (\rho\vec{q}\, q_n + p\vec{n})dA \qquad (31)$$

$$\vec{M} = -\int_{C.S.} \vec{r} \times (\rho\vec{q}\, q_n + p\vec{n})dA.$$

If we specialize further to incompressible potential flow, we have $\rho =$ const. and $\vec{\omega} = 0$. Then it follows that $\mathcal{P} = \frac{p}{\rho}$ and since H in the case of potential flow is a universal constant,

$$\frac{p}{\rho} + \frac{q^2}{2} = \text{const.}$$

or $p = -\frac{\rho q^2}{2} +$ const. We can neglect the const. as it will drop out of the integral of $p\vec{n}$. Hence

$$\vec{F} = -\int_{C.S.} \rho(\vec{q}\, q_n - \frac{q^2}{2}\vec{n})dA.$$

But $q_n = q\cos(n,s)$ and $\vec{q} = q\vec{s}$ where $\vec{s}$ is a unit vector along the velocity direction. So

$$\vec{F} = \frac{\rho}{2}\int_{C.S.} q^2(\vec{n} - 2\vec{s}\cos(n,s))dA$$

and

$$\vec{M} = \frac{\rho}{2} \int_{C.S.} q^2 \vec{r} \times (\vec{n} - 2\vec{s} \cos(n,s)) dA.$$

Let $\vec{n'} = \vec{n} - 2\vec{s} \cos(n,s)$. It is then seen from the figure that $\vec{n'}$ is a unit vector which is the image of the vector $\vec{n}$ reflected in the plane normal to the vector $\vec{s}$. This plane is tangent to the potential surface Φ = const. and passing through our point. Thus we have

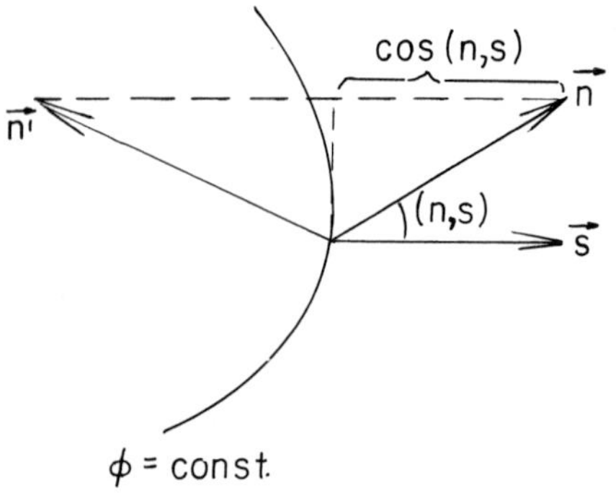

$$\vec{F} = \frac{\rho}{2} \int_{C.S.} q^2 \vec{n'} dA \tag{33}$$

$$\vec{M} = \frac{\rho}{2} \int_{C.S.} q^2 (\vec{r} \times \vec{n'}) dA \tag{34}$$

where $\vec{n'}$ is the unit vector symmetrical to $\vec{n}$ with respect to the potential surface. These two equations give the force and moment exerted by a perfect, incompressible, irrotational fluid upon material bodies inside the surface C.S. when the flow is steady and the effect of gravity can be neglected.

Problem 1. Given a steady motion with the velocity field

$$u = (y-z)(x^2+y^2+z^2-xy-yz-zx+a^2)$$
$$v = (z-x)(x^2+y^2+z^2-xy-yz-zx+a^2)$$
$$w = (x-y)(x^2+y^2+z^2-xy-yz-zx+a^2),$$

find the streamlines, the vortex lines, the surfaces on which H is a constant and the pressure distribution.

Problem 2. If $\vec{\Omega} = \frac{\vec{\omega}}{\rho}$ and $d\sigma$ is the differential element of length in the direction of $\vec{\omega}$, show that $\frac{d\vec{\Omega}}{dt} = \Omega \frac{\partial \vec{q}}{\partial \sigma}$ is equivalent to the two theorems of Helmholtz.

CHAPTER II

MOTION IN TWO DIMENSIONS - AIRWING OF INFINITE SPAN

If all quantities are independent of z, and $w = 0$ so that the motion is parallel to the xy-plane, we say that it is a plane motion or two dimensional. In such a case we may envisage a volume of fluid of unit thickness moving parallel to the xy-plane.

1. Steady Motion of an Incompressible Fluid.

If the motion is steady, all partial derivatives with respect to t will be zero and, if the fluid is incompressible, we will have $\rho = \text{const.}$ We make these two assumptions in this section.

The two theorems of Helmholtz and the equation of continuity are, as shown in Chapter I, equivalent to Euler's hydro-mechanical set up as far as the velocity distribution is concerned. The continuity equation $\operatorname{div} \vec{q} = 0$ becomes

$$\frac{\partial u}{\partial x} + \frac{\partial v}{\partial y} = 0.$$

This equation is identically satisfied if we assume a function ψ such that

$$u = \frac{\partial \psi}{\partial y} \qquad v = -\frac{\partial \psi}{\partial x}. \tag{1}$$

The function $\psi(x,y)$ is called the stream function since $\psi(x,y) = \text{const.}$ gives us a streamline for the motion. This follows from Equation (1) as we conclude from there that the directional derivative of ψ in any direction gives the component of the velocity in a direction making an angle of -90° with the direction of the derivative. Since $\vec{q}$ is tangent to a streamline, it has no component normal to the streamline; hence $\frac{\partial \psi}{\partial s} = 0$ where s is measured along the streamline, which means that $\psi = \text{const.}$ along each streamline. The value of ψ will vary from streamline to streamline. If we define as dn the element in a direction normal to ds and such that the angle from ds to dn is $+90^{\circ}$, we have

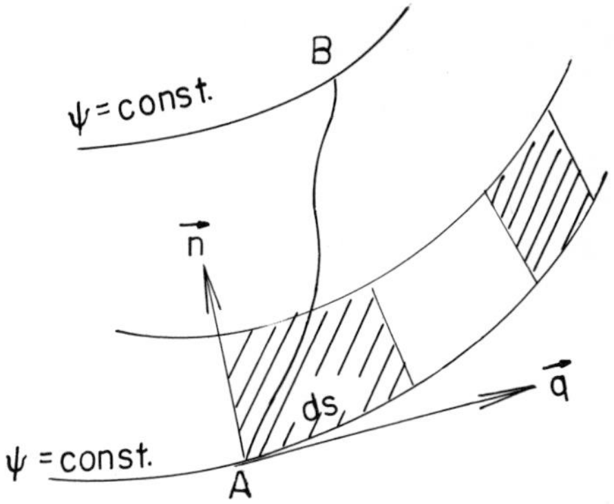

$$\frac{\partial\psi}{\partial n} = q, \quad d\psi = q\, dn.$$

Thus, the increment $d\psi$ in ψ from one streamline to the next equals the amount of fluid passing between these two streamlines per unit time. (dn is the cross section of the filament of width dn and of unit thickness.) It follows that the amount of fluid crossing any cruve AB, shown in the figure, in unit time is $\psi_B - \psi_A$, which is the value of ψ at B minus the value of ψ at A.

In order to apply the theorems of Helmholtz we need to bring in the vortex vector $\vec{\omega}$. Since the motion is two dimensional $w = 0$, $\frac{\partial u}{\partial z} = 0$ and $\frac{\partial v}{\partial z} = 0$, so

$$\omega_x = 0 \qquad \omega_y = 0 \qquad \omega_z = \frac{\partial v}{\partial x} - \frac{\partial u}{\partial y} .$$

Thus in two dimensional motion the vector $\vec{\omega}$ is normal to the xy-plane.

The first theorem of Helmholtz then says that each vortex line, normal to the xy-plane, must travel along a streamline. The second theorem states that $\omega_z ds\, dn$ is a constant. Notice in the foregoing figure the shaded area will move between the streamlines as shown. By the equation of continuity these two shaded areas are equal so ds dn remains constant too. Hence ω must stay constant along each streamline. But from Equation (1)

$$\omega_z = \frac{\partial v}{\partial x} - \frac{\partial u}{\partial y} = - \frac{\partial^2\psi}{\partial x^2} - \frac{\partial^2\psi}{\partial y^2} = -\Delta\psi$$

so the constancy of $\vec{\omega}$ gives us as equation of motion

$$\Delta\psi = F(\psi). \tag{2}$$

Each function $\psi(x,y)$ satisfying this equation can be the stream function of an incompressible perfect fluid; e.g., $\psi = (x^2 + y^2)^2$ with $\Delta\psi = 16(x^2 + y^2) = 16\sqrt{\psi}$ is a solution while $\psi = x^3 + y^3$ with $\Delta\psi = 6(x+y)$ contradicts (2).

We have now to determine the pressure distribution since thus far we have studied the equations from which p had been eliminated. We start with the relation connecting ω with H (Chapter I, Section 5). This is, as $\frac{\partial}{\partial t} = 0$,

$$\vec{q} \times \vec{\omega} = \text{grad } gH.$$

Here $\vec{q}$ is normal to $\vec{\omega}$. Thus the absolute value of $\vec{q} \times \vec{\omega}$ is equal to $q\omega$. The direction of the vector $\vec{q} \times \vec{\omega}$ is normal to $\vec{q}$ and to $\vec{\omega}$, therefore, parallel to the xy-plane and normal to the tangent of the streamline. According to the right hand rule the positive sense of $\vec{q} \times \vec{\omega}$ is opposite to the positive normal as defined above, if ω_z is positive. Hence

$$q\omega_z = -g \frac{\partial H}{\partial n}$$

or

$$\omega_z = -g \frac{\partial H}{q\partial n} = -g \frac{\partial H}{\partial \psi}$$

and since $F(\psi)$ is $-\omega_z$ we have

$$g \frac{\partial H}{\partial \psi} = F(\psi).$$

In this way the value of H for each streamline is found when we know $F(\psi)$:

$$H = \frac{1}{g} \int F(\psi) d\psi \tag{3}$$

(except for an additive constant, which is of no importance). Then from the definition of H

$$\frac{q^2}{2g} + \frac{p}{\rho g} + h = H$$

we find the pressure

$$p = (H - h - \frac{q^2}{2g})\rho g. \qquad (3a)$$

In some cases it is convenient to have the expression for ω or $\Delta\psi$ in intrinsic coordinates. We let the positive direction of the x-axis coincide with the direction of $\vec{q}$ and the direction of the positive y-axis with that of the normal $\vec{n}$ at the point considered. Then

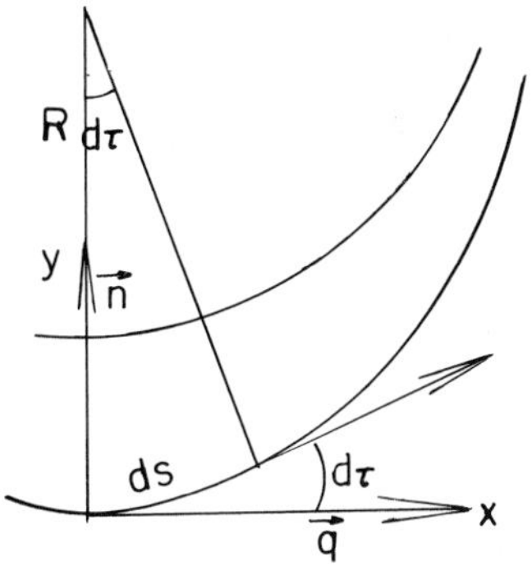

$$\omega_z = \frac{\partial q_n}{\partial s} - \frac{\partial q}{\partial n}$$

and while $q_n = 0$, $\frac{\partial q_n}{\partial s} \neq 0$. This is seen from the figure in which the radius of curvature $R = \frac{ds}{d\tau}$. There $dq_n = q d\tau$ and so

$$\frac{\partial q_n}{\partial s} = q \frac{d\tau}{ds} = q/R.$$

The expression for ω_z becomes

$$\omega_z = q/R - \frac{\partial q}{\partial n} .$$

Here R is positive when the center of curvature is on the same side of the streamline as the positive direction of dn.

Now our Equation (2) reads

$$F(\psi) = \frac{\partial q}{\partial n} - q/R. \tag{4}$$

This formula can be used for a kind of graphical integration. Suppose we have one streamline given and the q values along this line as shown in the figure. This enables us to plot a neighboring streamline s_2 corresponding to an increment $d\psi$ by using the equation $dn = \frac{d\psi}{q}$. Now from (4) we get

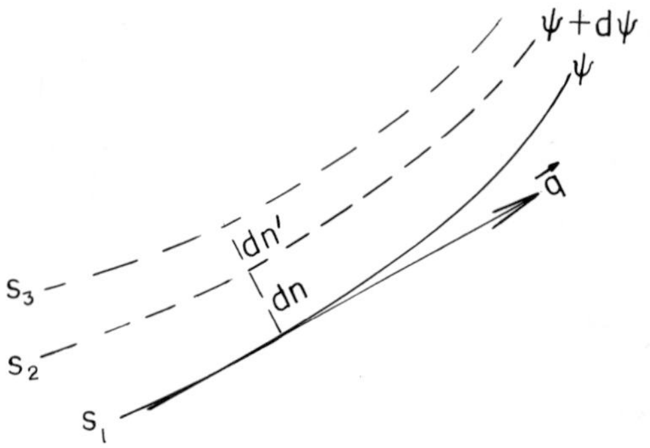

$$\frac{\partial q}{\partial n} = F(\psi) + q/R.$$

As q and R are known along the given streamline s_1 and $F(\psi)$ is supposed to be given too, we can compute the values $\partial q/\partial n$ and, therefore, the new values of q by adding $\frac{\partial q}{\partial n}\, dn$ to the original values. Then we find the new distances dn' equal to $d\psi$ divided by the new values of q, and in this way we obtain a third streamline s_3. We continue in the same manner by considering the pair of second and third lines s_2 and s_3 instead of the first and second lines s_1 and s_2. In this second step the radius of curvature along the second streamline has to be used. But it is obvious that, if only a finite section of each of the first and second lines (or a finite section of the first line with the q-values) is given, this procedure cannot be extended up to the ends of the given portions. Therefore, in order to really get a solution, we have to know the endpoints of all the streamlines independently. This is the case if along two cross sectional lines (C_1 and C_2 in the figure) the value of ψ is given. We have to mark the points on C_1 and C_2 that correspond to equal increments $d\psi$. These are the endpoints of our streamlines. Thus we see that a solution can be determined when, besides the function $F(\psi)$, two neighboring streamlines s_1 and s_2 (or what is the same, one streamline s_1 and the velocity q along s_1) and the ψ-values along two cross sectional lines are given.

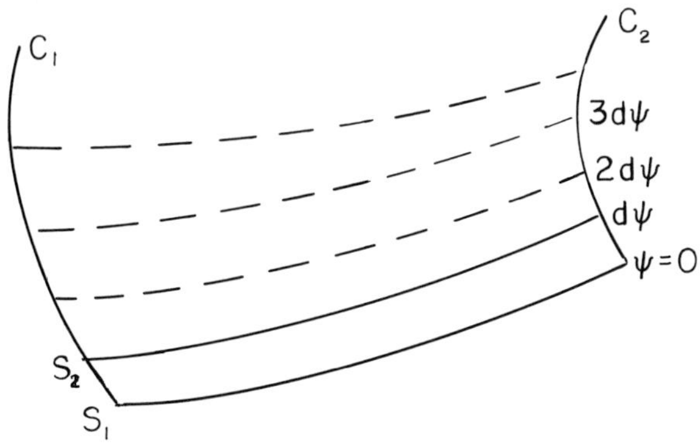

Actually, the practical problem will be a little different. We shall have given, besides the ψ values along the cross sectional curves, one streamline s_1 and another streamline s_n at a finite distance (the two walls of a channel). In this case we can say that ψ is given on the entire boundary of a closed area. In order to find the corresponding solution we assume tentatively a line s_2 and proceed in the above described manner. The result will be that we will get a final streamline passing through the endpoints of s_n but not coinciding with s_n. Now our task is to modify the assumption about s_2 so as to get the final streamline to coincide with the given s_n. This procedure can be considered as an approximate graphical integration of the equation $\Delta\psi = F(\psi)$ for given F and given ψ-values along the complete boundary of the area.

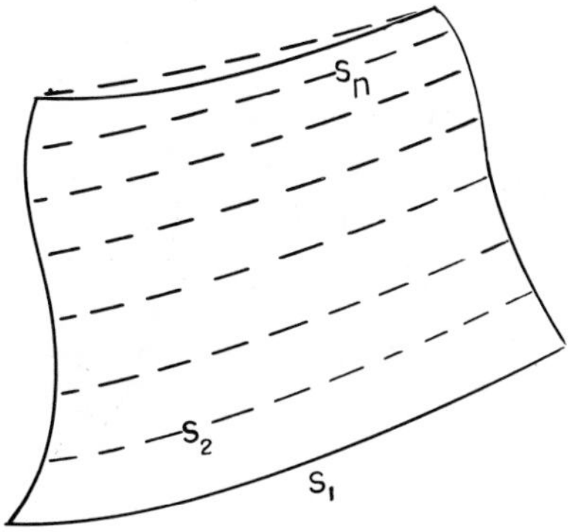

2. Steady, Irrotational Motion of an Incompressible Fluid.

The simpliest case of the problem in the preceding section is obtained by assuming $\omega = 0$ and hence $F(\psi) = 0$ everywhere in the fluid. Then (2) becomes

$$\Delta\psi = 0. \tag{5}$$

We also have $\frac{\partial v}{\partial x} = \frac{\partial u}{\partial y}$ since $\omega = 0$, which is identically satisfied by setting

$$u = \frac{\partial\phi}{\partial x}, \qquad v = \frac{\partial\phi}{\partial y}. \tag{6}$$

Then the function $\phi(x,y)$ has the property that its derivative in any direction is the component of the velocity in that direction, or grad $\phi = \vec{q}$. We again call ϕ the potential function. The equation of continuity gives, since div $\vec{q}$ = div grad $\phi = \Delta\phi$,

$$\Delta\phi = 0. \tag{7}$$

For the stream function ψ, we had $\frac{\partial\psi}{\partial n} = q$, $\frac{\partial\psi}{\partial s} = 0$ while now

$$\frac{\partial\phi}{\partial s} = q, \quad \frac{\partial\phi}{\partial n} = 0$$

so the lines along which ϕ = const. (called potential lines) must be everywhere orthogonal to the streamlines. If we construct a network of the two families of lines such that all $d\phi$ and $d\psi$ have the same value, then $dn = ds$ at each point since $q = \frac{\partial\phi}{\partial s} = \frac{\partial\psi}{\partial n}$. In differential geometry such a network of curvilinear squares is called an

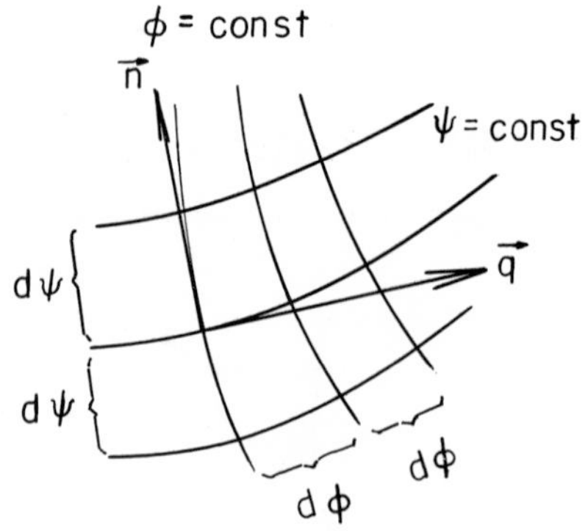

isothermal system.

The definition of ϕ and ψ gives us

$$u = \frac{\partial \phi}{\partial x} = \frac{\partial \psi}{\partial y}$$
$$v = \frac{\partial \phi}{\partial y} = -\frac{\partial \psi}{\partial x} \tag{8}$$

which are just the Cauchy-Riemann equations for $\phi + i\psi$ to be a function w of the complex variable $z = x + iy$ which possesses a derivative at each point z. That is, $\frac{dw}{dz} = w'$ exists if and only if the real and imaginary parts ϕ and ψ of w respectively satisfy these equations. Then

$$w' = \frac{\partial \phi}{\partial x} + i\frac{\partial \psi}{\partial x} = \frac{\partial \psi}{\partial y} - i\frac{\partial \phi}{\partial y} = u - iv. \tag{9}$$

Since $\vec{q} = u + iv$ is the velocity vector, then $u - iv$, the complex conjugate of q, is the reflection of the velocity vector in the x-axis.

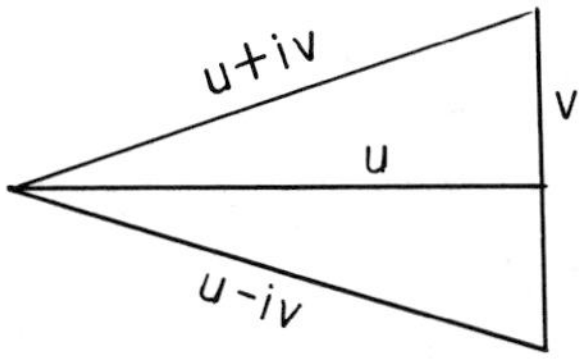

A simple example of a function of z is

$$w = z^2 = x^2 - y^2 + 2ixy.$$

Here $\phi = x^2 - y^2$ and $\psi = 2xy$ are two orthogonal families of hyperbolas.

If r and δ are polar coordinates of the point (x,y), we have $z = x + iy = re^{i\delta}$. As q is the amount of velocity and w' the reflected velocity vector we can write $w' = qe^{-i\theta}$, where θ is the angle $\vec{q}$ makes with the x-axis.

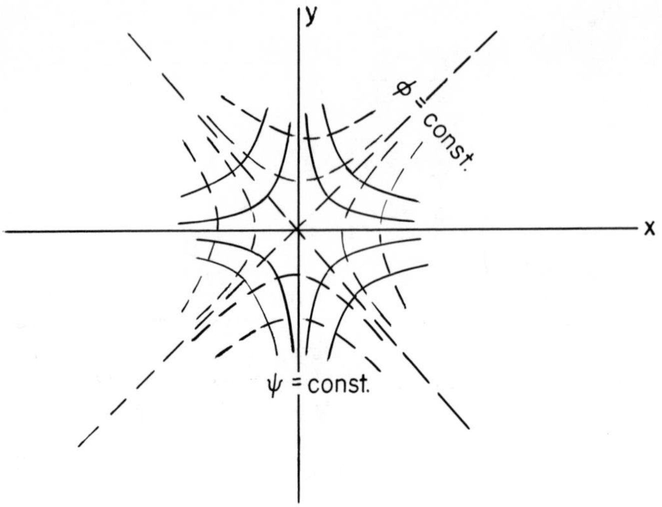

If $f(z)$ is regular inside a closed curve C we know by Cauchy's integral theorem that $\oint_C f(z)dz = 0$. In our case we consider a closed curve C filled with fluid and such that the velocity field inside this curve is regular. Thus by Cauchy's theorem $\oint w'dz = 0$. Now

$$\oint w'dz = \oint (u-iv)(dx+idy)$$

$$= \oint (udx+vdy) + i\oint (udy-vdx).$$

We note that $\oint(udx+vdy) = \Gamma$, the circulation. If we denote Λ, $\oint(udy-vdx)$, Cauchy's theorem states

$$\Gamma + i\Lambda = 0.$$

However, this can only happen if both $\Gamma = 0$ and $\Lambda = 0$. The fact that Γ, the circulation, is zero is known for vortex-free motion. Now

$$\Lambda = \oint (udy-vdx) = \oint q \sin(q,s)ds = \oint q_n ds = 0$$

states that the total quantity of fluid entering or leaving the circuit is zero. This means that there are no sources or sinks present.

In the case of a solid body immersed in a fluid the integral $\oint w'dz$ must not vanish for a curve surrounding the body, since we do not have a regular velocity field filling the entire area enclosed by the curve. However, if we consider another closed curve C_2 around the same body and draw lines AB and CD, as shown in the figure, then we can say $\oint w'dz = 0$ over the path (DCFBAED) which is a simple closed curve. Now let the lines AB and DC approach each other, and in the limit the integrals over these paths will cancel each other. Thus we get

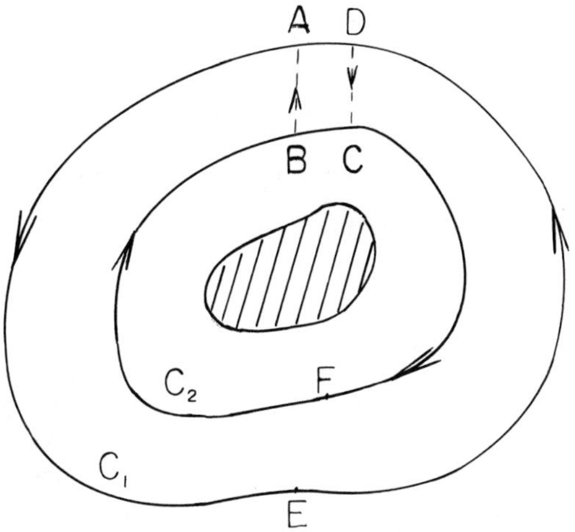

$$\oint_{C_1} w'dz + \oint_{C_2} w'dz = 0.$$

From this we see that if we call Γ_1 the circulation around C_1 and Γ_2 the circulation around C_2, both taken in the positive sense then $\Gamma_1 = \Gamma_2$. The case of several bodies in a fluid can be treated analogously to the one above. For example if we consider three bodies, then we will get three circulations Γ_1, Γ_2 and Γ_3 about each body by itself. Then we can deduce that the circulation Γ around a closed curve containing all the bodies on the inside is given by

$$\Gamma = \Gamma_1 + \Gamma_2 + \Gamma_3.$$

This can be done by breaking up the path as shown in the figure and proceeding as before.

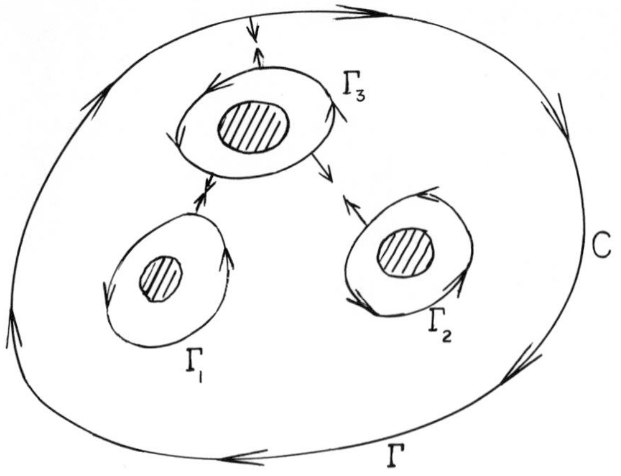

We now wish to examine the form which the momentum equations (33) and (34) of Chapter I take in plane motion. In two-dimensional flow we consider solid bodies to be cylinders of infinite length with generators normal to the plane under consideration. Since the force will be proportional to the thickness, we compute the force per unit span, i.e., we consider a layer of unit thickness.

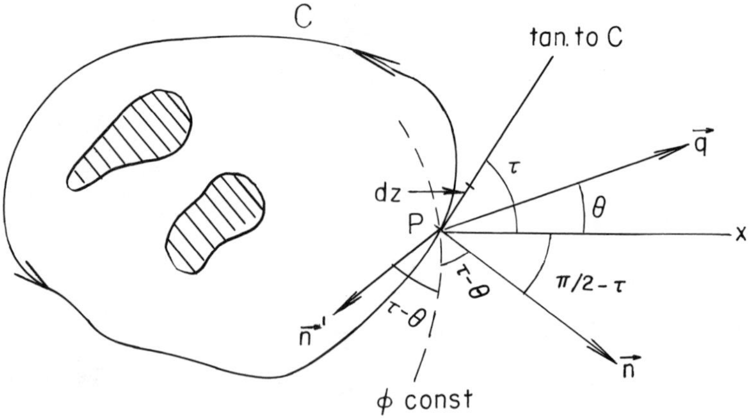

In the above diagram $\vec{q}$ is the velocity vector at the point P, θ is the angle $\vec{q}$ makes with the x-axis. Let τ be the angle the tangent to C at P makes with the x-axis, and let $\vec{n}$ be the outward normal to C at P. Then the vector $\vec{n}'$ will be the unit vector which is the reflection of $\vec{n}$ in the surface Φ = constant passing through P. Since dz is the increment of z it is a vector of length $d\ell$ in the direction τ tangent to the curve C. Thus we have $dz = d\ell\, e^{i\tau}$ and hence $dA = 1\cdot d\ell = e^{-i\tau}dz$. Now

$$w' = qe^{-i\theta}, \quad q = w'e^{i\theta}, \quad q^2 = w'^2e^{2i\theta}.$$

From the above figure it is clear that the angle $\vec{n}$ makes with the x-axis is $\frac{\pi}{2} - \tau$ in the negative sense, and also that $\vec{n}'$ makes an angle of $\frac{\pi}{2} - \tau + 2(\tau-\theta)$ $= \frac{\pi}{2} + \tau - 2\theta$ in the same way. Since $\vec{n}'$ is aunit vector we get

$$\vec{n}' = e^{-(\frac{\pi}{2} + \tau - 2\theta)i}.$$

Multiplying the equations $q^2 = w'^2e^{2i\theta}$ and $dA = e^{-i\tau}dz$, we get $q^2dA = w'^2e^{(2\theta-\tau)i}dz$, which becomes $i\,w'^2dz = q^2dA\,e^{(\tau - 2\theta + \frac{\pi}{2})i}$, since $i = e^{\frac{\pi}{2}i}$. Let us denote by F_x and F_y the components of the force vector $\vec{F}$ in the x- and y-directions respectively. Then we note that the expression $F_x - i\,F_y$ is the reflection of $\vec{F}$ in the x-axis. Now since $e^{(\tau - 2\theta + \frac{\pi}{2})i}$ is the reflection of $\vec{n}'$ in the x-axis, the equation for $\vec{F}$,

$$\vec{F} = \frac{\rho}{2}\int q^2\,\vec{n}'\,dA,$$

derived in Chapter I (Equation (33)), enables us to deduce, with the aid of the expression above for $i\,w'^2dz$ that

$$\frac{\rho}{2}\,i\int w'^2dz = F_x - i\,F_y. \tag{10}$$

Thus we see that the vector $\vec{F}$, like $\vec{q}$ is not expressible as a complex variable, but its complex conjugate is expressible in this manner.

For the vector $\vec{M}$, the moment, we proceed analogously. By (10) we have

$$\frac{\rho}{2}\, i\, w'^2 dz = dF_x - i\, dF_y .$$

If we take the moment with respect to the origin,

$$dM = x\, dF_y - y\, dF_x .$$

By using the identity

$$z(dF_x - idF_y) = (x+iy)(dF_x - idF_y) = x\, dF_x + y\, dF_y + (ydF_x - xdF_y)i$$

and denoting the quantity $x\, dF_x + y\, dF_y$ by dN, we have

$$-\frac{1}{i}\, z(dF_x - idF_y) = dM + i\, dN .$$

Hence

$$M + Ni = -\frac{\rho}{2}\oint w'^2 z\, dz . \tag{11a}$$

By equating the real parts of this last equation, we have

$$M = -\frac{\rho}{2}\,\mathscr{R}\left(\oint w'^2 z\, dz\right). \tag{11b}$$

We note that Equation (11a) gives the moment with respect to the origin. If we wish to take the moment with respect to the point P given by the position vector z_0, then (11a) becomes

$$M_0 + N_0 i = -\frac{\rho}{2}\oint(z-z_0)w'^2 dz. \qquad (11c)$$

The Equations (10) and (11b) are known as the equations of Blasius.

The quantity N derived above does not enter into the usual problems of statics. However, in so-called astatics, which is the study of the changes that occur when the forces acting on a body are rotated by a certain amount each about its point of application, the quantity N plays a certain role and is called the virial of the forces. This theory of astatics was developed by Hamilton; we shall make some additional remarks about this subject later.

Problem 3. Under what conditions will the streamlines of the plane motion of an incompressible fluid be similar conics with common axis, i.e., of the form $ax^2 + 2bxy + cy^2 = \text{constant}$?

Problem 4. Find the streamlines, potential lines, pressure distribution for the irrotational motion given by

(a) $w = e^z$

(b) $z = w + e^w$.

Problem 5. Prove the following statement. In irrotational motion in a horizontal plane the lines of constant velocity direction and the lines where the pressure is a constant are perpendicular.

Problem 6. Compute the force exerted upon the walls of the channel whose border is:

$$x^2 - y^2 = a^2$$
$$b > a \quad \text{for} \quad -c < y < c.$$
$$x^2 - y^2 = b^2$$

3. Effect of Uniform Flow upon Immersed Bodies.

We consider the irrotational plane motion of a fluid which we suppose fills the entire plane. Let there be an arbitrary but finite number of bodies immersed in the fluid. We say that we have uniform flow if no matter how z tends to infinity the velocity $\vec{q}$ approaches a constant vector which we denote by $\vec{V}$. This vector $\vec{V}$ is called the velocity vector at $z = \infty$. Obviously this kind of motion is the reverse of the motion produced in a fluid at rest by a body (or set of bodies moving with uniform velocity $-\vec{V}$).

We can represent $\vec{V}$ as $-Ve^{-\alpha i}$ where V is the magnitude of $\vec{V}$ and α is the angle $\vec{V}$ makes with the x-axis, taken in the negative sense (see figure). (This would correspond to a body moving to the right in a fluid at rest.) Thus w' which is the reflection of the velocity vector in the x-axis becomes $w' = -Ve^{\alpha i}$ at $z = \infty$. The angle α is often referred to as the angle of attack or angle of incidence.

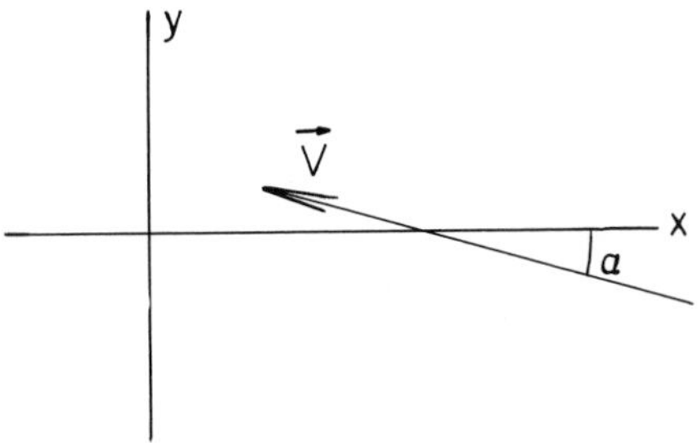

If we consider a circle $|z| = R$ where R is taken so large that all the solid bodies are contained in this circle, then w' is a regular function of z in the domain $|z| > R$. Hence we can expand $w'(z)$ in a power series about the point $z = \infty$ obtaining

$$w'(z) = A + \frac{B}{z} + \frac{C}{z^2} + \cdots\cdots . \tag{12}$$

The expansion (12) possesses the following three properties.

(1) A equals the given (reflected) velocity vector at $z = \infty$: $A = -Ve^{\alpha i}$. This is clear since from (12) $w' = -Ve^{\alpha i} = A$ at $z = \infty$.

(2) The quantity B is a pure imaginary number. This can be shown by the

elementary theorem of residues which states that for any analytic function $f(z)$, $\frac{1}{2\pi i}\oint_C f(z)dz$ equals the sum of the residues of $f(z)$ at the singular points interior to C. In our case of w' the residue is well known to be B.[1] Hence

$$\oint w'dz = \Gamma + \Lambda i = 2\pi i B. \qquad (13)$$

Now Γ, the circulation must not be zero since we have solid bodies inside the circle, while Λ must be zero in the case of a regular velocity field which has neither sources nor sinks present. Thus $B = \frac{\Gamma}{2\pi i}$, and property (2) is proved.

(3) <u>The coefficients A,B,C,... in the expansion (12) are linear functions of the velocity vector V.</u> That this is true for A is seen immediately by property (1). To prove this for all the coefficients we have to examine the conditions which have to be fulfilled by the velocity field $\vec{w}'$. We are given first $\text{grad}\ \phi = \vec{q} = \vec{V}$ for $z = \infty$ and then we have $\frac{\partial\phi}{\partial n} = 0$ at every point on the surface of each of the bodies. Incidentally the motion may not be determined uniquely by these conditions, but we can say that all possible motions must satisfy these conditions. Now let us decompose $\vec{V}$ into two parts: $\vec{V} = \vec{V}_1 + \vec{V}_2$. Then for $\vec{q} = \vec{V}_1$ at $z = \infty$ and $\frac{\partial\phi_1}{\partial n} = 0$ at every point on the surface of each body, we get a certain solution ϕ_1 for the potential and a corresponding w_1' for the (reflected) velocity field. Similarly, for $\vec{q} = \vec{V}_2$ at $z = \infty$ and $\frac{\partial\phi_2}{\partial n} = 0$ at the bodies we get a solution ϕ_2 and w_2'. Then the sum $\phi_1 + \phi_2 = \phi$ is a solution for $\vec{q} = \vec{V}$ at $z = \infty$ as

---

[1]Without using the residue theorem we may conclude as follows: On a circle of radius R, we have $z = Re^{i\delta}$ and, therefore, $dz = Rie^{\delta i}d\delta$. Thus the integral $\int z^n dz = R^{n+1}i\int_0^{2\pi} e^{(n+1)\delta i}d\delta$. If n is an integer not equal to -1, the integral on the right side is clearly equal to zero. And for $n = -1$, we get $\int z^{-1}dz = i\int_0^{2\pi} d\delta = 2\pi i$. Applying this to (12) and integrating term by term, we get (13).

$\frac{\partial\phi}{\partial n} = \frac{\partial\phi_1}{\partial n} + \frac{\partial\phi_2}{\partial n} = 0$ and at infinity $\vec{q} = \vec{V}_1 + \vec{V}_2 = \vec{V}$. As w_1' and w_2' are the derivatives of w_1 and w_2, the value w' corresponding to the resulting solution is $w' = w_1' + w_2'$. Thus the quantities $A, B, C, \ldots$ can be broken up such that $A = A_1 + A_2$, $B = B_1 + B_2$, $C = C_1 + C_2, \ldots$ corresponding to the decomposition of $\vec{V}$. This states that $A, B, C, \ldots$ are linear functions of $\vec{V}$.

Now using these three properties of the coefficients in the expansion (12) we can find with the aid of Equations (10) and (11c) definite expressions for $\vec{F}$ and $\vec{M}$. From (12) we have

$$w'^2 = A^2 + \frac{2AB}{z} + \frac{(B^2+2AC)}{z^2} + \cdots\cdots . \tag{12a}$$

By making use again of the theorem on residues, and of this last expression, Equation (10) becomes

$$F_x - i\,F_y = \frac{\rho}{2}\, i \oint w'^2 dz = \frac{\rho}{2}\, i\; 2AB \cdot 2\pi i,$$

since the residue of w'^2 is $2AB$. Now $B = \frac{\Gamma}{2\pi i}$ and $A = -Ve^{\alpha i}$ which give us

$$F_x - F_y i = -\rho\Gamma V i e^{\alpha i}. \tag{14a}$$

The magnitude F of $\vec{F}$ then is given by the absolute value of the right hand side and this is $\rho \cdot |\Gamma| \cdot V$. Let us introduce the (positive or negative) magnitude L

$$L = \rho\Gamma V, \tag{14b}$$

which generally is called the <u>lift force</u>. We note that we have a positive lift only in the case of a positive circulation around the bodies under consideration. Equation (14a) yields moreover the direction of $\vec{F}$. If L is positive the direction of the reflected force vector $F_x - F_y i$ is given by $-ie^{\alpha i}$, and hence the direction of $\vec{F}$ by $ie^{-\alpha i}$. This means that we rotate $+90^{\circ}$ from the direction $e^{-\alpha i}$, and we see (note figure) that $\vec{F}$ is perpendicular to $\vec{V}$. Thus if L or Γ is positive $\vec{F}$ will be in the upward direction as shown in the figure (i.e. turned from $\vec{V}$ by 90° in the negative sense). while if L and Γ are negative it will be in the opposite direction (dotted vector). From the above discussion we can conclude that a set of bodies immersed in a fluid of uniform flow is subject to a force of amount $\rho\Gamma V$ and normal to the direction of flow (i.e. normal to $\vec{V}$ at infinity which corresponds to: normal to the velocity of the bodies in a fluid at rest). In horizontal flight the direction of L is vertically upward and this is why we call L a lift.

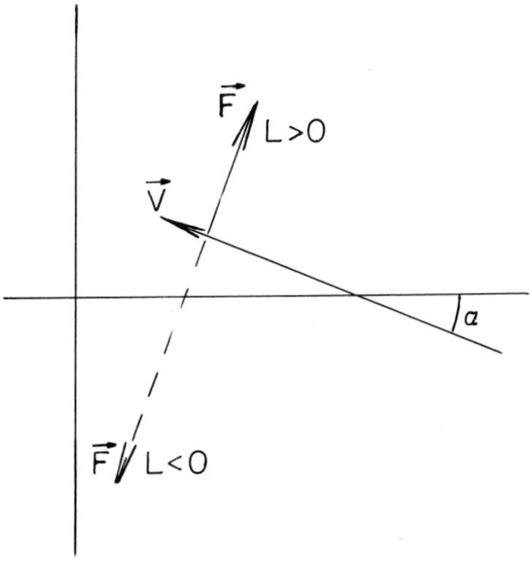

If we decompose $\vec{V}$ into its horizontal and vertical components V_x and V_y, then with the aid of properties (2) and (3) above, we can write $B = (b_1 V_x + b_2 V_y)i$, where b_1 and b_2 are real constants. Since $V_x = -V\cos\alpha$ and $V_y = V\sin\alpha$, we get $B = (-b_1\cos\alpha + b_2\sin\alpha)Vi$. Let us define $a = \frac{1}{2}\,|(b_1^2 + b_2^2)^{1/2}|$ and β by the relations

$$\sin\beta = \frac{b_1}{2a}\,,\ \cos\beta = \frac{-b_2}{2a}\,,$$

thus getting $B = -Vi\,2a\sin(\alpha+\beta)$. Since $L = \rho\Gamma V = 2\pi i\rho VB$, we conclude that

$$L = 4\pi a\rho V^2 \sin(\alpha+\beta) \tag{15}$$

where a and β are constants. Equation (15) was first given by Joukowski (1910) for a quite specified body, the so-called Joukowski wing. We found here that the same expression applies to each single body and even to an arbitrary set of bodies.

It is obvious that the angle α by itself cannot enter into a physical equation since it is the angle $\vec{V}$ makes with the x-axis, and the x-axis can be chosen arbitrarily. In our equation (15) not α but $\alpha + \beta$ appears, i.e., the angle $\vec{V}$ makes with a certain direction. This direction makes an angle β with the x-axis, and it is customary to call this direction the direction of the first axis of the immersed bodies. This direction in its relative position with respect to the set of bodies must be known (like the constant a). We shall speak about this later, however. At any rate we note that the angle $\alpha + \beta$ is independent of the coordinate system; it is the angle between $\vec{V}$ and the first axis. This angle is often called the effective angle of incidence. Thus Equation (15) states that

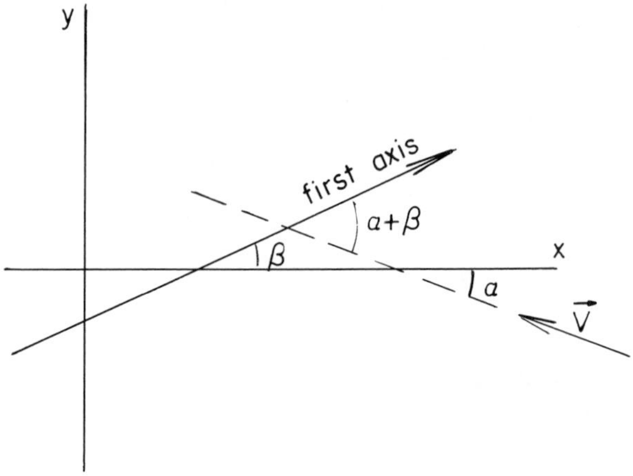

the magnitude of lift L is proportional to the density ρ, to the square of V, and to the sine of the effective angle of incidence. The factor a depends upon the magnitude and shape of the bodies and will be studied later.

We now proceed to find the form that Equation (11c) for the moment takes in uniform flow. Equation (11c) states

$$M_0 + N_0 i = -\frac{\rho}{2}\oint (z-z_0)w'^2 dz$$

$$= -\frac{\rho}{2}\oint zw'^2 dz + \frac{\rho}{2} z_0 \oint w'^2 dz.$$

From the expansion (12a) for w'^2 about $z = \infty$, we know

$$zw'^2 = A^2 z + 2AB + \frac{(B^2+2AC)}{z} + \cdots\cdots,$$

and the residue for this expression is clearly $B^2 + 2AC$. Thus, we get by using the theorem on residues

$$M_0 + N_0 i = -\frac{\rho}{2}(B^2+2AC)2\pi i + \frac{\rho}{2} z_0\, 2AB\cdot 2\pi i.$$

Now by using property (3) above we can represent C as $C_1'V_x + C_2'V_y$. Or we may write $C = c_1 V e^{\alpha i} + c_2 V e^{-\alpha i} = V(c_1 e^{\alpha i} + c_2 e^{-\alpha i})$ with c_1, c_2 arbitrary complex constants. Placing this last expression for C, and the expressions $A = -Ve^{\alpha i}$ and $B = -Vi\, 2a \sin(\alpha+\beta)$ which were previously obtained, in the above equation for $M_0 + N_0 i$, we get

$$M_0 + N_0 i = 2\pi i\, \rho V^2 [2a^2 \sin^2(\alpha+\beta) + e^{\alpha i}(c_1 e^{\alpha i} + c_2 e^{-\alpha i}) + 2z_0 a i e^{\alpha i} \sin(\alpha+\beta)]. \quad (16)$$

Now this is the expression for $M_0 + N_0 i$ when the moment is taken about the point z_0. This point z_0 which as yet was undetermined we fix to be the point $z_0 = -\frac{c_1}{a} e^{-i\beta}$. Now equating the real parts of (16) and substituting in this value of z_0, we get

$$M_0 = 2\pi\rho V^2 \mathscr{R} i \left[c_1 e^{2\alpha i} + c_2 - \frac{c_1}{a} e^{-\beta i} \cdot 2aie^{\alpha i} \frac{e^{(\alpha+\beta)i} - e^{-(\alpha+\beta)i}}{2i} \right].$$

Simplifying

$$M_0 = 2\pi\rho V^2 \mathscr{R} i (c_2 + c_1 e^{-2\beta i}). \quad (16a)$$

This last equation may be written in the form

$$M_0 = -2\pi\rho V^2 \operatorname{Im}(c_2+c_1 e^{-2\beta i}). \tag{16b}$$

We note that (16a) and (16b) give expressions for M_0 which are independent of α. It is customary to call the point z_0 (the relative position of which with respect to the bodies is determined by $z_0 = -\frac{c_1}{a} e^{-\beta i}$) the focus of the bodies under consideration. If the amount is taken with respect to this point, then the moment M_0 is independent of the angle of attack. It may happen, in a particular case, that the amount of the moment, M_0, is zero. Then we see that the lift force passes through the focus no matter what angle of attack we have.

Suppose that we know the position of the focus F and the direction β of the first axis for a set of bodies. We can write $L = L_0 \sin(\alpha+\beta)$ where $L_0 = 4\pi a\rho V^2$ is a positive constant dependent upon the still unknown constant a. A further constant is the value M_0 (positive or negative) of the moment of the lift force with respect to F. Draw the line ℓ_0 in the direction of the first axis at a distance $h_0 = \frac{M_0}{L_0}$ from F, for $h_0 > 0$ on the side shown in the figure. With the help of F, L_0 and the line ℓ_0 we can easily find the magnitude, the direction and the line of action of the lift force everytime the angle α of $\vec{V}$ is given. We draw a parallel to $\vec{V}$ through F which cuts ℓ_0 in S. Then we take the normal line to $\vec{V}$ passing through S as the line of action of L. First, the direction is correct, since it is normal to $\vec{V}$. Second,

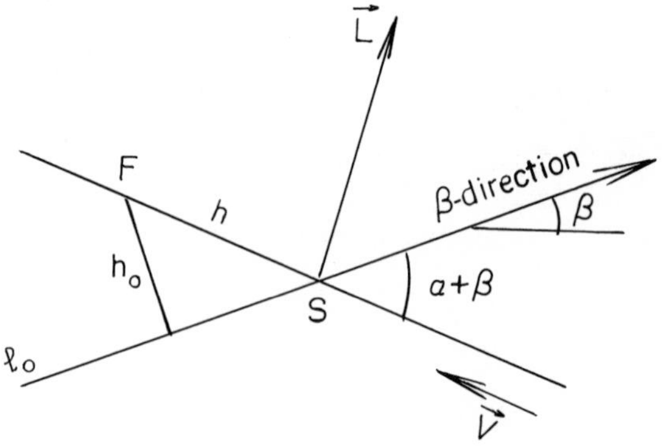

the moment of this force with respect to F is positive and of amount $L \cdot h$ where $h = \overline{FS}$. Now $h = \frac{h_0}{\sin(\alpha+\beta)}$ and $L = L_0 \sin(\alpha+\beta)$ so the moment is $L_0 h_0$ as it should be.

If the angle α of $\vec{V}$ changes, the point S moves along ℓ_0 and the line of action passes through S normal to $\overline{FS}$. A family of straight lines defined in this way (i.e., each line is normal to a ray through a fixed point F and passes through the intersection of the ray with a fixed line ℓ_0) envelopes, as we know, a parabola. The line ℓ_0 is the vertex tangent and F is the focus of the parabola.

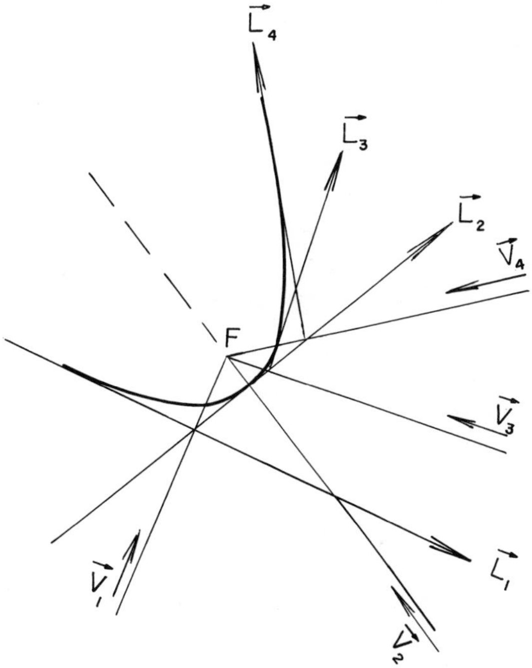

This parabola is called the metacentric parabola, since it is the focus of all metacenters. The metacenter, generally speaking, is the point where the line of action of the lift L is tangent to its envelope, i.e., to the parabola. If the angle of flow changes by a small amount $d\alpha$ then the line of action of L turns about the metacenter.

To find the metacenter P for a given direction of flow $\vec{V}$, i.e., the point of the parabola, elementary geometry supplies the following procedure. We have L normal to $\vec{V}$ at S. Find P' so that $\overline{SP'}$ is parallel to the axis of the parabola or normal to ℓ_0 while $\overline{FP'}$ is parallel to $\vec{L}$ or normal to $\vec{V}$. Then P is the point in which a line through P' parallel to V intersects the line of action of L.

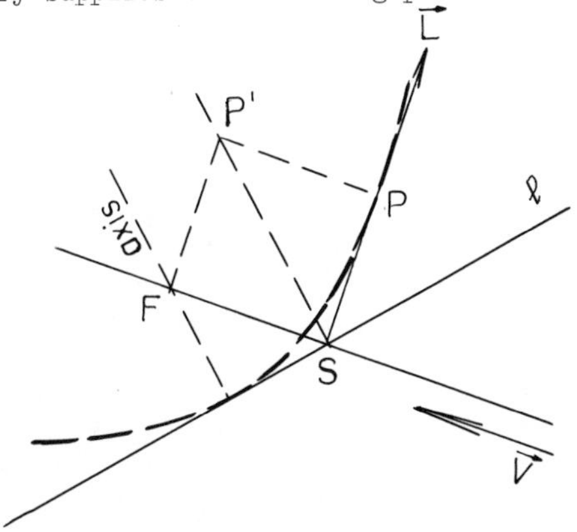

The idea of a metacenter was originally used in ship construction.

In ships the metacenter must lie above the center of gravity for stability in rolling. Here stability, as usual, is only considered in the sense of small disturbances. The figure shows the cross section of a ship with longitudinal stability. If the ship rolls through a small angle $d\alpha$, the resulting lift will turn by the same angle relative to the body, and the position of the line of action of the lift will intersect the old position (which is the symmetry axis in the figure) at a point which we call the metacenter M. The gravity force passes constantly through the center of gravity c.g. When M lies above the c.g. the couple of lift force and gravity force after the disturbance counteracts the disturbance as seen in the figure. Otherwise, the couple would increase the distrubance and the ship would roll over.

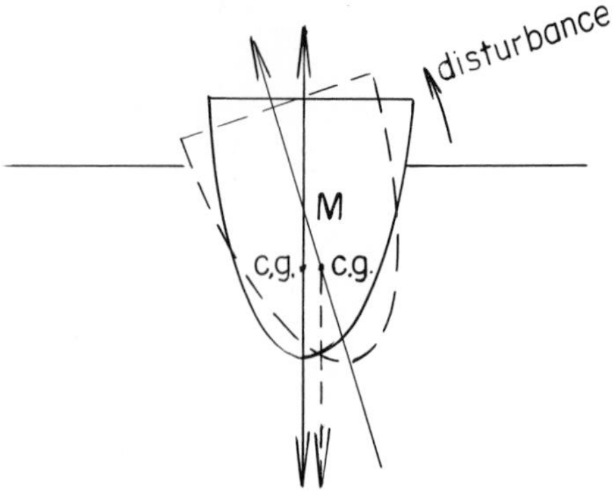

We have similar conditions for bodies in airflow. The point P must lie above the center of gravity for stability.

Thus far in our discussion the moment M_0 about F was assumed to be positive so that h_0 was greater than zero. Since F was above ℓ_0 in all the figures we can say that h_0 is measured positive downwards from F. For an airwing together with a tail wing we have, in general, the above case of positive h_0. In the case of a single airwing we often have $h_0 < 0$. The metacentric parabola then opens downwards as shown in the figure.

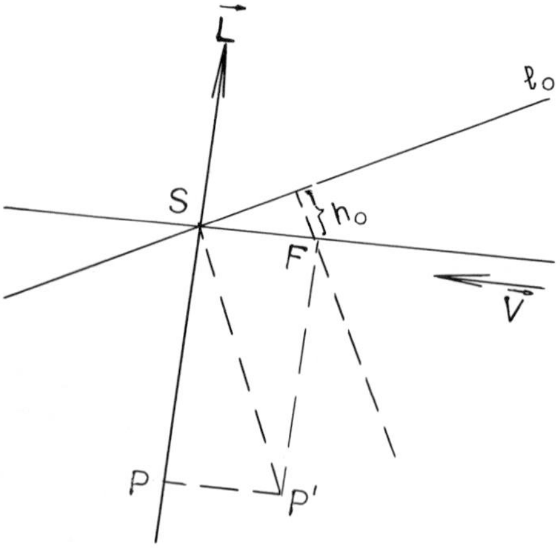

A very important but special case is when $h_0 = 0$. Then the moment M_0

is zero and the lift force passes through F for all angles of attack. We no longer have a parabola but the lines of action form a pencil through F. We then say that F is the center of lift.

In the present argument we have used three complex constants A, B and C to determine the lift forces for all angles α. Thus, the action of the flow upon the bodies depends upon five real constants, since one of the six components of A, B, C accounts for the arbitrary orientation of the coordinate system. The location of F takes two constants, the location of ℓ_0 two more and the fifth constant is needed to determine the value of a in $L_0 = 4\pi a\rho V^2$. These five constants give the whole distribution of forces. The metacentric curve is always a parabola whatever the shape of the bodies may be.

It is incorrect to say that the equilibrium is stable, unstable or indifferent according as $h_0 > 0$, $h_0 < 0$ or $h_0 = 0$. Unless, we know the position of the center of gravity nothing can be said about stability. A correct statement would be for instance that the equilibrium is stable if the c.g. is below the point P. Now we saw that in the case of $h_0 > 0$ the entire parabola or the locus of all P's lies higher than in the case of $h_0 < 0$. Thus it is easier to reach a stable condition with $h_0 > 0$ than with $h_0 < 0$. If when $h_0 = 0$ the positions of F and the c.g. coincide and only if they do is the equilibrium indifferent.

We have so far calculated M_0 and now wish to find the value of N_0, the virial of the forces. We had

$$M_0 + N_0 i = 2\pi\rho V^2 i\left[c_1 e^{2\alpha i} + c_2 + 2a^2\sin^2(\alpha+\beta) + 2aiz_0\sin(\alpha+\beta)e^{\alpha i}\right]. \tag{16}$$

Before, we introduced $z_0 = -\frac{c_1}{a}e^{-i\beta}$ but now we wish to study the value of this expression at another point

$$z_0' = -\frac{c_1}{a}e^{-i\beta} + ae^{i\beta}.$$

Then

$$N_0 = 2\pi\rho V^2 \mathcal{R}\Big[c_1 e^{2\alpha i} + c_2 + 2a^2 \sin^2(\alpha+\beta)$$

$$- 2ic_1 e^{(\alpha-\beta)i} \frac{e^{(\alpha+\beta)i} - e^{-(\alpha+\beta)i}}{2i}$$

$$+ 2ia^2 e^{(\alpha+\beta)i} \sin(\alpha+\beta)\Big]$$

$$N_0 = 2\pi\rho V^2 \mathcal{R}\left[c_2 + c_1 e^{2\beta i}\right] \tag{16c}$$

which is also independent of α. The new point z_0' is at G which is at a distance a from F in the direction of β. For G the virial N_0 is constant, independent of the angle of attack α.

We now wish to discuss the significance of the quantity N, the virial, which we obtained in Equation (11a). Let us review the simple case of two-dimensional statics. Let $\vec{F}_\nu$ be a set of forces with components X_ν, Y_ν, and let the points of application be (x_ν, y_ν). Then the moment, M, about the origin is given by

$$M = \sum_\nu (x_\nu Y_\nu - y_\nu X_\nu).$$

If we let $X = \sum_\nu X_\nu$; $Y = \sum_\nu Y_\nu$, then we get

$$M = xY - yX. \tag{17}$$

If we consider x and y as variables this is the equation of a straight line and this line is the line of action of the resultant force.

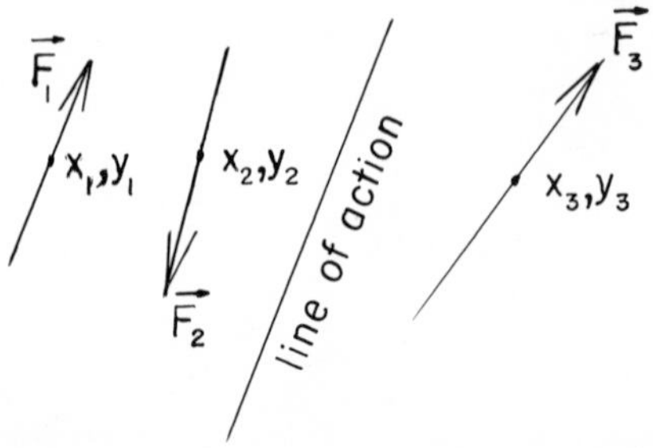

As we said previously, the quantity N arises in astatics which is the study of the changes that occur, if each force is turned about its point of application by an amount α without changing its magnitude. When this happens we will get a new line of action and a new resultant force. This resultant force will have the same magnitude as the original resultant, but will be turned by the same angle α. As we change α from 0 to 2π, we will get a family of straight lines which has a certain curve as its envelope. In order to find the point where a line of action touches the envelope, we must consider an infinitesimal change $d\alpha$ or take the derivative of the Equation (17) for the line of action with respect to α. Thus we get.

$$\frac{dM}{d\alpha} = x\frac{dY}{d\alpha} - y\frac{dX}{d\alpha}. \tag{17a}$$

This equation is also a straight line, and solving this simultaneously with the Equation (17), we get the desired point. Since $X_\nu = F_\nu \cos\alpha_\nu$, $Y_\nu = F_\nu \sin\alpha_\nu$, as we rotate each force by the same amount $d\alpha_\nu = d\alpha$, we get, since F_ν remains the same

$$dX_\nu = -F_\nu \sin\alpha_\nu d\alpha, \quad dY_\nu = F_\nu \cos\alpha_\nu d\alpha.$$

Now by summing

$$dX = \sum dX_\nu = -Y d\alpha, \quad dY = \sum dY_\nu = X\, d\alpha.$$

The expression for the moment M about the origin can be written in the form

$$M = \sum F_\nu r_\nu \sin(F_\nu, r_\nu).$$

Since the point of application and the amount of the force are unchanged as we rotate by an amount $d\alpha$, we get

$$dM = \sum F_\nu r_\nu \cos(F_\nu, r_\nu) d\alpha = \sum (x_\nu X_\nu + y_\nu Y_\nu) d\alpha = N\, d\alpha.$$

The quantity $N = \sum (x_\nu X_\nu + y_\nu Y_\nu)$ is what we called the virial of the forces. Thus we see that the two equations

$$\begin{aligned} M &= xY - yX \\ N &= xX + yY \end{aligned} \tag{18}$$

which form two perpendicular lines have as their point of intersection, the point where the line of action under consideration meets the envelope of the family of all lines of action. This point is called the Hamilton center of the given system of forces. If the Hamilton center coincides with the origin we have $xY - yX = 0$, $xX + yY = 0$, or $M = 0$ and $N = 0$. In other words: the Hamilton center is the point for which both the moment and virial vanish.

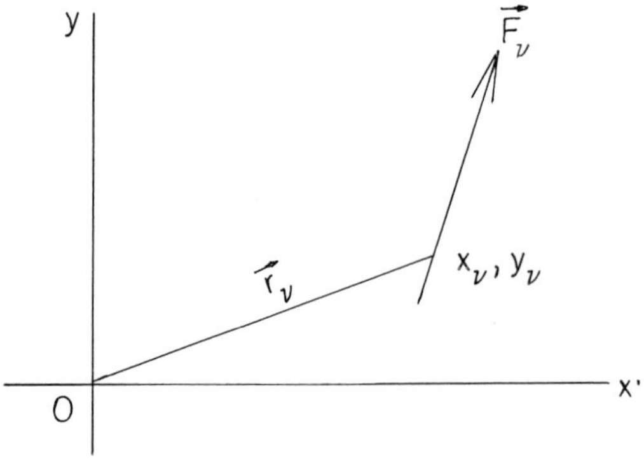

The above equations hold for the case where the moment was taken with respect to the origin. If we take the moment with respect to an arbitrary point (x_0, y_0), then Equations (18) become

$$\begin{aligned} M_0 &= (x-x_0)Y - (y-y_0)X \\ N_0 &= (x-x_0)X + (y-y_0)Y. \end{aligned} \tag{18a}$$

Thus M_0 can be considered as a function of the point (x_0, y_0) and we get a moment distribution throughout the plane. We see that since this is a linear function, the moment will be a constant along any straight line parallel to the resulting force. Thus we get a system of parallel lines such that the moment is constant along each of these lines, and there is one line for which $M_0 = 0$. This line is the line of action.

We can apply precisely the same argument to the quantity N_0. We will get a system of parallel lines where N_0 will be constant and one line where $N_0 = 0$. Clearly this system will be orthogonal to the system of lines for M_0 obtained above. The line $N_0 = 0$ contains all points for which the virial vanishes. The point of intersection of the lines $M_0 = 0$, $N_0 = 0$ is the point we called the Hamilton center; it is the point about which the resultant force turns.

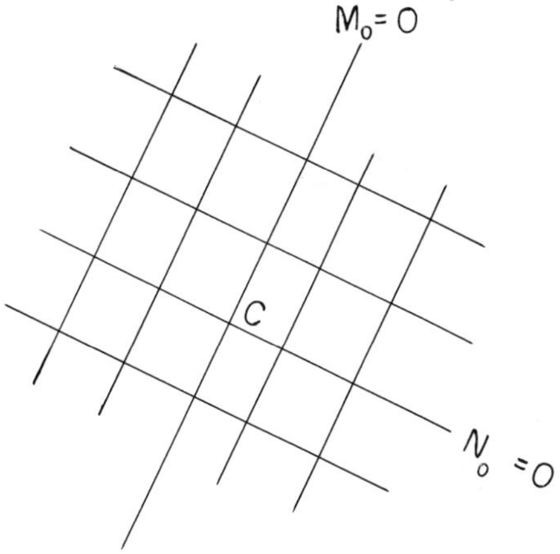

This point C, the Hamilton center, has a significance similar to that of the metacenter. The metacenter P was also found to be a point where the line of action meets its envelope. However, in determining this point P we considered the lines of action for each value α in accordance with the hydrodynamic conditions. This means that if α changes by an amount $d\alpha$, then by the hydrodynamical equations not only the direction of the pressure force at each element of surface changes, but simultaneously the magnitude of p changes also in a certain way. The effect of both change in direction and in magnitude was the new line of action as considered before. The metacenter P was the point of intersection of <u>this</u> new line of action with the original one. On the other hand, we now consider the case where only the direction and not the magnitude of pressure force at any element changes. The effect is then the new line of action as considered in the last paragraph, and the Hamilton center C is the point of intersection of the original line of action with this new line of action.

In general, there is no way of deciding which holds true in the case of a real disturbance under flying conditions. It may be that at the very moment a small angular disturbance occurs the pressure at a point changes in no way or partly or completely into the value which would correspond to the new α-value. Our result is that the point on the line of action that has to be assumed as the turning point is either C or P or any point in between. If we accept this point of view, we have to insure that the center of gravity lies below both points.

We now wish to show how the point C can be found when we know the constants A, B, C in the expansion (12). As was shown above the point C is the point of intersection of the lines $M_0 = 0$ and $N_0 = 0$. Now the line $N_0 = 0$ can be found in the same way as we previously found the line $M_0 = 0$ (line of action). To do this we use the point G: $z_0' = -\frac{c_1}{a} e^{-i\beta} + ae^{\beta i}$ for which the virial is

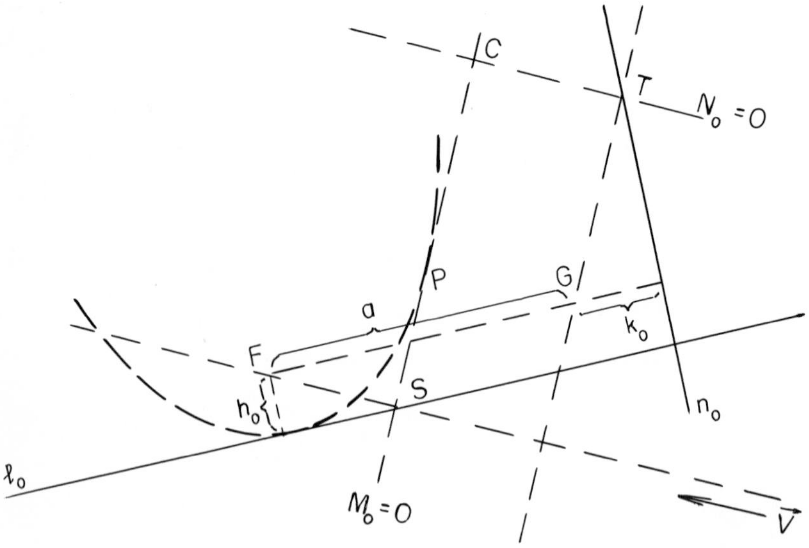

independent of the angle of attack. If N_0 is the constant value of the virial, we compute $k_0 = \frac{N_0}{L_0}$ and plot the straight line n_0 perpendicular to ℓ_0 at a distance k_0 from G. Then in the same way as before we get the line $N_0 = 0$ by drawing a normal to $\vec{V}$ through G which intersects n_0 at the point T. The desired line $N_0 = 0$ is the line passing through T parallel to $\vec{V}$. This gives the Hamilton center C as the point of intersection of the lines $M_0 = 0$, $N_0 = 0$.

From this construction we see that the quantities A, B and C in the expansion (12) for w' not only determine, for each α, the magnitude and line of action of the lift, but also the quantities determining the Hamilton center. Thus the problem of determining equilibrium and stability reduces to the problem of computing the three constants A, B and C for a given set of bodies or a single body.

4. Circular Cross-Section and Theorem of Joukowski.

We consider now the particular case of a single circular cylinder moving uniformly through a fluid at rest. Or, what is the same thing, the uniform flow of the fluid in the opposite direction with the body immersed in it. We let the center of the circular cross-section be the origin of our coordinate system, and let the radius of the circle be a. The conditions of our motion are that the velocity at infinity must be the constant vector $\vec{V}$, and that on the boundary of the cylinder the normal component of the velocity vector must be zero. First we may consider the motion where $\alpha = 0$. We can verify that the function

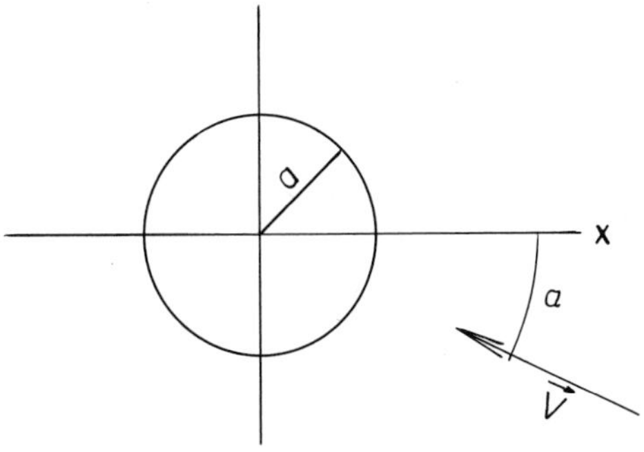

$$w = -V\left(z + \frac{a^2}{z}\right) \tag{19}$$

satisfies our conditions. Since

$$w' = -V\left(1 - \frac{a^2}{z^2}\right) \tag{19a}$$

we have at $z = \infty$, $w' = -V$ which is one condition. Also any point on the circle is given by $z = ae^{\delta i}$, and $w' = -V(1-e^{-2\delta i})$. Hence $w'e^{\delta i} = -V(e^{\delta i} - e^{-\delta i})$ $= -2iV \sin \delta = -2V \sin \delta e^{\frac{\pi}{2} i}$. Thus $w' = |w'| \cdot e^{(\frac{\pi}{2} - \delta)i}$, and we see that the velocity at any point on the circle is in the direction of the tangent to the circle at that point, which is the second requirement. Thus for the special case $\alpha = 0$ we have a solution to our problem in (19).

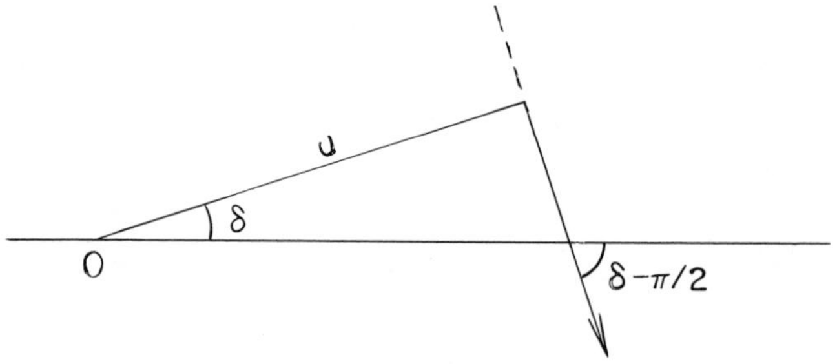

For the case where α is not zero we consider the transformation $z = z_1 e^{\alpha i}$. This transformation changes each point P into P' in such a way that P is rotated about the origin by an amount $-\alpha$. The function $w = -V(z + \frac{a^2}{z})$ is transformed into $w = -V(z_1 e^{\alpha i} + \frac{a^2}{z_1} e^{-\alpha i})$, and this gives us the solution for the case where

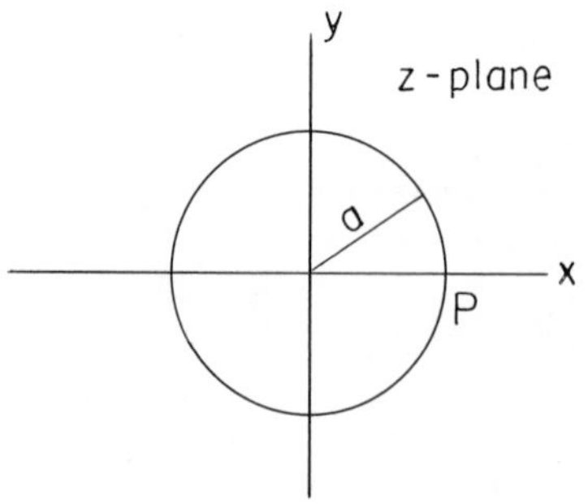

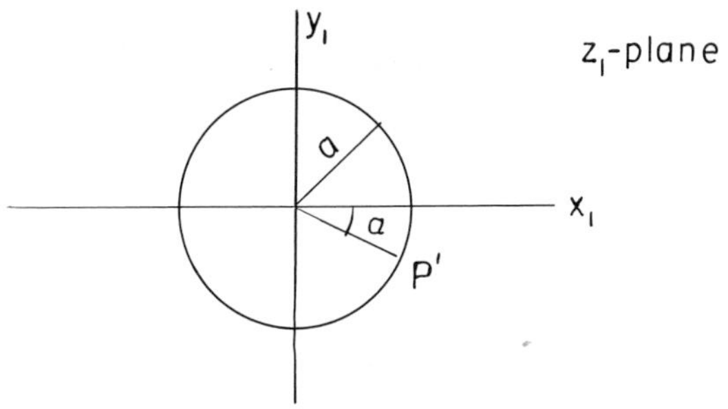

$\vec{V}$ makes an angle α with the x-axis. If we now write z instead of z_1 we have

$$\left.\begin{aligned} w &= -V(ze^{\alpha i} + \frac{a^2}{z} e^{-\alpha i}) \\ w' &= -Ve^{\alpha i}(1 - \frac{a^2}{z^2} e^{-2\alpha i}) \end{aligned}\right\} \tag{19b}$$

and the new conditions for our motion are clearly satisfied. However, this solution is not unique. If we consider the function

$$w = -Ve^{\alpha i}(z + \frac{a^2}{z} e^{-2\alpha i}) + \frac{\Gamma}{2\pi i} \log z \tag{20}$$

where Γ is a real constant we get a more general solution. We can verify this easily, for

$$w' = -Ve^{\alpha i}(1 - \frac{a^2}{z^2} e^{-2\alpha i}) + \frac{\Gamma}{2\pi i} \frac{1}{z} \tag{20a}$$

and $w' = -Ve^{\alpha i}$ at $z = \infty$ as before. Also, on the circle, the additional contribution in velocity is a real constant times $\frac{1}{i} \cdot \frac{1}{z}$. This is of the form $\frac{1}{ia} e^{-\delta i}$ on the circle, and we get the direction of this additional velocity as being that of the tangent to the circle at the point. Thus we see that we did find a solution for the problem of the translational motion of a circular cylinder, but this solution is not unique. It is known except for a real constant Γ which is arbitrary.

Let us first study the significance of the different terms in Equation (20).

We use the graphical method of super-position to plot the streamlines of our motion. According to this method, if we have two families of streamlines plotted with constant and equal values of $\Delta\psi$, as shown in the figure, the resultant flow will be along the dashed curves which are the diagonals of the small curvilinear rectangles. The value of $\Delta\psi$ will be the same for the resulting family.

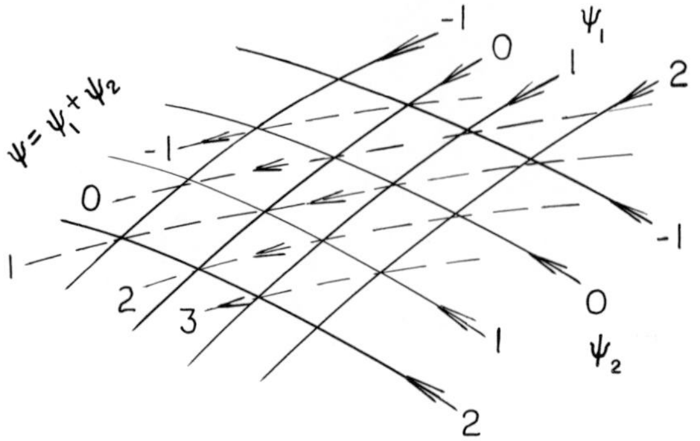

In our case we first have the linear term $w_1 = -Ve^{\alpha i}z = -V(x \cos \alpha - y \sin \alpha) - iV(x \sin \alpha + y \cos \alpha)$. The stream function $\psi_1 = -V(x \sin \alpha + y \cos \alpha)$ gives for constant values of ψ_1 the straight lines

$$x \sin \alpha + y \cos \alpha = \text{const.}$$

parallel to the direction $\vec{V}$ of flow as shown.

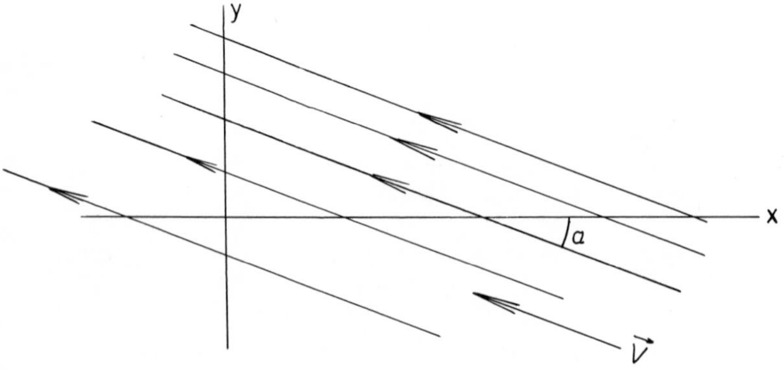

The second term is $w_2 = -\frac{Va^2}{ze^{\alpha i}}$. Now a term of the form $\frac{c}{z}$ where c is a real constant can be written

$$\frac{c}{z} = \frac{c}{x+iy} = \frac{cx}{x^2+y^2} + i \frac{-cy}{x^2+y^2}$$

or $\psi_2 = -\frac{cy}{x^2+y^2}$. For ψ_2 constant we then have the family of circles

$$x^2 + y^2 = \text{constant } y$$

with centers on the y-axis and passing through the origin. The factor $e^{\alpha i}$ in the denominator rotates the circles through the angle $-\alpha$ giving the circles shown in the figure. The superposition of the two families of lines and circles are shown in the next figure. This shows the flow in the case when $\Gamma = 0$. The flow is symmetric to both axes shown in the figure.

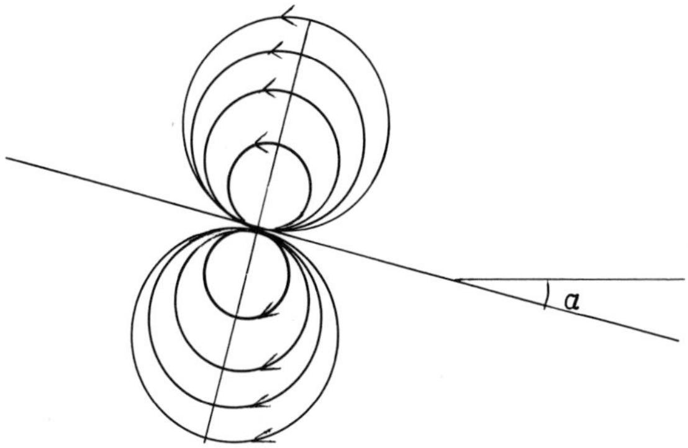

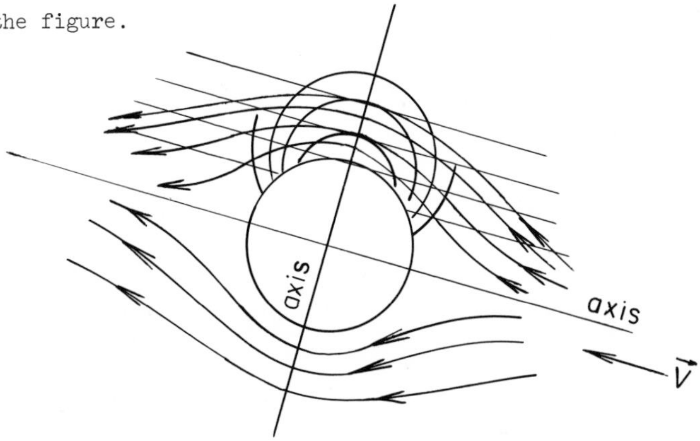

The third and last term in the expression for w is a circular motion. Setting $z = re^{\delta i}$

$$w_3 = \frac{\Gamma}{2\pi i} \log z = \frac{\Gamma}{2\pi} (\delta - i \log r)$$

and

$$\psi_3 = - \frac{\Gamma}{2\pi} \log r.$$

For ψ_3 constant we have log r = const. or r = const. which gives a family of circles with centers at the origin. If Γ is positive the motion is counterclockwise as shown. Γ is the circulation of this motion since we see immediately that

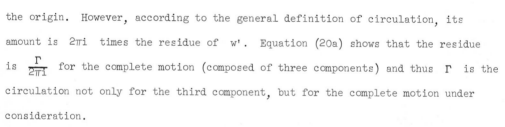

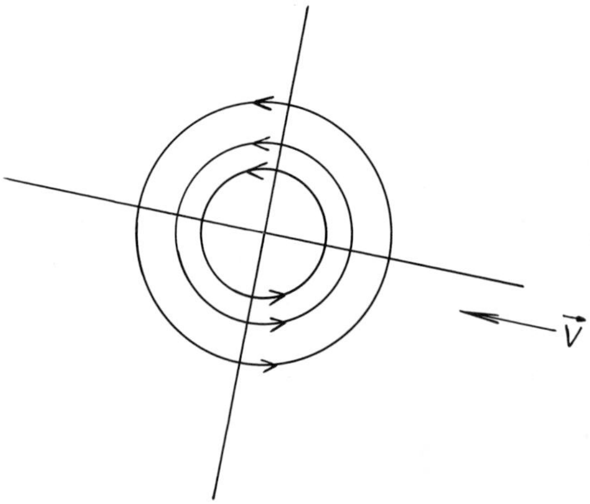

$$\oint w' dz = \frac{\Gamma}{2\pi i} \oint \frac{dz}{z} = \Gamma$$

when the contour of integration is any circle about the origin. However, according to the general definition of circulation, its amount is $2\pi i$ times the residue of w'. Equation (20a) shows that the residue is $\frac{\Gamma}{2\pi i}$ for the complete motion (composed of three components) and thus Γ is the circulation not only for the third component, but for the complete motion under consideration.

The super-position of this third type of motion onto the other two will leave the symmetry with respect to the axis through the center of the circle normal to the direction of $\vec{V}$, but will destroy the symmetry with respect to the line through the center parallel to $\vec{V}$. For above this second line, the velocities will add, but below it they will subtract. Hence, the velocities at the points of the cylinder are smaller below than above, and we see that the presence of circulation leads by Bernoulli's theorem to higher pressure values on the lower side of the cylinder than on its upper side. This explains the presence of a lift force. For not too large values of Γ, we get upon super-position streamlines as shown in the accompanying figure.

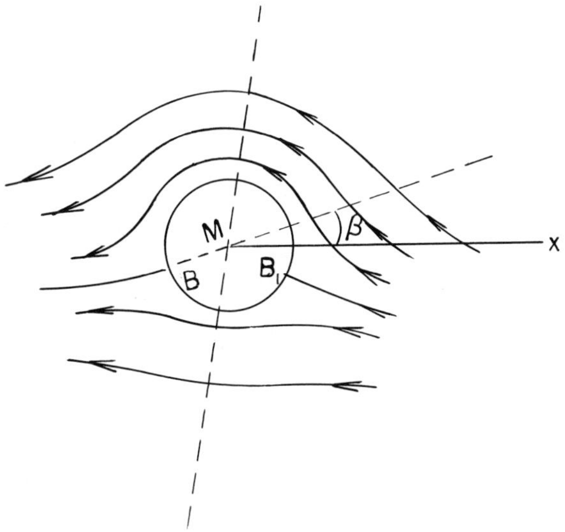

We can find two points B and B_1 on the circle where the velocity is zero. For on the circle $z = ae^{\delta i}$, the velocity there is

$$w' = -Ve^{\alpha i}(1-e^{-2(\alpha+\delta)i}) + \frac{\Gamma}{2\pi i}\frac{1}{a}e^{-\delta i} .$$

Then w' is zero if

$$\Gamma = 2\pi i\ a\ V(e^{(\alpha+\delta)i} - e^{-(\alpha+\delta)i}) = -4\pi a\ V \sin(\alpha+\delta)$$

or

$$\sin(\alpha+\delta) = -\frac{\Gamma}{4\pi aV} .$$

If $|\Gamma| < 4\pi aV$, we then have two values for δ and two points on the circumference where the velocity is zero. They lie symmetrically to the symmetry axis. If Γ is too large, we have no such "stagnation points" and if $\Gamma = 0$ we have $\delta = -\alpha$,

$\delta = \pi - \alpha$ or the stagnation points B and B_1 are directly opposite each other as shown in the figure we had for this case.

Let β be the angle which $\overline{BM}$ makes with the x-axis. Then we have $\delta = \beta + \pi$ and $\alpha + \delta = \alpha + \beta + \pi$ so

$$\Gamma = 4\pi aV \sin(\alpha+\beta). \tag{21}$$

At any rate Γ is known if the location of the stagnation points is known.

Now let us apply what we learned in Section 3 about the force exerted upon a body in a uniform flow. We found that the resulting force is normal to the velocity vector and by Equations (14b) and (15) has magnitude $L = \rho V\Gamma$ with $\Gamma = 4\pi a\ V \sin(\alpha+\beta)$ where β was the angle of the first axis and a was a constant. In the present case of a circle we find (21); this shows: (1) The direction of the first axis is determined by BM (where B is the stagnation point and M is the center). (2) The constant a equals the radius of the circle. As to the moment of the force and its line of action there is no question. By symmetry L must pass through the center M for all angles of attack α. Hence M is the focus and $h_0 = 0$.

However, the whole theory is not complete in the case of a circular cylinder since we have no way to determine the stagnation points or, what is the same, the value of Γ. On the other hand the situation is different in the case of an airwing section where we can use the essential idea of Joukowski, which we are going to discuss now.

From the beginning of the development of flying it has been known that a body with a definite lift and definite Γ cannot have as cross section a circle or a similar curve, but must be a body with sharp trailing edge, i.e., a body whose cross section has a corner or a cusp (see figure). Such a cusp or corner with a finite angle, as shown in the figure, is mathematically a singular point of the boundary. If we apply the theory we have just developed for the case of a circle, to such a cross section, we shall get again an undetermined solution, i.e., a function w including one unknown constant Γ. But, as Joukowski, a Russian mathematician, discovered about 1910, all these solutions in general involve an infinite velocity at the singular point. Velocity values beyond a certain finite limit are physically impossible, since, according to the Bernoulli equation, the sum of $\frac{q^2}{2}$ and p/ρ is a constant and to high values of q, therefore, would correspond negative p. Moreover, Joukowski also found that for one and only one particular value of Γ the velocity at the singular point remains finite. It is then obvious that the real flow corresponds to the function w with this particular Γ-value. We have to see how this Γ can be determined.

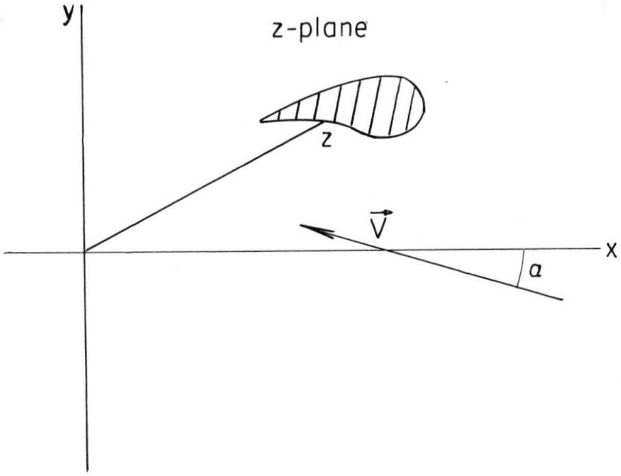

Let us use the idea of conformal transformation. Denote by f(z) an analytic function of z. Then $z' = f(z)$ determines such a transformation. Let us assume that we are able to determine f(z) in such a way that z' traces out a circle as z takes on all values along the given profile. A complex function $w(z') = \phi(z') + i\psi(z')$ determines, as we know, an irrotational flow in the z' plane. Suppose we have found the w(z') corresponding to the flow around the circle, i.e., a function w(z') whose imaginary part ψ has a constant value for all points of the circle (so as to have the derivative of ψ zero in the direction of the tangent, or normal component of velocity zero). If now we introduce into the expression w(z') the function $z' = f(z)$ we obtain a function w of z: $w[f(z)]$. This function determines a flow in the z-plane and the imaginary part ψ has a constant value when $f(z) = z'$ is on the circle of z is on the profile. In this way, by using the mapping function f we deduce from the flow around the circle a flow around the given profile.

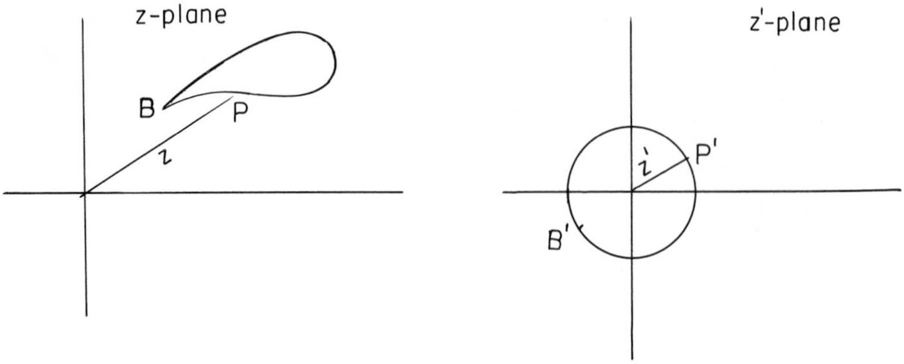

Now the velocity of flow in the profile plane is determined by

$$w' = \frac{dw}{dz} = \frac{dw}{dz'} \cdot \frac{dz'}{dz} = \frac{dw}{dz'} \cdot \frac{df}{dz} .$$

The first factor on the right hand side is the (reflected) velocity of the flow in the circle-plane. The second factor is the derivative of the mapping function. It is easy to see that this derivative cannot have a finite value at the singular point of the boundary, $\frac{df}{dz}$ must be infinite, when we take for z the point B. It follows that, if we want to have a finite velocity w' at B, it is necessary that $\frac{dw}{dz'}$ vanishes at B', the image of B. Thus we come to the conclusion that B' must be one of the stagnation points on the circle.

In choosing the mapping function $f(z)$ we have to take care that the conditions at infinity are maintained. They must be the same in the z-plane as in the z'-plane. Thus the conception of Joukowski leads to the following formulation: If a profile with a singular point is given, we try to find the transformation $f(z)$ which transforms this profile into a circle, leaving unchanged the region at infinity. To the singular point B may correspond by this transformation the point B' on the circle. Then we take the solution

$$w(z') = -Ve^{\alpha i}(z' + \frac{a^2}{z'} e^{-2\alpha i}) + \frac{\Gamma}{2\pi i} \log z' \tag{20b}$$

with that value of Γ which makes this B' a stagnation point. By introducing $z' = f(z)$ into this Equation (20b) we obtain the definite unique solution $w(z)$.

Problem 7. Let AB and CD be two streamlines in irrotational plane motion. If AC and BD are potential lines normal to the streamlines, show that the force due to the dynamic pressure (i.e., to the term $\rho \frac{q^2}{2}$) along AB and CD is equal in magnitude, direction and line of action to that along AC and BD.

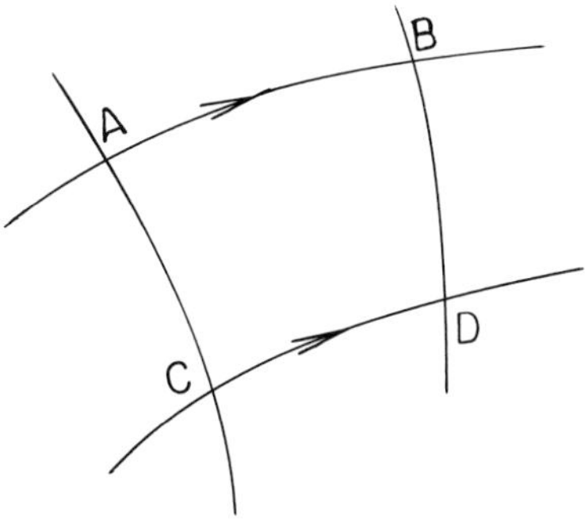

5. Solution of the Problem for a Simple Wing Profile.

Let us consider now a wing of infinite span, i.e., a cylinder with a cross-section containing one singular point. Our problem is, as just stated, to find a transformation $z' = f(z)$ which transforms this profile into a circle such that

(1) to all points of the profile correspond points of the circle, and

(2) $z \to z'$ at infinity, i.e., nothing is changed at infinity.

If we consider the inverse transformation $z = g(z')$, the second of these conditions states that $g(z')$ must be of the form

$$z = z' + \frac{k}{z'} + \frac{k_2}{z'^2} + \cdots\cdots . \tag{22}$$

This expansion must be valid for all points $|z'| \geq a$ except for one point on the circle $|z| = a$ (the singular point). Solving (22) for z', we get a development valid for sufficiently large $|z|$:

$$z' = z + \frac{k_1'}{z} + \frac{k_2'}{z^2} + \cdots\cdots . \tag{22a}$$

This transformation has the following properties:

(1) The first coefficient k_1' of (22a) equals $-k$ of (22). This can be seen as follows: Equation (22) can be written

$$z' = z - \frac{k}{z'} - \frac{k_2}{z'^2} - \cdots\cdots = z - \frac{k}{(z - \frac{k}{z'} \cdots\cdots)} - \frac{k_2}{z'^2} \cdots\cdots .$$

Thus $z' = z - \frac{k}{z}$ + terms of higher order. Comparing this with (22a), we get $k_1' = -k$.

(2) If the axes are translated to any other point, the expression maintains the same form with the same k. For, if we translate by an amount A, then (22) becomes

$$z + A = z' + A + \frac{k}{z'+A} + \cdots\cdots ,$$

and then

$$z = z' + \frac{k}{z'(1 + \frac{A}{z'})} + \cdots\cdots = z' + \frac{k}{z'} + \cdots\cdots .$$

(3) If the coordinate axes are rotated by an amount θ, then the constant k becomes $ke^{-2\theta i}$. This is easily seen as

$$ze^{i\theta} = z'e^{i\theta} + \frac{k}{z'e^{i\theta}} + \cdots\cdots ,$$

or

$$z = z' + \frac{ke^{-2\theta i}}{z'} + \cdots\cdots .$$

(4) If $z = z' + \frac{k}{z'} + \cdots$, and $z' = z'' + \frac{k_1}{z''} + \cdots$, are two such transformations, then the composite transformation of z'' to z again has the same form, and has for its coefficient of $\frac{1}{z''}$ the sum of the coefficient k and k_1. This follows by straightforward substitution. We get

$$z = z'' + \frac{k_1}{z''} + \cdots\cdots + \frac{k}{z'' + \frac{k_1}{z''} + \cdots} + \cdots\cdots ,$$

$$= z'' + \frac{k_1}{z''} + \cdots\cdots + \frac{k}{z''(1 + \frac{k_1}{z''^2} + \cdots)} + \cdots\cdots ,$$

$$= z'' + \frac{k+k_1}{z''} + \cdots\cdots .$$

Now we examine more closely the significance of k. For sufficiently large values of $|z'|$ the transformation (22) behaves like the transformation

$$z = z' + \frac{k}{z'} . \tag{23}$$

Here we let

$$k = c^2 e^{2\gamma i}, \tag{24}$$

and consider the transform of the circle $z' = Re^{i\delta}$ for a large value of R. This gives us

$$z = Re^{\delta i} + \frac{c^2}{R} e^{(2\gamma-\delta)i},$$

or

$$ze^{-\gamma i} = Re^{(\delta-\gamma)i} + \frac{c^2}{R}e^{-(\delta-\gamma)i}.$$

The real and imaginary parts of $ze^{-\gamma i}$ are, as the figure shows, coordinates x_1, y_1 of the point z in a coordinate system rotated by the angle γ. It follows from the last equation that

$$x_1 = (R + \frac{c^2}{R}) \cos (\delta-\gamma),$$

$$y_1 = (R - \frac{c^2}{R}) \sin (\delta-\gamma).$$

Now squaring and adding, we get

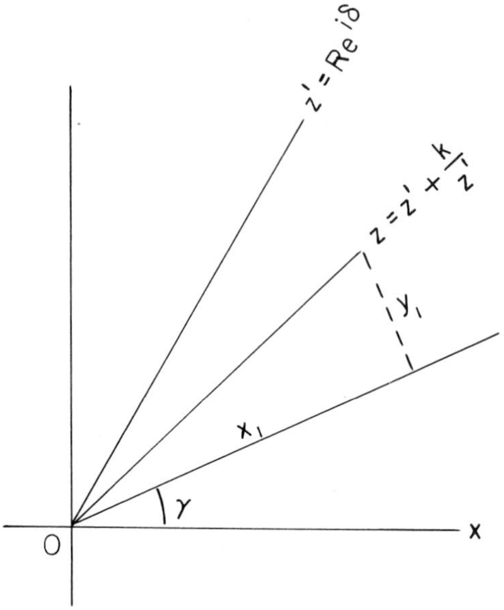

$$\frac{x_1^2}{(R + \frac{c^2}{R})^2} + \frac{y_1^2}{(R - \frac{c^2}{R})^2} = 1.$$

Thus the circle $|z'| = R$ goes into an ellipse which has its center at the origin and its major axis along the line through the origin making an angle γ with the x-axis. Since R is large, $\frac{c^2}{R}$ is small, and we see that the eccentricity of the ellipse is small and will approach zero as R tends to infinity; both semi-axes approach the radius R. The major axis of this ellipse is often called the <u>second axis of the profile</u>. The major and minor axes of the ellipse can be considered as the <u>principal axes of the profile</u>.

Now we have to find out what happens to our solution $w(z')$ for the circle when we apply a transformation (22). Equation (20b) reads

$$w(z') = -Ve^{\alpha i}(z' + \frac{a^2}{z'} e^{-2\alpha i}) + \frac{\Gamma}{2\pi i} \log z' . \qquad (20b)$$

Thus $w(z') = w[f(z)]$ will fulfill our conditions for the motion. At infinity nothing will be changed, and since the circle is a streamline in the z'-plane it will be transformed into a streamline coinciding with the profile. Since $w' = \frac{dw}{dz} = \frac{dw}{dz'}\frac{dz'}{dz}$, we have

$$w' = \left[-Ve^{\alpha i}(1 - \frac{a^2}{z'^2} e^{-2\alpha i}) + \frac{\Gamma}{2\pi i} \cdot \frac{1}{z'}\right](1 + \frac{k}{z^2} + \cdots\cdots).$$

Now $\frac{1}{z'} = \frac{1}{z(1 - \frac{k}{z^2} + \cdots)} = \frac{1}{z}(1 + \frac{k}{z^2} + \cdots) = \frac{1}{z}$ except for terms of order higher than the second. Similarly, $\frac{1}{z'^2} = \frac{1}{z^2}$ except for terms of higher order. Thus if we only write down terms up to $\frac{1}{z^2}$, the above expression for w' becomes

$$w' = \left[-Ve^{\alpha i}(1 - \frac{a^2}{z^2} e^{-2\alpha i}) + \frac{\Gamma}{2\pi i} \cdot \frac{1}{z}\right](1 + \frac{k}{z^2} + \cdots)$$

$$= -Ve^{\alpha i} + \frac{\Gamma}{2\pi i} \cdot \frac{1}{2} + (Va^2e^{-\alpha i} - Ve^{\alpha i}k)\frac{1}{z^2} + \cdots \qquad (25)$$

If we compare this with Equation (12) which we used as a basis for computing the lift and moment, we have $A = -Ve^{\alpha i}$ and $B = \frac{\Gamma}{2\pi i}$ as was obtained previously. Finally,

$$C = V(c_1e^{\alpha i} + c_2e^{-\alpha i}) = -V(ke^{\alpha i} - a^2e^{-\alpha i}),$$

and this gives us

$$c_1 = -k, \quad c_2 = a^2. \tag{26}$$

Thus if we know the center and the radius a of the circle, and the complex constant k (which means knowing c and γ), and the circulation Γ, we can find, according to Section 3, the force and the moment of the force exerted upon the airwing. To find Γ we have to know the point B' into which the singular point B of the profile is mapped. Then the angle determined by the x-axis and $B'M$ is β. We then find Γ by the formula obtained:

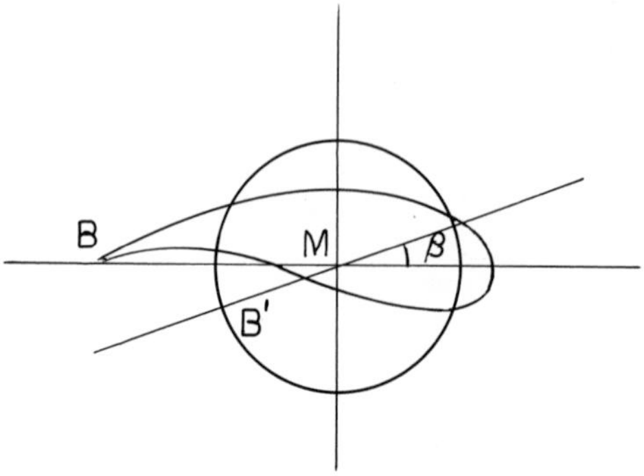

$$\Gamma = 4\pi a V \sin(\alpha+\beta).$$

The direction β was called in Section 3 the direction of the first axis. We see now that the first axis of a simple profile with one singular point B is parallel to the line through B' (mapping point of B) and M, (the center of the circle). The point M is usually called the <u>center of the profile</u>, the radius a of the circle the <u>radius of the profile</u>, and the line $B'M$ itself the <u>first axis of the profile</u>.

Once we know Γ we can obtain the lift force by using Equation (15):

$$L = \rho\Gamma V = 4\pi a \rho V^2 \sin(\alpha+\beta). \tag{15}$$

To obtain the moment M_0, we first show how we find the focus, F, of the profile. The point F is given by $z_0 = -\frac{c_1}{a} e^{-\beta i}$. In our case

$z_0 = \frac{k}{a} e^{-\beta i} = \frac{c^2}{a} e^{(2\gamma-\beta)i}$. To reach this point we draw the first axis through $\overline{MB'}$. Then draw the line MA making the angle γ (half the argument of k) with the x-axis. If we draw ME so that $< EMD$ is twice $< AMD$, and then go out along it a distance $\frac{c^2}{a}$ we will reach the desired point F. We can state this simply by saying that the focus F is at a distance $\frac{c^2}{a}$ from the point M along the line which is the reflection of the first axis in the second axis of the profile. It follows, by the way, that a profile with center of lift, i.e., F lying on the first axis, is determined by the condition that the first and second axes have the same direction, or $\gamma = \beta$.

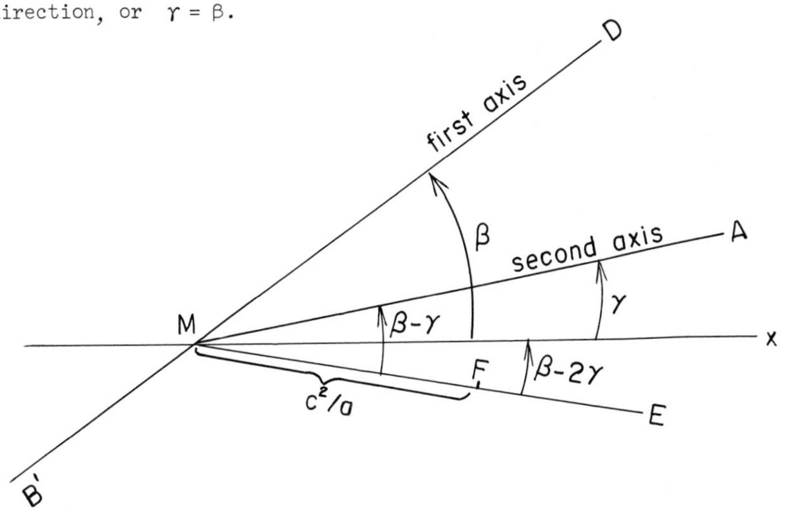

The moment M_0 with respect to this point F was found to be (in (16a)) independent of α and equal to

$$M_0 = 2\pi\rho V^2 \mathscr{R} i(c_2 + c_1 e^{-2\beta i}). \tag{16a}$$

Now $c_2 = a^2$, a real constant, and $c_1 = -k$. Thus

$$M_0 = 2\pi\rho V^2 \mathscr{R}(-ike^{-2\beta i}),$$

and since $k = c^2 e^{2\gamma i}$

$$M_0 = 2\pi\rho V^2 c^2 \sin 2(\gamma-\beta).$$

The distance h_0 of the line ℓ_0 (parallel to the first axis) from F is (see Section 3)

$$h_0 = \frac{M_0}{L_0} = \frac{c^2}{2a} \sin 2(\gamma-\beta),$$

while the distance of F from the first axis is, according to the above construction,

$$\frac{c^2}{a} \sin 2(\beta-\gamma).$$

Hence ℓ_0 is halfway between F and the first axis, or the first axis is the directrix of the metacentric parabola.

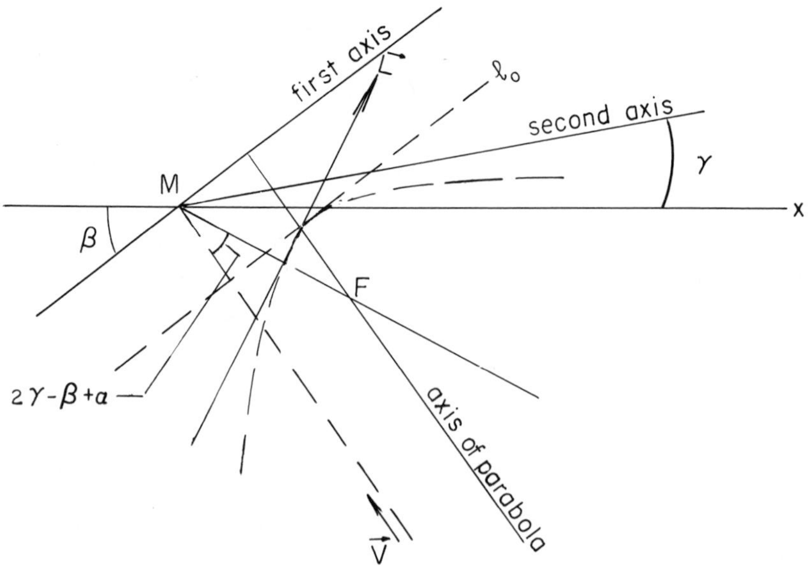

center M of the circle. By the principles of statics M is the moment M_0 about F plus the moment of the force $\vec{L}$ with respect to the point M, when $\vec{L}$ is imagined as acting through F. In our case we have by simple trigonometric formulas

$$\begin{aligned} M &= M_0 + L\frac{c^2}{a}\cos(2\gamma-\beta+\alpha) \\ &= 2\pi\rho c^2 V^2[\sin 2(\gamma-\beta) + 2\sin(\alpha+\beta)\cos(2\gamma-\beta+\alpha)] \\ &= 2\pi\rho c^2 V^2 \sin 2(\alpha+\gamma). \end{aligned}$$

Thus the two formulas which determine the action upon the wing are

$$\begin{aligned} L &= 4\pi\rho a V^2 \sin(\alpha+\beta) \\ M &= 2\pi\rho c^2 V^2 \sin 2(\alpha+\gamma) \end{aligned} \tag{28}$$

where $\alpha + \gamma$ is the angle between $\vec{V}$ and the second axis, $\alpha + \beta$ is the angle between $\vec{V}$ and the first axis, a is the radius of the mapping circle and c^2 is a positive constant.

6. Example of Airwing Sections.

(a) The Joukowski Profiles.

Joukowski used the transformation

$$z = z' + \frac{c^2}{z'} \tag{29}$$

where c^2 is a real positive constant. If we add and subtract $2c$ to both sides of this equation and divide the results we get

$$\frac{z+2c}{z-2c} = \left(\frac{z'+c}{z'-c}\right)^2 . \tag{29a}$$

If we let z' vary in such a way that the argument $\frac{z'+c}{z'-c}$ remains constant = η (see figure), we get the part of the circle passing through c and $-c$ with center on the y-axis which is above the x-axis. For this argument η is equal to the angle between the two radii vectors from $-c$ and c to z'. The argument of $\frac{z+2c}{z-2c}$ is according to (29a) just twice the argument of $\frac{z'+c}{z'-c}$ (modulo 2π). Hence the point z corresponding to z' traces out an arc of a circle through $2c$ and $-2c$ again with center on the y-axis and with angle 2η (for $\eta < \frac{\pi}{2}$) or $2\pi - 2\eta$ (for $\eta > \frac{\pi}{2}$) at z.

When z' traces out the portion of the circle through c and $-c$ which is below the x-axis the angle at z' becomes $\pi - \eta$ so the angle at z becomes $2\pi - 2\eta$ (or 2η) and z, therefore, lies on the same arc of circle through $2c$ and $-2c$. Thus the circle in the z'-plane is transformed into an arc of a circle doubly described or into a slit in the form of a circular arc in the z-plane, giving us a sort of degenerate airwing.

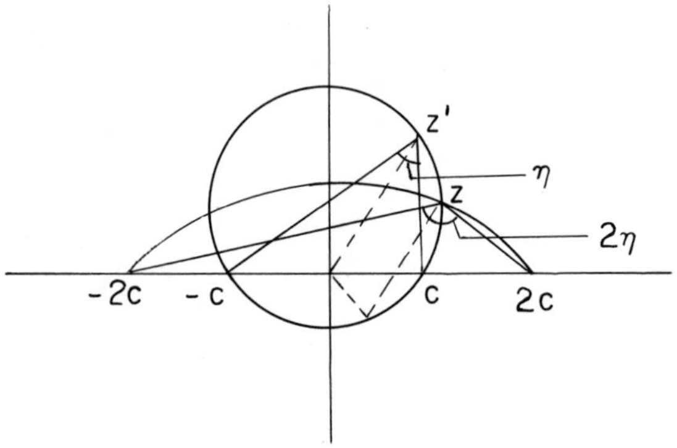

The next figure shows how one point z of the arc corresponds to two points z' and z_1' of the original circle, since for the two points marked we have

$$z' = \frac{c^2}{z_1'}, \qquad z_1' = \frac{c^2}{z'} .$$

The point A' at which the z' circle cuts the positive y'-axis has the coordinate $z' = bi$ where $b = \overline{OA'}$. The corresponding point in the z-plane is

$$z = bi + \frac{c^2}{bi} = i(b - \frac{c^2}{b})$$

$$= i(\overline{OA'} - \overline{OA_1'})$$

since $\overline{OA'} \cdot \overline{OA_1'} = c^2$. This is the point A in which the z circle cuts the positive y-axis. It follows that M', the center of the original circle, is half way between 0 and A. Moreover, we easily see that the center M_1 of the arc through 2c, -2c lies on the normal line to B'M' through B'.

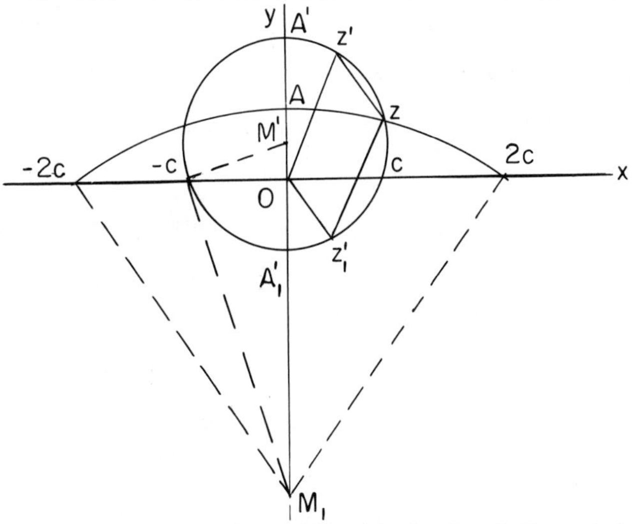

If b = c = a, the center of the circle is at the origin. The circle is then doubly mapped onto the portion of the x-axis joining 2a and -2a by the transformation (29). A point on the iδ' circle has the coordinate $z' = ae^{i\delta'}$ and

$$z = ae^{i\delta'} + \frac{a^2}{a} e^{-i\delta'} = 2a \cos \delta'$$

is real, taking on the values from +2a to -2a as δ' varies from 0 to π or from π to 2π.

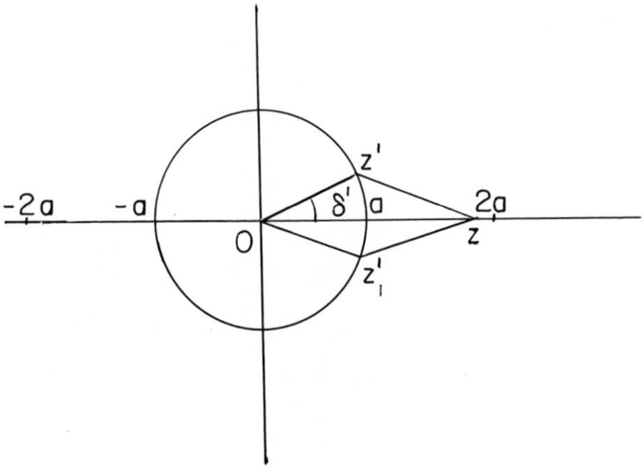

Now in order to obtain a suitable airwing section by conformal transformation from a circle, Joukowski used the following idea. We take the same transformation formula (29) and the same coordinate system, but we apply the transformation to a circle whose center does not lie on the y-axis, but is shifted to the right along the line B'M' in our former figure. Let us call M_1 this former point M' and denote by M the new center. Let the distance $\overline{M_1B'}$ be a_1. We choose the point M near to M_1 only such that $a > a_1$. With M as center we draw a circle of radius a passing through B' which will be tangent to the former circle at B'. Then the point B' is still mapped onto the point B at -2c, but the new circle as a whole is mapped onto a <u>curve which surrounds the slit</u> we found above, as shown in the figure. The nearer M is to M_1 the closer the new transfrom will be to the slit. At B the contour of the new profile must be tangent to the first one, the circular arc, since both circles have a common tangent at B'. Thus the profile or the map of the bigger circle has a cusp at its left end, with tangent coinciding with the tangent of the circular slit. On the other hand, the point +c is now an inner point of the circle mapped. It follows that the image includes the point +2c in its interior. We have thus only <u>one</u> singular point instead of the two at $\pm 2c$ in the case of the slit.

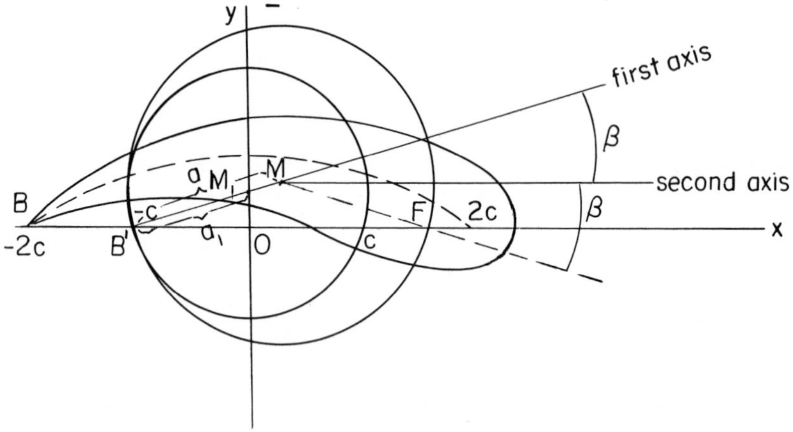

We can get various forms of profiles by varying the thickness and camber. As $M \to M_1$ the profile approaches the circular slit so we get greater thickness by taking larger $\overline{MM_1}$. The ratio $\frac{c}{a_1}$ $(\sim \frac{c}{a})$ which equals $\cos\beta$ determines the concavity or camber. If $a_1 = c$ so that $\frac{c}{a_1} = 1$, we have the case of no camber or the profile is symmetric to the x-axis as shown. For $\frac{c}{a_1} < 1$ we have camber which becomes larger as $\frac{c}{a_1}$ decreases.

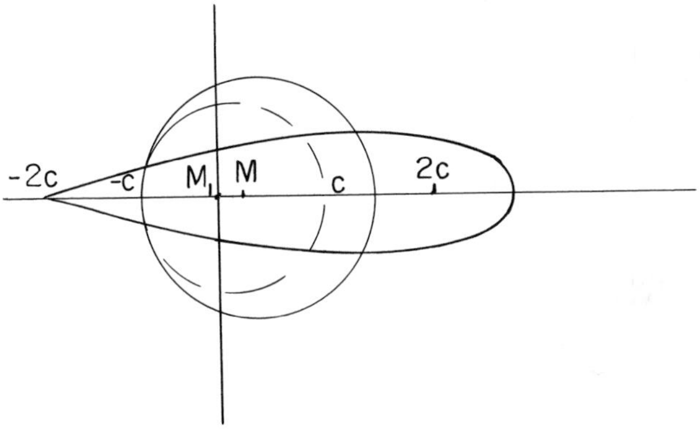

The point B' immediately gives us β and the first axis. Since $k = c^2$ is real, $\gamma = 0$ and the second axis is parallel to the x-axis. We know $\frac{c^2}{a}$ which is $\overline{MF}$ and the angle which $\overline{MF}$ makes with the horizontal, $2\gamma - \beta$, is now $-\beta$. Since $c < a_1 < a$, $\frac{c^2}{a} < a$ so F lies inside the circle. We can then find the metacentric parabola and everything is known about the force and moment.

These results concerning Joukowski profiles can be used for obtaining some approximate estimates for all forms of airwing sections. Actual airwings are similar to thin Joukowski profiles where the distance $\overline{MM_1}$ is small. On the other hand, the camber is usually not large, i.e., $\frac{c}{a}$ is near to 1. Then F is close to the periphery of the circle since $\overline{MF} = \frac{c^2}{a} \sim a$, and the length of the profile is only slightly different from four times the radius of the circle. Hence, we can say that the radius a of the profile is about one quarter of its length and its center lies approximately in the middle of the profile section. As β is small the first axis is approximately parallel to the x-axis, i.e., to the longitudinal axis of the profile. F is the "one-quarter point" about one-quarter of the length of the profile from its leading edge. With β small, the focus and directrix of the metacentric parabola almost coincide so we can say approximately that we have a center of lift.

All these facts hold true precisely when the profile degenerates into a straight slit, which is as we know a limiting case of a Joukowski profile. They are first approximations in all cases where the actual airwing section is not very much different from a straight slit. (Compare the remarks at the beginning of Section 7.)

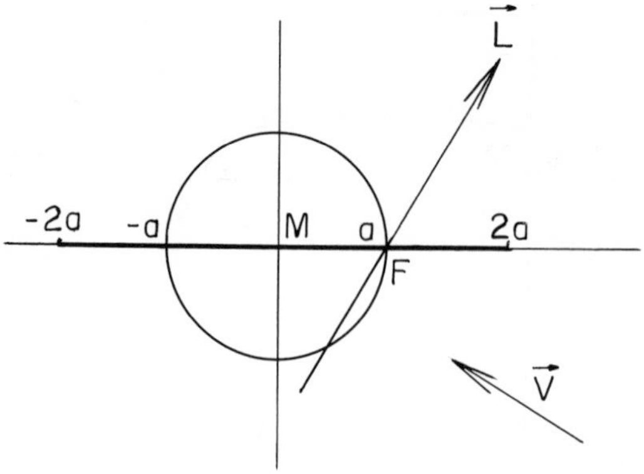

The geometrical construction of a Joukowski profile can be performed in the following way. Let C, the circle through B' with center at M, be the one we desire to map. We can find another circle C' such that the points on it are of the form $\frac{c^2}{z'}$ where z' is a point on C. (Inversion of a circle.) We obtain C' by first drawing OM' so that angle M'OM is twice angle M_1OM, and then using M' as center of the circle through B'. Now to obtain the image of a point P' on C we first find the point P_1' on C' by making angle $P_1'OP'$ twice xOP'. The image of P', the point P is obtained by completing the parallelogram $P'OP_1'P$.

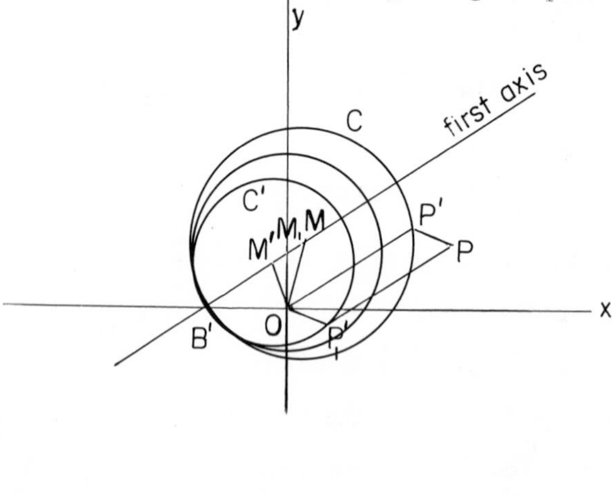

(b) Karman-Trefftz Profiles.

Since the actual construction of airwings with a cusp at the singular point B is not feasible, the Karman-Trefftz profiles get around this difficulty. In this type of profile, instead of a cusp at the point B, the boundaries make a finite angle at the left end point B. To obtain this kind of profile, instead of the transformation (29a), we consider the following, more general transformation

$$\frac{z+c_1}{z-c_1} = \left(\frac{z'+c_2}{z'-c_2}\right)^n , \tag{30}$$

where c_1, c_2 and n are constants. We will consider this transformation for values of n close to 2, and write $n = 2 + \varepsilon$ where ε is small.

As before, if we consider values of z' such that the lines from z' to $-c_2$ and c_2 form the constant angle η, we will get an arc of a circle through points $-c_2$ and c_2 with center on the y-axis. The complementary arc of this circle will subtend the angle $\pi - \eta$ at the points $-c_2$ and c_2. Thus the transform of those points of the circle subtending the angle η, according to (30) will consist of the points z such that the lines from z to $-c_1$ and c_1 form the angle $n\eta$. This again will be an arc of a circle through $-c_1$ and c_1 with center on the y-axis. Now corresponding to those points z' subtending the angle $\pi - \eta$ we get, by the transformation (30), points z subtending the angle $n\pi - n\eta$ (mod 2π). This will also be an arc of a circle through the points $-c_1$ and c_1. However, as $n = 2 + \varepsilon$, $n\pi - n\eta = 2\pi + \varepsilon\pi - n\eta$. Hence

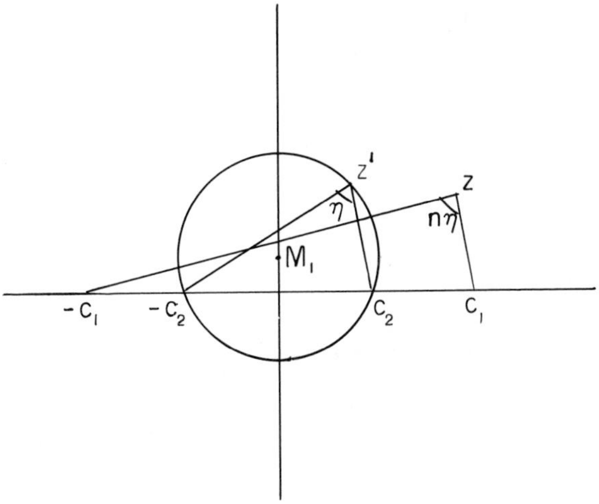

the new arc of a circle through $-c_1$ and c_1 subtends an angle different by $\varepsilon\pi$ (since 2π has no effect) from the angle $n\eta$ subtended by the first arc. Thus as transform of the complete circle through $-c_2$ and c_2 we get a circular sickle, i.e., two circular arcs through $-c_1$ and c_1 making the angle $\varepsilon\pi$ at the points $-c_1$ and c_1.

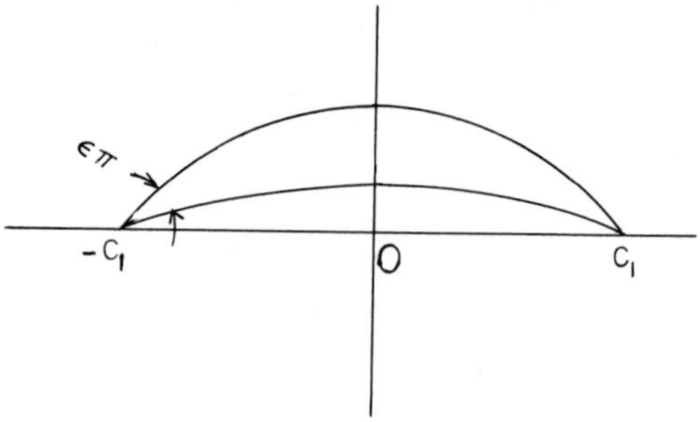

If we now apply the transformation (30) to the circle with center at M going through $-c_2$, (as we did in the case of the Joukowski profile), where M is at a small distance from M_1 along the line $B'M_1$, we will get a profile with a finite angle (of amount $\varepsilon\pi$) at the trailing edge. For, as we saw before, at the point B, the transforms of both circles C and C_1 must have the same tangents

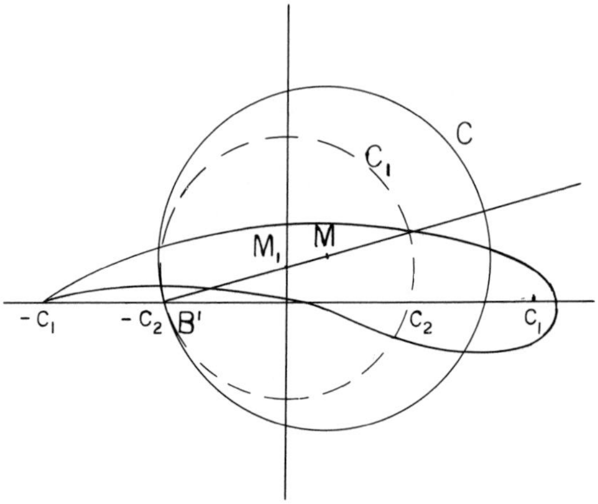

So far we have not yet examined whether the transformation (30) now used fulfills our condition at infinity. This was immediately clear in (29). Now we must see that in (30) also the condition that nothing changes at infinity is satisfied and we must find the value of the constant k (the coefficient of the $\frac{1}{z'}$ term in the expansion). To do this we write the transformation (30) in the form

$$\frac{1 + \frac{c_1}{z}}{1 - \frac{c_1}{z}} = \left(\frac{1 + \frac{c_2}{z'}}{1 - \frac{c_2}{z'}} \right)^n .$$

Expanding this we get

$$1 + \frac{2c_1}{z} + \cdots = 1 + \frac{2nc_2}{z'} + \cdots .$$

Thus to have $z = z'$ at ∞, we must have

$$c_1 = nc_2. \tag{31}$$

To find the value of k, we write (30) in the form

$$\frac{z}{c_1} = \frac{\left(1 + \frac{c_2}{z'}\right)^n + \left(1 - \frac{c_2}{z'}\right)^n}{\left(1 + \frac{c_2}{z'}\right)^n - \left(1 - \frac{c_2}{z'}\right)^n}$$

$$= \frac{2 + n(n-1)\frac{c_2^2}{z'^2} + \cdots}{2n\frac{c_2}{z'} + 2^n \frac{(n-1)(n-2)}{6}\frac{c_2^3}{z'^3} + \cdots} .$$

Thus

$$z = c_1 \left[\frac{z'}{nc_2} + \frac{n(n-1)c_2^2}{2nc_2} \cdot \frac{1}{z'} - \frac{(n-1)(n-2)}{6nc_2}\frac{c_2^2}{z'} + \cdots \right]$$

$$= z' + \frac{1}{3}(n^2-1)\frac{c_2^2}{z'} + \cdots .$$

Hence the coefficient k of $\frac{1}{z'}$ is

$$k = c_2^2 \frac{n^2 - 1}{3} = \frac{1}{3} c_1^2 \frac{n^2 - 1}{n^2} . \tag{32}$$

Therefore, for any assumed value of c_2, we get c_1 and k immediately by (31) and

(32). Then we find the lift L and the moment M with respect to the center, M, which becomes

$$L = 4\pi a\rho V^2 \sin(\alpha+\beta)$$

$$M = 2\pi c_2^2 \frac{n^2-1}{3} \rho V^2 \sin 2\alpha.$$

Here we used the fact that from the value of k we have $\gamma = 0$. The value of ρ is immediately given by the line B'M. A profile with a center of lift is determined by $\gamma = \beta$. Now, if γ and β are both zero we have the case of a symmetric profile. Actually, symmetric profiles (camber zero) give too small a lift force. We thus see that neither the Joukowski nor Karman-Trefftz profiles gives us an example for a cross-section with center of lift.

(c) General Profiles

In the case of the Joukowski profile we saw that the camber and thickness could be varied, but all other quantities (like having a cusp at the singular point) were already determined. For the Karman-Trefftz profile we had one additional parameter which we could control. That turned out to be the size of the angle at the singular point. In the case we are about to consider, we will investigate transformations which allow us many more degrees of freedom. For example, we will, in addition to being able to vary the camber, thickness, and angle at the singular point, be able to have a profile with a center of lift, etc.

We consider a transformation of the form

$$z = z' + \frac{k}{z'} + \frac{k_2}{z'^2} + \cdots + \frac{k_n}{z'^n}. \tag{33}$$

We wish to consider the transform of a circle about the origin. It is evident that the transformation must not possess any singular points outside this circle, and on the other hand we know it must have exactly one singular point on the circle. Let $-c'$ (c' positive, real) be this singular point on the boundary, and $a_1, a_2, \ldots a_n$

be the singular points interior to the circle. Taking the derivative of (33), we obtain, since the singular points are zeros of $\frac{dz}{dz'}$

$$\begin{aligned}\frac{dz}{dz'} &= 1 - \frac{k}{z'^2} - \frac{2k_2}{z'^3} - \cdots - \frac{nk_n}{z'^{n+1}} \\ &= (1 + \frac{c'}{z'})(1 - \frac{a_1}{z'}) \cdots (1 - \frac{a_n}{z'}).\end{aligned} \tag{34}$$

Since the expression for $\frac{dz}{dz'}$ has no term in $\frac{1}{z'}$, we immediately have the condition

$$c' - \sum_{i=1}^{n} a_i = 0, \qquad c' = \sum_{i=1}^{n} a_i. \tag{35}$$

According to this equation, the center of gravity of the n points a_ν is on the positive real axis at a distance $\frac{c'}{n}$ from the origin. The center of gravity of the whole system of singular points including $-c'$ is at the origin. Furthermore, we see from (34), which is an algebraic equation of degree n, that k is minus the sum of all the various products of the zeros of (34) taken two at a time, or

$$\begin{aligned}k &= c'(a_1 + \cdots + a_n) - \sum_{\mu \neq \nu}^{1..n} a_\mu a_\nu \\ &= c'^2 - \sum_{\mu \neq \nu} a_\mu a_\nu.\end{aligned} \tag{36}$$

Thus the procedure of finding a general wing section will be as follows. We choose a set of a_ν values and a circle of radius a passing through $-c'$, large enough so that all singular points a_ν are interior to it. Then we have the center M, the radius a and the angle β (between $\overline{B'M}$ and the x-axis), and by means of (36) we find k. The values a_ν determine the right hand side of (34) which on being integrated gives us the function z of z'. In this way the points z of the profile derived from the points z' of the circle can be found successively. The result is that we have a profile with its corresponding constants M, a, β, k. For $n = 1$ this procedure leads back to the Joukowski profiles, since $a_1 = c$ according to (35) and $k = c^2$ according to (36).

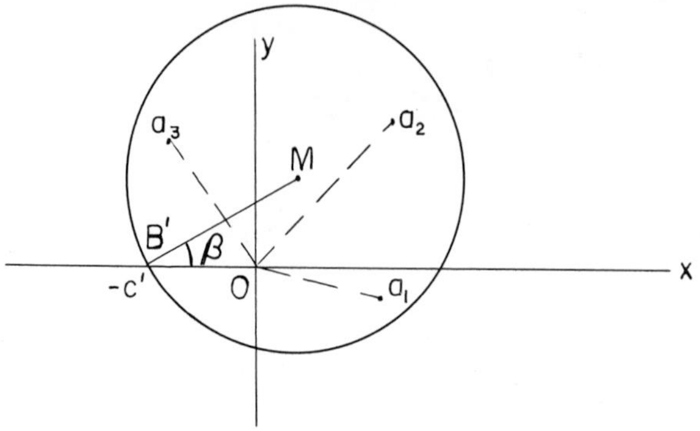

In particular, this procedure can be used in order to find cross sections with a center of lift, the metaccentric parabola being reduced to a point. We have only to satisfy the condition $\beta = \gamma$, where γ is half the argument of the expression (36). Take, for instance, $n = 3$ and select for a_1, a_2, a_3 the very special values

$$a_1 = c' \qquad a_2 = c''e^{\theta i} \qquad a_3 = -c''e^{\theta i}.$$

Then (35) is automatically fulfilled while (36) gives us

$$k = c'^2 + c''^2 e^{2\theta i}.$$

The number

$$\frac{k}{c'} = c' + \frac{c''^2}{c'} e^{2\theta i}$$

has the same argument 2γ as k. We get the angle 2γ (see figure) by making $\overline{OG} = \frac{c''^2}{c'}$ and $< xOG$ twice the angle xOa_2. Bisecting

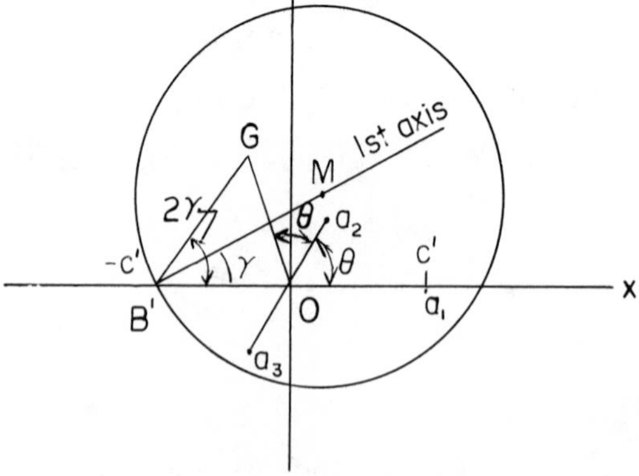

the angle OB'G we get γ, and we will have $\gamma = \beta$, if we take this bisector for the first axis on which the center M of the circle lies. Using (34) we obtain

$$\frac{dz}{dz'} = \left(1 - \frac{c'^2}{z'^2}\right)\left(1 - \frac{c''^2 e^{2\theta i}}{z'^2}\right)$$

$$= 1 - \frac{c'^2 + c''^2 e^{2\theta i}}{z'^2} + \frac{c'^2 c''^2 e^{2\theta i}}{z'^4} .$$

Integrating this equation gives us the transformation required:

$$z = z' + \frac{c'^2 + c''^2 e^{2\theta i}}{z'} - \frac{(c'c''e^{\theta i})^2}{3z'^3} .$$

This checks the value for k and gives besides

$$k_2 = 0 \qquad k_3 = -\frac{1}{3}(c'c''e^{\theta i})^2 \qquad k_4 = \cdots = 0.$$

Thus we get a certain type of profile with center of lift. It turns out that such profiles for which $\beta = \gamma$ have in general the so-called S-form. That is the lower sides of such profiles as well as their middle line have a point of inflection near to the trailing edge. With this slight lift in the sharp tail the wing is more stable, as was well known even before the theory was developed. The Joukowski and Karman-Trefftz profiles all had $\beta > \gamma$ so their metacentric parabolas all open downwards. Such a situation is unfavorable to stability, since it is then harder to make the metacenter lie above the center of

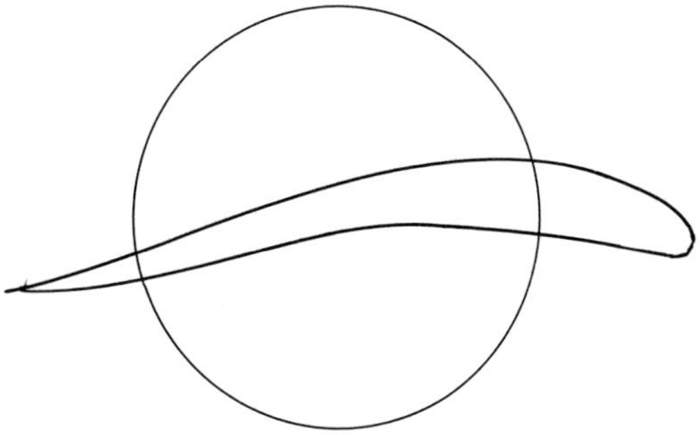

gravity for all angles of attack. The direction of opening of the parabola can in these cases be reversed by adding a tail surface.

7. The Direct Problem - Theory of Thin Wings.

Now we want to turn to the direct problem in the theory of airwings. Up to now we assumed a transformation of a circle and sought the profile into which the circle was mapped. The real problem is of course the opposite: We have to consider a given cross-section and to find the lift force acting on it. This can be done if we find the values of a and M for the circle into which the profile can be transformed as well as the point B' and the coefficient k of the transformation. Such a solution can only be worked out by an approximation method. However, some precise statements can be made about these quantities we are looking for, which we present in the form of three theorems without proof.

Theorem I. If any region of the complex plane bounded by a closed curve is given, and we map the outside of this curve onto the outside of a circle by a schlicht (simple) transformation in such a way that the infinity is unchanged, then the radius a of the circle satisfies the inequality $a \geq \frac{d}{4}$ where d is the diameter (the maximum of all distances joining any two points) of the closed region. Moreover, we have $a = \frac{d}{4}$ only for a straight slit; in all other cases $a > \frac{d}{4}$. This is a consequence of the distortion theorem of function theory. In the case of an airfoil we usually find d as the radius of a circle whose center is at the singular point B and is tangent to the nose of the profile. This theorem is included in the next.

Theorem II. If the circle onto which the profile is mapped has the radius a, then a circle of radius $2a$ but with the same center includes the whole of the profile. Thus the center of the circle is somewhat near to the middle of the profile. This theorem again is included in the following.

Theorem III. The circle of radius 2a includes, not only the whole profile, but also the double of any circle which is completely interior to the profile. Thus the only point of the profile which can lie on the circle of radius 2a is the singular point B. The circle must pass beyond the leading end by a distance which increases with the thickness of the profile at this end. Hence, the center of the mapping circle is actually shifted toward the leading edge from the middle of the profile.

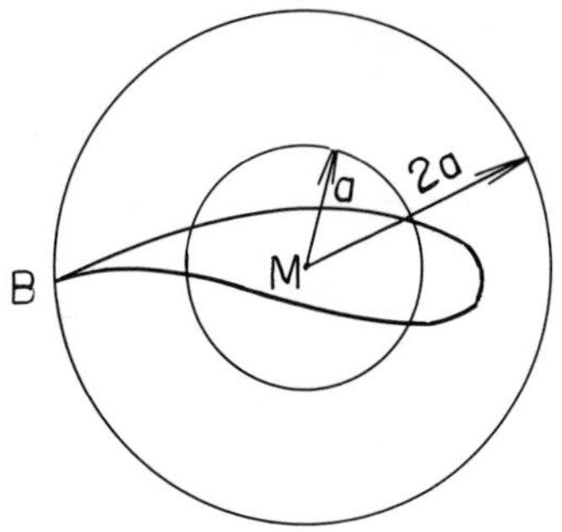

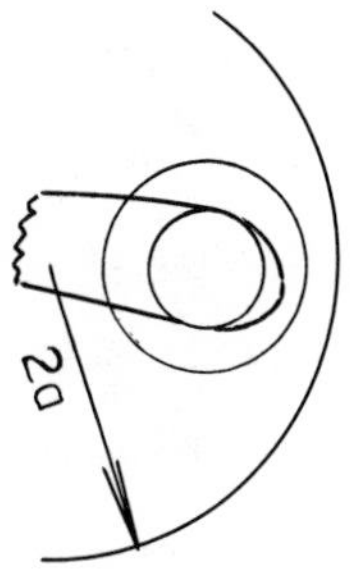

Another theorem could be stated which indicates that the focus F is interior to the mapping circle of radius a and not too far away from its periphery.

Real airwings are thin without very much camber so do not differ greatly from a straight slit. Thus a is not much different from $\frac{d}{4}$. If ℓ is the chord or width of a profile of this type then the lift force is almost

$$\pi\ell\rho V^2 \sin(\alpha+\beta)$$

where 4a is replaced by ℓ, and this lift acts approximately at the "one-quarter" point F.

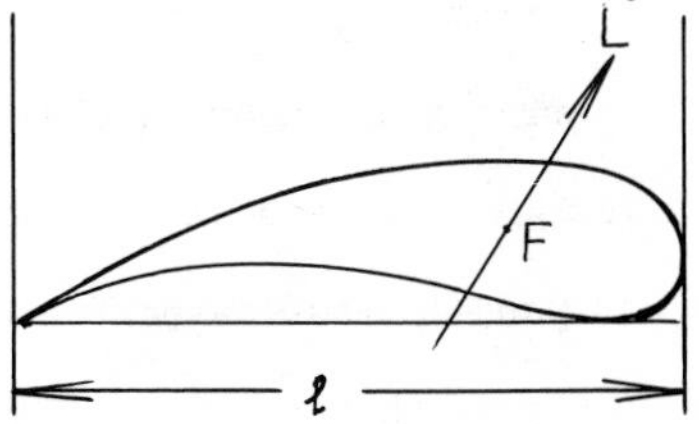

The problem of finding the mapping function for a given region is equivalent to the Riemann problem (the first problem of potential theory) for the outside of the domain. We will use here a special method of solution adapted to our case, profiting by the fact that all profiles are somewhat similar to each other. If a definite profile is given, we can easily choose another profile from among all those already considered in the inverse problem for which we know the solution. E.g., if the given profile has a cusp we use as an auxiliary profile a Joukowski profile, chosen so that at B the cusps will coincide and both profiles have as nearly the same shape and size as possible. (If the angle at B is finite we can use instead a Karman-Trefftz profile.) We know the mapping circle and the transformation function for this auxiliary profile. When we apply the transformation which carries the auxiliary profile into a circle, the given profile will be mapped onto a curve whose shape is nearly that of a circle. In the figure P is the given profile and P_1 the auxiliary profile; the transformation which maps P_1 onto the circle C_1 will map P onto the curve K.

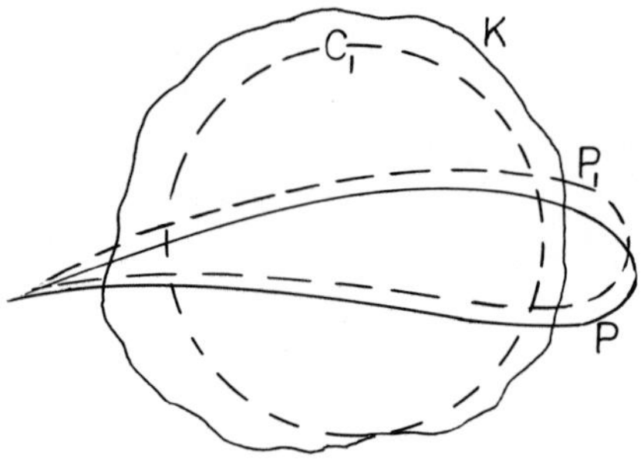

The problem thus reduces to the mapping of this near circle K onto a circle. The latter need not be the same as the mapping circle of the auxiliary profile; in general, it cannot if the conditions at infinity are to be satisfied. The product of the two transformations, the first carrying the profile into a near circle K and the second carrying the near circle K into a circle, give the transformation of the direct problem. In actual practice where we have thin, flat profiles, a straight slit is used in most cases for the auxiliary profile. (Thin-wing theory.)

Let us assume that the nearly circular curve K into which our given profile is mapped (by the transformation corresponding to the auxiliary profile) has as its polar equation

$$r = a'[1 + \eta(\phi)]. \tag{37}$$

Here a' is the radius of the circle C_1 which maps the auxiliary profile and $|\eta|$ is small. r and ϕ are the coordinates shown in the figure.

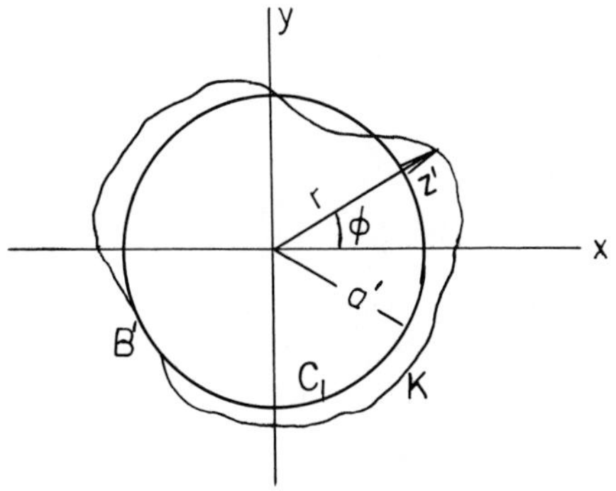

Since $\eta(\phi)$ is periodic of period 2π we express it in the form of a Fourier series

$$\eta(\phi) = a_0 + a_1 \cos \phi + a_2 \cos 2\phi + \cdots$$
$$+ b_1 \sin \phi + b_2 \sin 2\phi + \cdots$$

The singular point plays no role for $\eta(\phi)$ as the profiles coincide at their singular points and both neighborhoods are mapped onto the same element of the circle.

Our first problem is to find a transformation which maps the outside of the circle $r = a'$ onto the outside space of the curve (37). This transformation will, in general, not satisfy our condition at infinity; we will take care of that later. Now we are going to prove that the transformation (which we will denote by S) defined by

$$S: \quad z - z' = z'[a_0 + (a_1+b_1 i)\frac{a'}{z'} + (a_2+b_2 i)\left(\frac{a'}{z'}\right)^2 + \cdots] \tag{38}$$

has approximately (i.e., for small η) the required property. First, the right hand side is convergent for all $|z'| > a'$. Next multiply this equation by $e^{-\phi i}$, and we will get for the transform of the circle, on which $z' = a'e^{\phi i}$ (see last figure),

$$ze^{-\phi i} - a' = a'[(a_0 + a_1 \cos\phi + b_1 \sin\phi + a_2 \cos 2\phi + \cdots) - i(a_1 \sin\phi - b_1 \cos\phi + a_2 \sin 2\phi - b_2 \cos 2\phi \cdots)]. \tag{38a}$$

Hence, secondly $\mathscr{R}(ze^{-\phi i} - a') = a'\eta(\phi)$ or

$$\mathscr{R}(ze^{-\phi i}) = a'[1+\eta(\phi)] = r \tag{38b}$$

for points mapping the circle. On the other hand

$$\begin{aligned} \mathrm{Im}(ze^{-\phi i} - a') &= \mathrm{Im}(ze^{-\phi i}) \\ &= -a_1 \sin\phi - a_2 \sin 2\phi - \cdots \\ &\quad + b_1 \cos\phi + b_2 \cos 2\phi + \cdots \quad . \end{aligned} \tag{38c}$$

We see that the transformation (38) is such that the real part of the vector z when turned by an amount ϕ equals the radius vector of the curve (37). This means that if P is a point z of the transform of the circle corresponding to $z' = re^{i\phi}$, then the projection of P upon the ray making an angle ϕ with the x-axis is a point of the curve (37). In other words, if

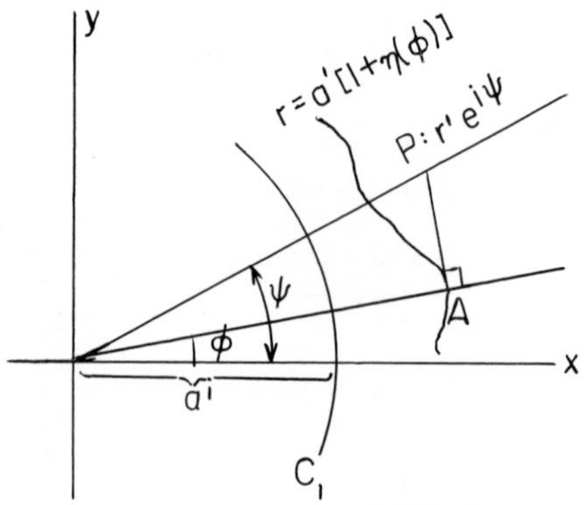

$$z = r'e^{i\psi} \quad \text{then} \quad ze^{-i\phi} = r'e^{i(\psi-\phi)}$$

and

$$r = \mathscr{R}(ze^{-i\phi}) = r' \cos(\psi - \phi).$$

As we assume that the curve (37) is close to the circle and has tangents close to those of the circle, (i.e., $\eta'(\phi)$ as well as $\eta(\phi)$ is small), then the distance of the point P from the curve (37) (see figure) will be of the second order. Thus we can say that (38) maps the circle into the curve (37) approximately, i.e., except for terms of higher order in η and η'.

We can reformulate the "direct problem" in the following symbolic manner. Let P be the given profile. Then we desire to find a circle C and a "schlicht" (simple) transformation T of the form

$$z = z' + \frac{k}{z'} + \cdots \tag{39}$$

such that

$$T(C) = P. \tag{40}$$

We denote the auxiliary profile by P_1 and since P_1 is similar to P, we write $P_1 \sim P$. This P_1 was selected from among those profiles for which we know a transformation T_1 of the form

$$z = z' + \frac{k'}{z'} + \cdots ,$$

and a circle C_1, such that

$$T_1(C_1) = P_1. \tag{40a}$$

Or denoting the inverse of T_1 by T_1^{-1}, we may write

$$T_1^{-1}(P_1) = C_1.$$

Since C_1 is known we represent the center M by the position vector m' and let the radius be a'. The process described above says that we apply the transformation T_1^{-1} to P, obtaining a curve K which was represented by (37). Symbolically

$$T_1^{-1}(P) = K, \ T_1(K) = P.$$

If we denote the transformation (38) by S, we have

$$S(C_1) \sim K,$$

but we note that S is not of the form (39). To get around this difficulty we shall introduce a new transformation S' such that

$$S'^{-1}(C_1) = C_2, \ S'(C_2) = C_1 \tag{41}$$

where C_2 is again a circle. The object of introducing this transformation is to have the transformation

$$T_1' = SS'$$

be of the form (39). If this is done, then

$$T_1'(C_2) = SS'(C_2) = S(C_1) \sim K.$$

Applying the transformation T_1 to this last equation and denoting by T_2 the product of the transformations T_1, T_1', we get

$$T_2(C_2) = T_1T_1'(C_2) \sim T_1(K) = P.$$

Hence the transformation T_2 and the circle C_2 give an approximate solution to the direct problem.

If we desire greater accuracy we can denote $T_2(C_2)$ by P_2 and, taking P_2 instead of P_1 as the auxiliary profile, repeat the above process. We will get a T_3 and C_3 which will supply a closer approximation. However, in actual computation it turns out that this process applied once gives sufficient accuracy in most cases.

In order to carry out the process described the next step is to find the transformation S' transforming a circle C_2 into the given circle C_1 so that it satisfies the conditions stated, namely that $SS' = T_1'$ is of the form (39). As S' takes the circle C_2 into the circle C_1 it may be of the form

$$S': \quad z' = \frac{a'}{a}(z''-m''),$$

where the circle C_2 has the radius a and the center given by the position vector m''. Since the transformation S is given by (39), T_1' or the product of S and S' becomes

$$z - \frac{a'}{a}(z''-m'') = \frac{a'}{a}(z''-m'')\left[a_0 + (a_1+b_1 i)\frac{a'}{\frac{a'}{a}(z''-m'')} + (a_2+b_2 i)\frac{a'^2}{\frac{a'^2}{a^2}(z''-m'')^2} + \cdots\right]. \tag{42}$$

$$T_1': \quad z = \frac{a'}{a}(1+a_0)z'' + a'(a_1+b_1 i) - \frac{a'}{a}(1+a_0)m'' + aa'\frac{a_2+b_2 i}{z''} + \cdots .$$

In order for (42) to be of the form (39), the following two conditions must hold:

(1) $\frac{a'}{a}(1+a_0) = 1$

(2) $a'(a_1+b_1 i) = \frac{a'}{a}(1+a_0)m''$ (= m'' by (1)).

From the first of these two conditions we find that the radius a of C_2 is given by

$$a = a'(1+a_0) = a' + \frac{a'}{2\pi}\int_0^{2\pi} \eta(\phi)d\phi \tag{43a}$$

since a_0 is the constant term of the Fourier expansion of $\eta(\phi)$. If $\eta(\phi)$ is small, then a_0 will be small, and a is not much different from a'. The second condition states that the center M of C_2 is given by

$$m' + m'' = m' + a'(a_1+b_1 i) = m' + \frac{a'}{\pi}\int_0^{2\pi} \eta(\phi)e^{i\phi}d\phi. \tag{43b}$$

If $\eta(\phi)$ is known, we see in this way that the circle C_2 is determined.

We now investigate the quantity k of the transformation T_2. By property (4) of page 79 we saw that the product of two such transformations has as its coefficient of $\frac{1}{z}$ the sum of the corresponding coefficients for each of the transformations. If we let k be this coefficient for T_2, then $k = k' + k''$, where k' is the coefficient in T_1 and k'' that in T_1'. Then we have

$$\begin{aligned} k &= k' + k'' = k' + aa'(a_2+b_2 i) \\ &= k' + \frac{aa'}{\pi}\int_0^{2\pi} \eta(\phi)e^{2\phi i}d\phi. \end{aligned} \tag{43c}$$

Thus we see that everything we need for determining the lift and moment is known except the direction of the first axis, i.e., the corrected value of β. Let β' be the direction of the first axis for the transformation T_1 and the circle C_1, and let β'' be the additional correction needed to get the β-value for T_2 and C_2 so as to have $\beta = \beta' + \beta''$. The transformation T_1 takes the singular point B of both P and P_1 into the point B' of C_1 since the profiles coincide at that point. Moreover, K and C_1 have the same tangents at this point since P_1 and P coincide in direction at B. The line B'M gives the value of β'. Now the transformation S' is a pure translation plus contraction so as to preserve angles. However, the transformation S (given by (38)) rotates the point B'. The amount of this rotation is obtained by taking the imaginary part of (38a) (see figure on page 102):

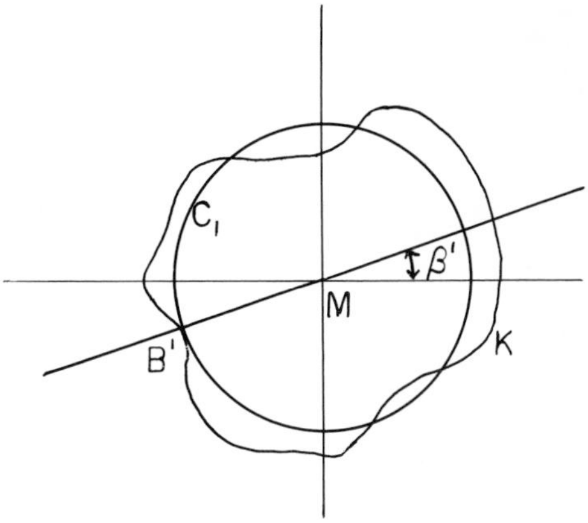

$$a'(\psi-\phi) \sim \mathrm{Im}(ze^{-i\phi} - a') = \mathrm{Im}(ze^{-i\phi}).$$

The quantity $\psi - \phi$ is the angle by which a radius vector to a point of the circle is turned if it is to become a radius vector of a point of K. Thus in the transformation of the curve K into the circle, the radius vector rotates by $-(\psi-\phi)$. It follows that the correction β'' is

$$\beta'' = -\frac{1}{a'}\,\mathrm{Im}(ze^{-\phi i} - a') \quad \text{for} \quad \phi = \pi + \beta',$$

as the argument for the point B' is given by $\pi + \beta'$. According to (38c) this last expression for β'' can be written

$$\beta'' = \sum_{n=1}^{\infty} a_n \sin n\phi - \sum_{n=1}^{\infty} b_n \cos n\phi \quad \text{for} \quad \phi = \pi + \beta'.$$

In order to find the value of this series for β'' without having to find all the integrals for a_ν and b_ν we resort to a transformation of the series expansion of η. Since

$$\eta(\phi) = a_0 + \sum_{\nu=1}^{\infty} a_\nu \cos \nu\phi + \sum_{\nu=1}^{\infty} b_\nu \sin \nu\phi$$

we have

$$\frac{1}{2}[\eta(\phi+\xi) - \eta(\phi-\xi)] = -\sum_\nu a_\nu \sin \nu\phi \sin \nu\xi$$
$$+ \sum_\nu b_\nu \cos \nu\phi \sin \nu\xi .$$

Multiply both sides of this equation by $\cot \frac{\xi}{2} \, d\xi$ and integrate from 0 to π term by term. Using the formula

$$\int_0^\pi \sin n\xi \cot \frac{\xi}{2} \, d\xi = \pi$$

we then have

$$\frac{1}{2}\int_0^\pi [\eta(\phi+\xi) - \eta(\phi-\xi)] \cot \frac{\xi}{2} \, d\xi = \pi \left[-\sum_\nu a_\nu \sin \nu\phi + \sum_\nu b_\nu \cos \nu\phi \right] .$$

The integral on the left is defined if $\eta(\phi)$ has a derivative at the point ϕ, which we assume. On the other hand the existence of the single integrals

$$\int_0^\pi \eta(\phi \pm \xi) \cot \frac{\xi}{2} \, d\xi$$

requires in addition that $\eta(\phi) = 0$ at the point ϕ considered. This is so at the point B' since the curve K cuts the circle C_1 at B'. Thus we are justified in using the following transformations

$$\int_0^\pi \eta(\phi-\xi) \cot \frac{\xi}{2} \, d\xi = \int_0^{-\pi} \eta(\phi+\xi) \cot \frac{\xi}{2} \, d\xi$$
$$= -\int_{-\pi}^{0} \eta(\phi+\xi) \cot \frac{\xi}{2} \, d\xi = -\int_{\pi}^{2\pi} \eta(\phi+\xi) \cot \frac{\xi}{2} \, d\xi .$$

The limits $-\pi, 0$ can be replaced by $\pi, 2\pi$ since the integrand is periodic of period 2π. Thus if $\eta(\phi) = 0$ we have

$$-\sum_\nu a_\nu \sin \nu\phi + \sum_\nu b_\nu \cos \nu\phi = \frac{1}{2\pi}\int_0^{2\pi} \eta(\phi+\xi) \cot \frac{\xi}{2}\, d\xi .$$

Set $\phi = \pi + \beta'$, which is the value of ϕ at the point B', and replace $\xi + \pi$ by x. Then we finally have for β'' the single integral

$$\beta'' = \frac{1}{2\pi}\int_0^{2\pi} \eta(\beta'+x) \tan \frac{x}{2}\, dx. \tag{43d}$$

We summarize as follows: The circle C_2, with radius a and center at m, is the second approximation of the map of the given profile P. The transformation T_2 has as its first coefficient k, while the point on C_2 into which the singularity of P is transformed determines the angle β with the x-axis. These four quantities a, m, k and β are given by the four equations:

$$\left.\begin{aligned} a &= a' + \frac{a'}{2\pi}\int_0^{2\pi} \eta(\phi)d\phi \\ m &= m' + \frac{a'}{\pi}\int_0^{2\pi} \eta(\phi)e^{\phi i}d\phi \\ k &= k' + \frac{aa'}{\pi}\int_0^{2\pi} \eta(\phi)e^{2\phi i}d\phi \\ \beta &= \beta' + \frac{1}{2\pi}\int_0^{2\pi} \eta(\beta'+\phi) \tan \frac{\phi}{2}\, d\phi \end{aligned}\right\} \tag{43}$$

It must be remembered that in (43) the primed quantities refer to the circle C_1 and the transformation T_1 which carries C_1 into the auxiliary profile P_1, while the function $\eta(\phi)$ is defined by the polar equation (37) of the curve K into which the profile P is carried by T. If the profile P is given by means of its graph, the curve K can be found point by point and then the six integrals in (43) are evaluated by numerical or graphical methods.

The simplest way in which this method can be applied is that used in the so-called thin-wing theory. Here one assumes that the auxiliary profile P_1 is nothing else than a straight line from the trailing edge to the leading edge of the profile. The following assumptions are made (see figure)

(1) The profile is thin with little camber.

(2) The angle of attack α is small.

(3) The contour of the profile can be represented by a double-valued function $y(x)$, in the range $-\ell/2 \leq x \leq \ell/2$ such that $|y|$ has only small values.

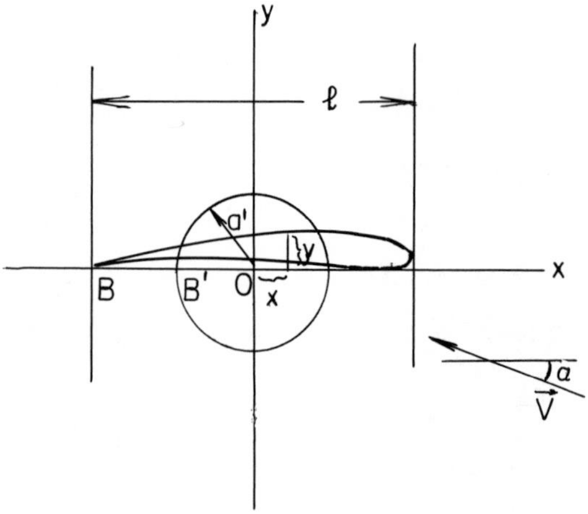

As the figure shows, we let the x-axis from $-\ell/2$ to $\ell/2$ be the auxiliary profile. Then with the axes taken as shown, the transformation

$$z' = z + \frac{c^2}{z}, \quad c = \frac{\ell}{4}$$

transforms the straight line into a circle C_1 with radius $a' = c$ and center at the origin so that $m' = 0$. Thus $a' = \frac{\ell}{4}$ while $k' = (\frac{\ell}{4})^2$ (so $\gamma' = 0$) and $\beta' = 0$. We substitute these values into the expressions (28)

$$L = 4\pi\rho a V^2 \sin (\alpha+\beta)$$

$$M = 2\pi\rho c^2 V^2 \sin 2(\alpha+\gamma)$$

for the lift and moment. At the same time using our second assumption that α is small, neglecting terms of higher order we get

$$L = \pi\ell\rho V^2 (\alpha+\beta'')$$

$$M = \frac{\pi\ell^2}{4} \rho V^2 (\alpha+\gamma'').$$

Here $\beta = \beta''$ and $\gamma = \gamma''$ since $\beta' = \gamma' = 0$. Since the center of the circle C_2 about which we take the moment M is shifted by an amount of the first order from the center of C_1, the origin, this moment differs from the moment about the origin by an amount of the second order, L times the shift. Thus except for terms of higher order than the first, we can assume that the above value of M holds for the moment about the origin.

In this way, of the six integrals in (43), we need only evaluate the two for β'' and γ''. For β'' we have

$$\beta'' = \frac{1}{2\pi} \int_0^{2\pi} \eta(\xi) \tan \frac{\xi}{2}\, d\xi$$

since $\beta' = 0$. Now $k' = c^2$, and k'' is a small complex number so

$$2\gamma'' = 2\gamma = \frac{1}{c^2} \operatorname{Im} k''$$

and

$$\gamma'' = \frac{\operatorname{Im} k''}{2c^2} .$$

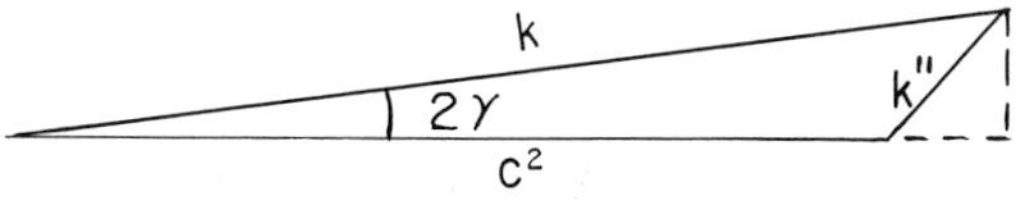

With $c^2 = a'^2$ this gives us

$$\gamma'' = \frac{1}{2\pi} \int_0^{2\pi} \eta(\xi) \sin 2\xi\, d\xi .$$

To find $\eta(\phi)$ we have to use the transformation which carries the slit into the circle and hence the profile into the curve K near to the circle. To the point with angle ϕ on the circle corresponds the point $x = 2a' \cos \phi$ on the x-axis. (See figure) Now in a conformal transformation angles are preserved and each infinitesimal region is carried over into a similar region. Hence a short line element normal to the slit at x will be carried over into another short element normal to the circle at z', and the ratio of these two elements will be the value of $\frac{dz}{dz'}$ for $z' = a'e^{\phi i}$ or

$$\frac{dz}{dz'} = 1 - \frac{c^2}{z'^2} = 1 - e^{-2\phi i}$$

$$= 2 \sin \phi \cdot ie^{-\phi i}.$$

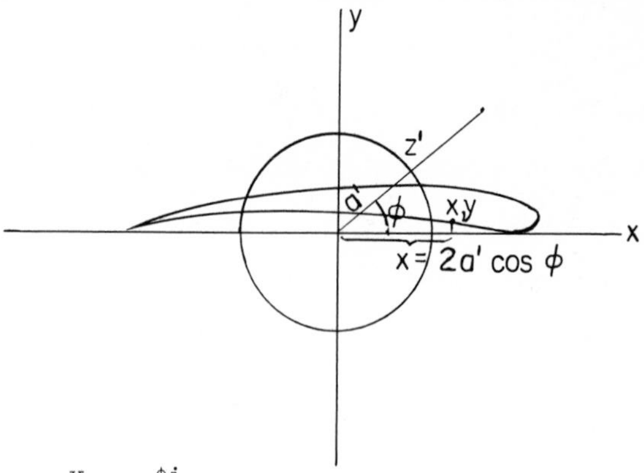

The small element yi is transformed into the segment

$$yi \cdot \frac{dz'}{dz} = \frac{y}{2 \sin \phi} e^{\phi i}$$

whose direction is ϕ, as we know, and its length $\frac{y}{2 \sin \phi}$ is $a'\eta(\phi)$. Since $x = 2a' \cos \phi$, this gives us for η the expression

$$\eta(\phi) = \frac{y(2a' \cos \phi)}{2a' \sin \phi} .$$

Then

$$\beta'' = \frac{1}{4\pi a'} \int_0^{2\pi} \frac{y(2a' \cos \xi)}{\sin \xi} \tan \frac{\xi}{2} \, d\xi$$

$$\gamma'' = \frac{1}{2\pi a'} \int_0^{2\pi} y(2a' \cos \xi) \cos \xi \, d\xi$$

giving for a thin profile the result

$$\left.\begin{aligned} L &= \pi \ell \rho V^2 \left(\alpha + \frac{1}{4\pi a'} \int_0^{2\pi} y(2a' \cos \xi) \frac{\tan \frac{\xi}{2}}{\sin \xi} \, d\xi \right) \\ M &= \frac{1}{4} \pi \ell^2 \rho V^2 \left(\alpha + \frac{1}{2\pi a'} \int_0^{2\pi} y(2a' \cos \xi) \cos \xi \, d\xi \right) \end{aligned}\right\} \qquad (44)$$

The first two assumptions which we have just used can be considered sufficiently correct. The third assumption, however, introduces errors which may cause the value of L to be wrong by as much as 15 to 20 per cent. This is due to the fact that we have used as auxiliary profile the straight line running from the trailing edge to the end of the nose. It would be correct to take the line joining the trailing edge to a point just inside the nose, which would correspond to our second singular point B_1. Then to find the part of the curve K into which the portion of the nose to the right of the point B_1 is mapped, we cannot use a function $y(x)$ but must resort to a direct point by point carrying out of the transformation, then obtaining $\eta(\phi)$ from the resulting curve K.

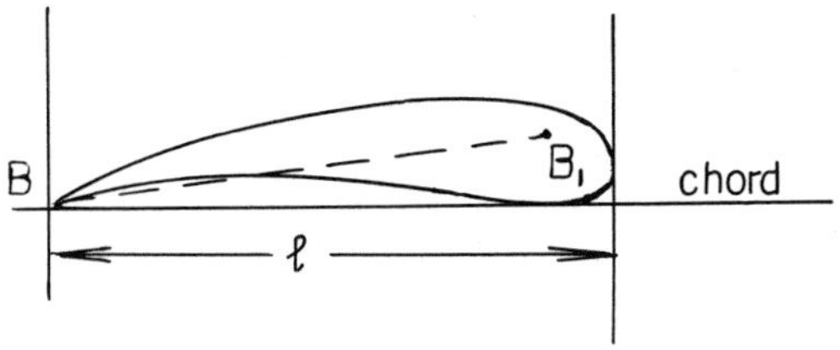

Problem 8. Plot the Joukowski profile corresponding to

$$a - a_1 = \frac{a_1}{5}$$

$$\frac{c}{a_1} = \cos\beta = \frac{4}{5} .$$

Find the focus F and plot the metacentric parabola. Find L, its line of action and M for $\alpha = 5^{\circ}$.

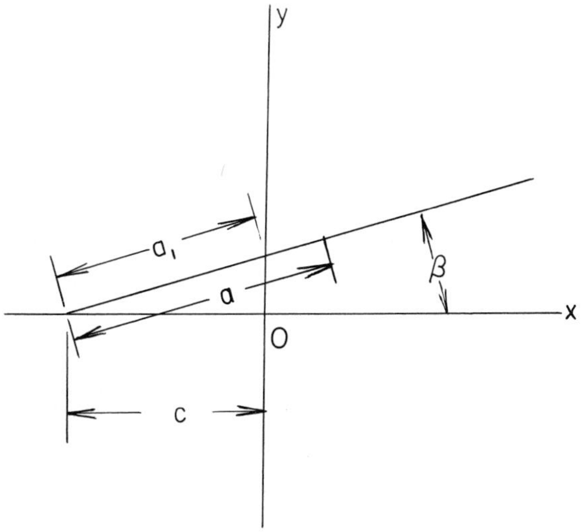

Problem 9. Try to find a suitable expression for the pressure distribution (or the distribution of q^2) for a Joukowski profile.

Problem 10. Given that a circle through $z' = -c_1$ ($c_1 > 0$ is real) is transformed into a profile by

$$z = z' + \frac{b^2 + c_1^2}{z'} - \frac{b^2 c_1^2}{3z'^3}$$

where b is a complex number.

(a) Show that $z' = -c_1$ is a singular point of the transformation and find all other singular points.

(b) If $m = \kappa b$, κ real constant, is the center of the circle show that $(m = x_0 + iy_0)$

$$x_0^2 + y_0^2 + c_1(1-\kappa^2)x_0 = \kappa^2 c_1^2$$

is the condition for the profile to have a center of lift.

(c) Find the metacentric parabola when $\kappa = \frac{1}{3}$, $x_0 = y_0 = \frac{c_1}{6\sqrt{2}}$.

(d) Find (by computation or graphically) some of the points on the profile and plot its general course for the values given in (c).

8. Final Remarks.

(a) Experimental Verification of the Lift Formula

In actual wind tunnel experiments we must use a finite wing-section, i.e., a wing model of finite span (see figure). However, by placing walls (w_1 and w_2 in the figure) at the ends of the model, we can make the flow go along parallel curves, and thus enforce a motion which approximates two dimensional flow to a high degree.

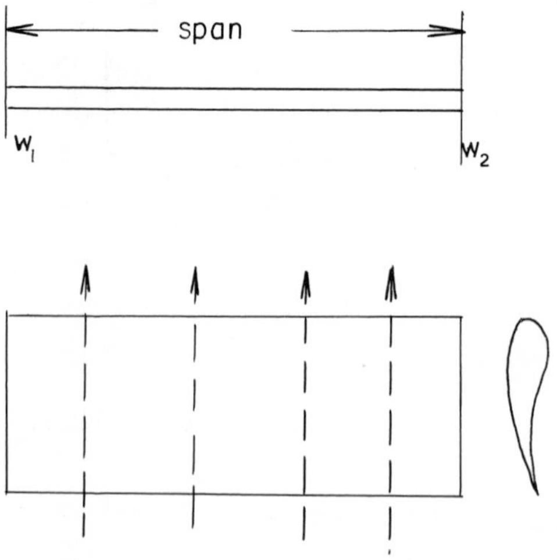

It is customary to present the results of testing

wings in terms of dimensionless coefficients. We introduce first a left coefficient, C_L, which is defined as:

$$C_L = \frac{L}{\frac{\rho}{2} V^2 A}, \tag{45}$$

where L is the lift force and A is the area of the top surface of the wing. This coefficient C_L is dimensionless since $\frac{\rho}{2} V^2$ is a pressure, and thus both numerator and denominator of (45) has the dimension of force. The quantity $\frac{\rho}{2} V^2$ is known as the stagnation pressure (i.e., the value of the pressure at the stagnation point when the pressure at infinity is set equal to zero). The quantity C_L compares the lift force to the force which the stagnation pressure would exert if it were acting on the whole wing.

Since the formula

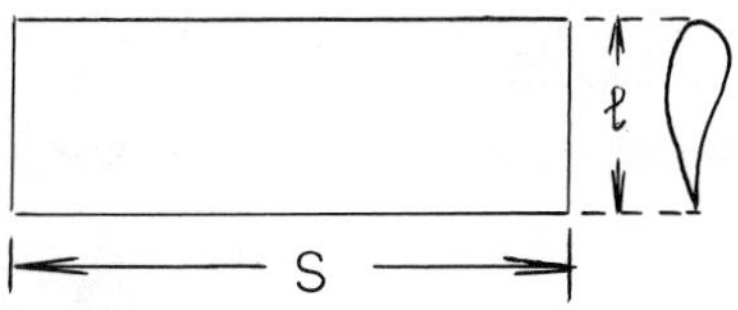

$$L = 4\pi a \rho V^2 \sin (\alpha+\beta)$$

gave the lift force per unit thickness, i.e., per unit span, we must take the area A per unit span too, i.e., $\frac{S\ell}{S} = \ell$. Thus we get

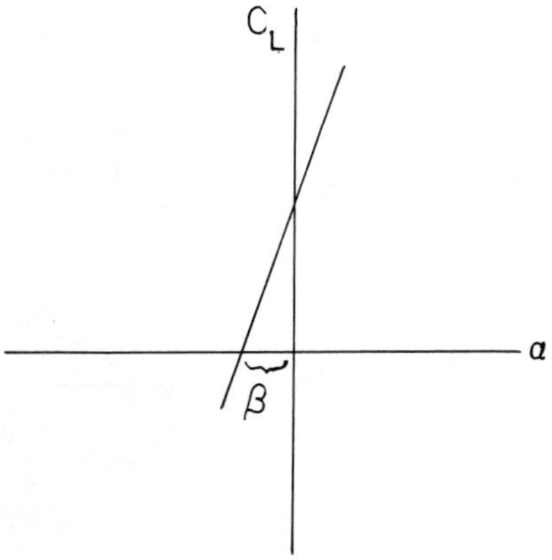

$$C_L = \frac{8\pi a \sin (\alpha+\beta)}{A}$$

Now for a thin wing we had $a \sim \frac{\ell}{4}$, and if we assume effective angle of incidence $\alpha + \beta$, which we denote by α', is small, we get

$$C_L \sim 2\pi\alpha'. \tag{46}$$

If we plot this curve taking α as horizontal axis and C_L for the ordinate, we get a straight line with slope 2π, intersecting the x-axis at the point $-\beta$.

Experimental results with Joukowski profiles gave a good approximation to this for values of α' up to 10 or 15 degrees. Naturally the slope of the line will vary slightly with the shape of the profile since the ratio $\frac{a}{\ell}$ varies through values slightly greater than $\frac{1}{4}$ as the shape of the profile changes.

Another method of verification of the results for the lift is one where we make use of the results obtained in the theory of three dimensional motion. These results give a relation between C_L, the lift coefficient for a wing of infinite span, and C_L', the lift coefficient for a finite wing in three dimensional motion. Thus experiments made on ordinary wing models in the wind tunnel give values of C_L' which in turn can be transformed into values for C_L.

(b) Moment Formula

Besides a verification of the results for the lift force, we inquire as to the line of action or the moment. In experimental work, one takes the moment with respect to a point M_1 at the leading edge of the profile. Previously, we found that the moment M with respect to the point M is given by

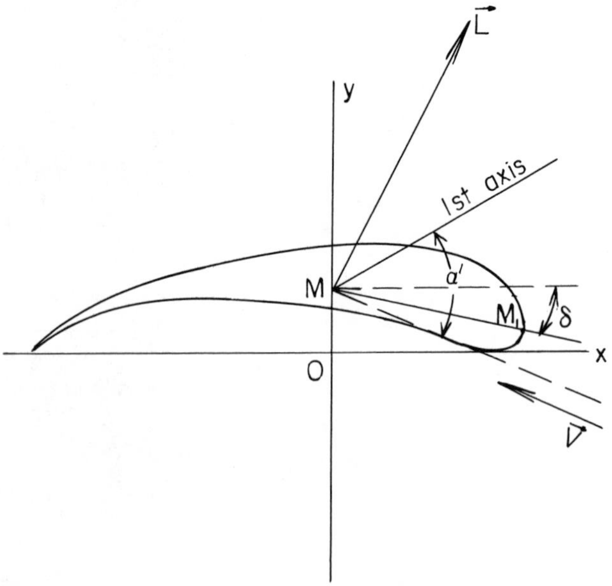

$$M = 2\pi\rho c^2 V^2 \sin 2(\alpha+\gamma).$$

Thus the moment M_1 with respect to the point M_1 is

$$M_1 = M - L\, d \cos(\alpha'-\delta),$$

where d is the distance $\overline{MM_1}$ and δ is the angle the line MM_1 makes with the x-axis (see figure). If we denote $\gamma - \beta$ by γ', we get for M_1,

$$M_1 = 2\pi\rho c^2 V^2 \sin 2(\alpha' + \gamma') - Ld \cos(\alpha' - \delta).$$

Now as in the case for measuring the lift, we put the results in the form of the dimensionless quantity C_M, the <u>coefficient of moment</u> which we define by

$$C_M = \frac{-M_1}{\frac{\rho}{2} V^2 A\ell} . \tag{47}$$

(The negative sign gives positive values for C_M.) By substituting the last expression for M_1 in this, and with some manipulation, we can express C_M as a function of C_L as follows:

$$C_M = - \frac{2\pi\rho c^2 V^2 \sin 2(\alpha' + \gamma') - Ld \cos(\alpha' - \delta)}{\frac{\rho}{2} V^2 A\ell}$$

$$= \frac{-2\pi\rho V^2}{\frac{\rho}{2} V^2 A\ell} [2c^2 \cos(\alpha' + \gamma')\{\sin\alpha' \cos\gamma' + \cos\alpha' \sin\gamma'\}$$

$$- 2ad \cos(\alpha' - \delta) \sin\alpha']$$

$$= - \frac{2L}{a\rho V^2 A\ell} [c^2 \cos(\alpha' + \gamma') \cos\gamma' - ad \cos(\alpha' - \delta)]$$

$$- \frac{2\pi\rho V^2}{\frac{\rho}{2} V^2 A\ell} 2c^2 \cos(\alpha' + \gamma') \cos\alpha' \sin\gamma'$$

$$= - \frac{C_L}{\ell} [\frac{c^2}{a} \cos(\alpha' + \gamma') \cos\gamma' - d \cos(\alpha' - \delta)] - \frac{8\pi c^2}{\ell^2} \cos(\alpha' - \gamma') \cos\alpha' \sin\gamma' .$$

Now since the angles are considered as small, we can replace their cosines by 1 and the sines by the angles themselves. Then the last expression for C_M becomes

$$C_M \sim C_L\left(\frac{d}{\ell} - \frac{c^2}{a\ell}\right) - \frac{8\pi c^2}{\ell^2}\,\gamma' .$$

Again, approximating, we may say that $\frac{d}{\ell} \sim \frac{1}{2}$ and $\frac{c^2}{a\ell} \sim \frac{c}{\ell} \sim \frac{1}{4}$. This gives us the expression for C_M in its reduced form:

$$C_M \sim \frac{1}{4} C_L + \text{constant}. \tag{47a}$$

This last equation states that C_M is a linear function of C_L, and that the slope has approximately the value $\frac{1}{4}$. If we plot C_M on the horizontal axis and C_L as ordinate, (as is customary in the diagrams issued by the experimental workers), we get a straight line as shown in the figure. It may be argued that all such approximations when higher degree terms are dropped always give a linear relation. However, if we denote by C_D the coefficient of drag (which will be discussed later) and plot this against C_L we get very distinctly a parabolic curve (see figure). Moreover, it is quite striking how closely all C_M lines have the same slope.

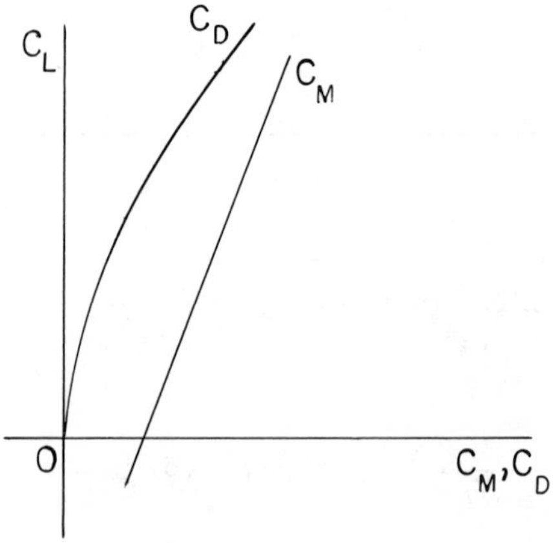

The relation (47a) has the property that it holds also for three dimensional motion, i.e., the slope of about $\frac{1}{4}$ is unaffected. For, in the case of three dimensional motion, the formula for lift L in two dimensions must be multiplied by a certain factor (depending upon the ratio of span to chord). This factor will enter in both C_M and C_L, and only the constant term in (47a) may be affected. Since the quantity γ' is small, the constant term in (47a) will be small and hence difficult to analyze.

(c) <u>Compressible Fluid</u>

In this chapter we always considered air as an incompressible fluid. In fact for a velocity V of 100 m.p.h., we have the stagnation pressure not much more that 1 per cent of the atmospheric pressure. Hence in these cases the density varies by not much more than $\pm$ 1 per cent. However, in recent developments of aircraft, we occasionally have velocities of 400 to 500 m.p.h., and then the variation in density becomes 16 to 25 times as much. We wish to add here a remark about how the results for incompressible fluids may be used, in such cases, to give an approximate solution which takes into account the compressibility.

In the case of compressibility, instead of the simple $\Delta\Phi = 0$, we have to use equation (22) of Chapter I:

$$\Delta\Phi = \frac{1}{c^2} q \frac{\partial}{\partial s} \left(\frac{1}{2} q^2\right), \tag{48}$$

where $c = \frac{dp}{d\rho}$ is the velocity of sound. If we consider the motion to be adiabatic, then we have

$$c^2 = c_0^2 - \frac{1}{2} (\kappa-1)(q^2 - q_0^2), \tag{48a}$$

where c_0 is the value of c for $q = q_0$.

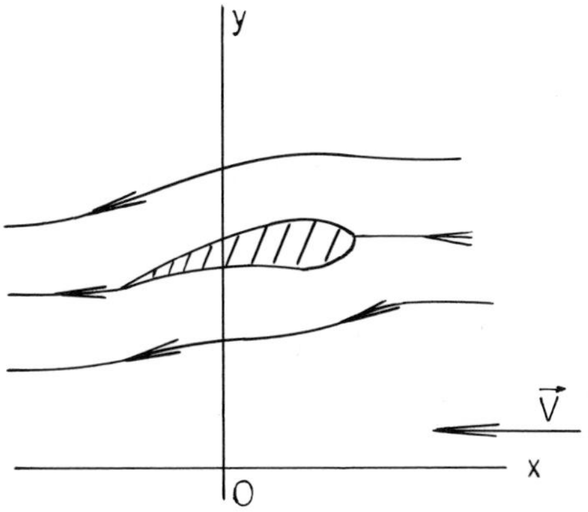

Let us take the x-axis in the direction of $\vec{V}$ so as to have the angle of attack α zero. As a first approximation we may assume that the velocity field $\vec{q}$ does not differ much from the constant velocity $\vec{V}$. This means that $\vec{q} = \vec{V} + \vec{q}'$ where $\vec{q}'$ is small, and also that all derivatives of $\vec{q}$ are small quantities. Similarly $q^2 - q_0^2$ in (48a) is small, and as the second factor in (48) is a derivative of $\vec{q}$ it will be small. Thus in (48) we can replace c^2 by c_0^2 neglecting terms of second order. If we let the x and y components of $\vec{q}'$ be u' and v', we have

$$q^2 = (-V+u')^2 + v'^2 = V^2 - 2Vu' + u'^2 + v'^2.$$

Now u'^2 and v'^2 are of second order, and $\frac{\partial}{\partial s}(V^2) = 0$ as V is constant. Thus, neglecting terms of higher order, (48) becomes

$$\Delta\Phi = \frac{1}{c_0^2}\left[u\frac{\partial}{\partial x} + v\frac{\partial}{\partial y}\right](-u'V) = -\frac{V}{c_0^2}\left(u\frac{\partial u'}{\partial x} + v\frac{\partial u'}{\partial y}\right).$$

As we have $u = -V + u'$ and $v = v'$, we obtain, again neglecting terms of higher order

$$\Delta\Phi = -\frac{V}{c_0^2}\left(-V\frac{\partial u'}{\partial x}\right) = \frac{V^2}{c_0^2}\frac{\partial u'}{\partial x}.$$

Since we know $u = \frac{\partial\Phi}{\partial x}$ the equation for compressible fluid takes the form

$$\frac{\partial^2\Phi}{\partial x^2}\left(1 - \frac{V^2}{c_0^2}\right) + \frac{\partial^2\Phi}{\partial y^2} = 0. \tag{49}$$

If we make the affine transformation $y' = y\sqrt{1 - \frac{V^2}{c_0^2}}$ assuming first that $V < c_0$, then (49) is transformed into

$$\frac{\partial^2\Phi}{\partial x^2} + \frac{\partial^2\Phi}{\partial y'^2} = 0 \qquad (\text{for } V < c_0). \tag{49a}$$

However, for $V > c_0$, the transformation $y' = y\sqrt{\frac{V^2}{c_0^2} - 1}$ takes (49) into

$$\frac{\partial^2\Phi}{\partial x^2} - \frac{\partial^2\Phi}{\partial y'^2} = 0, \qquad (\text{for } V > c_0). \tag{49b}$$

Equation (49a) is of the elliptic type and is the same as the one we had for incompressible fluids except for the transformation of y into y'. Thus we see that whenever we have a solution for the incompressible case, we get an approximate solution for the compressible fluid by using an affine transformation as long as the velocity V is less than the velocity of sound. Clearly, we do not get a solution for the same profile, but for one which is thicker by the amount given by the transformation

$$y' = y\sqrt{1 - \frac{V^2}{c_0^2}} .$$

On the other hand, equation (49b) of hyperbolic type, which holds when the velocity of flow is greater than the velocity of sound, gives an altogether different kind of solution with real characteristic lines. This case will be discussed later.

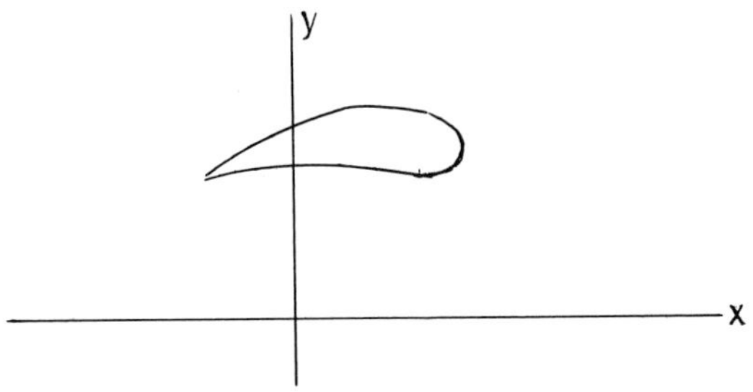

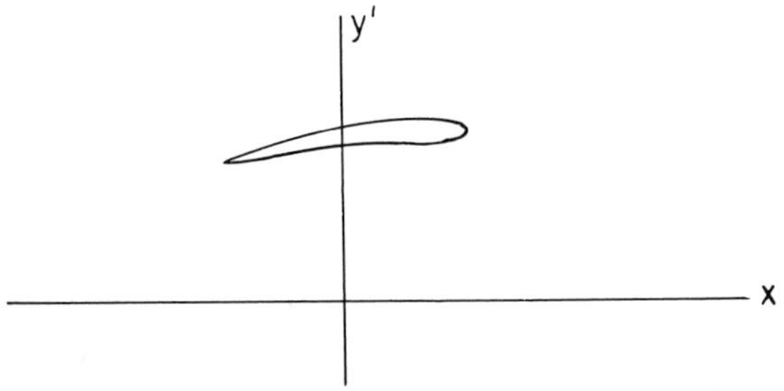

CHAPTER III

MOTION IN THREE DIMENSIONS

1. Vortex Lines and Vortex Sheets.

Consider the equation for the potential function of an irrotational, incompressible fluid in three dimensions

$$\Delta\Phi = 0$$

with the velocity vector $\vec{q} = \text{grad}\ \Phi$. The components of $\vec{q}$ are

$$u = \frac{\partial\Phi}{\partial x} \qquad v = \frac{\partial\Phi}{\partial y} \qquad w = \frac{\partial\Phi}{\partial z} .$$

This is not incompatible with the title of this section as all motions we will consider will be vortex free motions, i.e., the particles of fluid themselves will have no rotation. However, Φ will be discontinuous at certain points or on certain lines and surfaces. We will call these singularities in Φ under certain conditions, vortex points or vortex lines and vortex sheets respectively. We also have H a universal constant and shall neglect gravity so

$$\frac{\rho q^2}{2} + p = g\rho H = \text{const. throughout the motion.}$$

We first discuss briefly the notion of vortex lines in two dimensions. Here the irrotational flow is determined by a complex function $w = \phi + i\psi$. However, if we take

$$w = c \log z$$

we have a singular point at the origin. Outside any circle of small radius about the origin this motion is clearly irrotational. For c a real number the stream lines are the radii and indicate that the origin is either a point source or a point sink, depending on the sign of c. But if c is pure imaginary, say $\frac{\Gamma}{2\pi i}$ with Γ real, we have

$$w = \frac{\Gamma}{2\pi i} \log z. \qquad (1)$$

Then $\phi = \frac{\Gamma}{2\pi} \delta$ where δ is the argument of z. For δ = const. we then have ϕ = const.; the potential lines are straight lines through the origin and we have an irrotational flow in circles about the origin, which we may call a circulating flow. The velocity vector is tangent to a circle and of magnitude

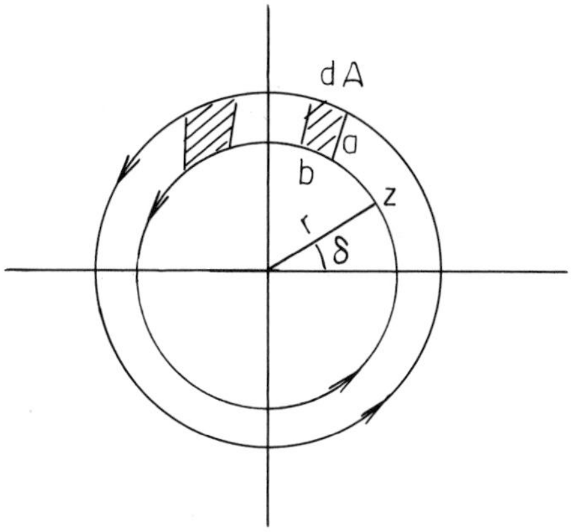

$$q = \frac{1}{r}\frac{\partial \phi}{\partial \delta} = \frac{\Gamma}{2\pi}\frac{1}{r}$$

Hence q is inversely proportional to the distance of the point from the origin or $\frac{1}{r^2}$ times the velocity in a rotating flow. Due to this variation in q we see that for a small element dA the mean rotation is zero. For while the side b (see figure) rotates counterclockwise, the side a rotates clockwise by the same amount. The sum of the two or the mean rotation of the element is zero.

The above solution is physically possible as long as r is not taken so small as to give a negative pressure. The region whose radius is this smallest value of r we can take as a circular fluid disc, rotating as a whole, like a solid, with constant angular velocity. It is customary to disregard the physical condition of positive pressure, and to consider the motion as irrotational throughout the plane except for the one point $z = 0$. This single point or the infinitesimal region around it then has an infinite angular velocity.

As we defined vorticity as the product of the vortex by the area, which equals the circulation (according to Stoke's theorem), we have for the vorticity of the infinitesimal area

$$\lim_{A \to 0} \int_{(A)} \operatorname{curl} \vec{q} \cdot d\vec{A} = \lim \oint_{(C)} w' dz = \lim \frac{\Gamma}{2\pi i} \oint_{(C)} \frac{dz}{z} = \Gamma .$$

As the area approaches zero the magnitude of the vortex vector approaches infinity in such a way that the vorticity remains finite. We can say that the irrotational motion given by (1) is connected with a vortex point or with a vortex line normal to the plane of vorticity Γ. The vortex line is to be considered as the limiting case of a cylinder with finite cross-section within which the fluid motion has a constant vortex. It is customary to say that the flow given by (1) is induced by a vortex line of vorticity Γ normal to the plane at the origin. As a matter of fact, a velocity distribution is uniquely determined by the conditions, that it is irrotational except for one point, that the velocity at infinity is zero and that the circulation about the singular point has the value Γ.

Instead of singularity being at the origin, it may be at the point ζ. Then

$$w = \frac{\Gamma}{2\pi i} \log (z-\zeta). \tag{1a}$$

We can add different functions of type (1a) for several vortices $\zeta_1, \zeta_2, \ldots$ to find a more general type of flow. Finally, instead of a finite number of distinct vortex points, we may assume a continuous vortex distribution along a curve C (see figure), in which case we could have as complex potential

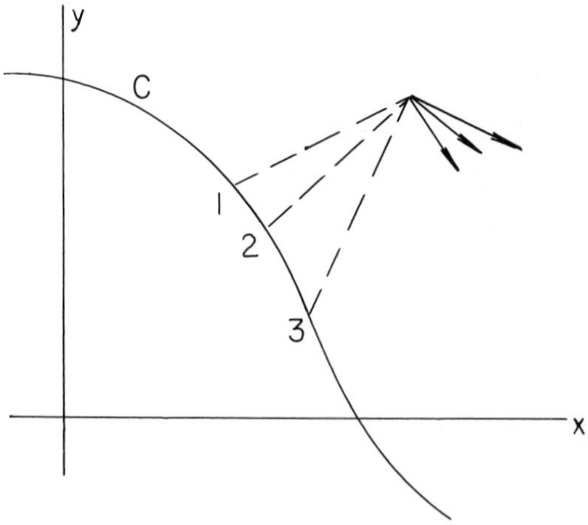

$$w(z) = \frac{1}{2\pi i} \int_{(C)} \log (z-\zeta) d\Gamma(\zeta).$$

The cylindrical surface of which the curve C is a cross-section is then called a <u>vortex sheet</u>.

Turning now to the three-dimensional case we take as a particular solution of $\Delta\Phi = 0$ the quantity

$$\Phi(x,y,z) = \frac{1}{r}, \quad \text{where} \quad r = \sqrt{(x-x')^2 + (y-y')^2 + (z-z')^2}$$

is the distance between the two points (x,y,z) and (x',y',z'). Then

$$\frac{\partial r}{\partial x} = \frac{x - x'}{r} = - \frac{\partial r}{\partial x'} = \cos (x,r).$$

This is also seen from the figure which is a cross-section plane through r and the x-axis. We also see that $r = 0$ is a singular point of the solution.

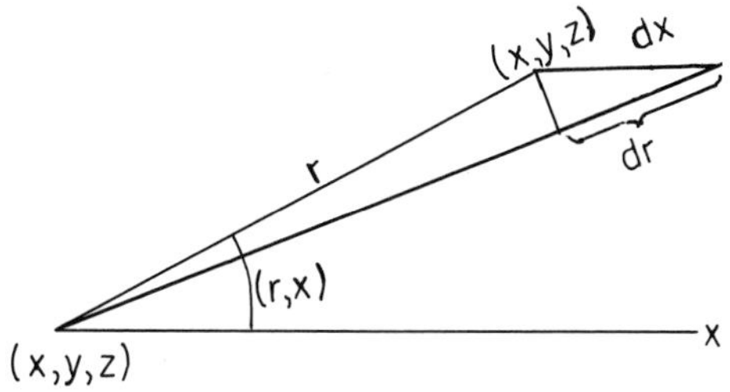

For several points (x_1',y_1',z_1'), (x_2',y_2',z_2'),... with associated quantities C_ν (which can be denoted by "mass" but may be negative) the solution is

$$\Phi = \sum_{\nu=1}^{n} \frac{C_\nu}{r_\nu}, \qquad r_\nu = [(x-x_\nu')^2 + (y-y_\nu')^2 + (z-z_\nu')^2]^{\frac{1}{2}}.$$

For a line or surface of such points this sum becomes an integral.

We may consider another type of generalization. Suppose we have a surface in space as shown in the figure. Draw a line normal to this surface at some point

P'; let the solid arrow denote the direction of the positive normal and the dotted arrow that of the negative normal. Then the boundary is taken in the positive sense shown. On the normal on opposite sides of the surface take two points at distance ϵ and associate with each a constant $+C$ or $-C$ as shown. For these two "mass" points the solution of $\Delta\Phi = 0$ is then

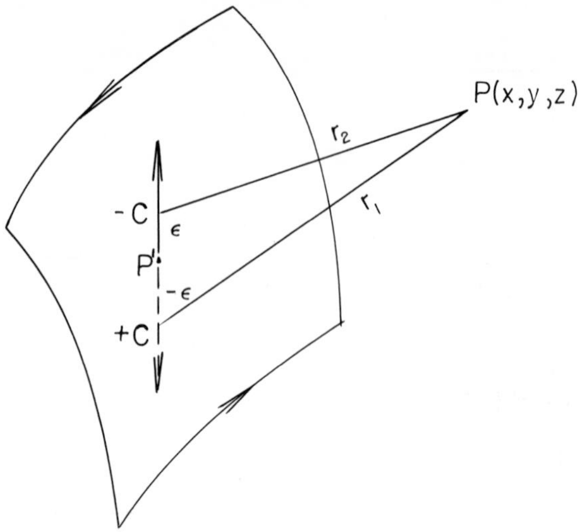

$$\Phi = C\left(\frac{1}{r_1} - \frac{1}{r_2}\right).$$

When $\epsilon \to 0$ the difference goes to the product of -2ϵ by the derivative of $\frac{1}{r}$ with respect to n:

$$\Phi \to -C \frac{\partial}{\partial n}\left(\frac{1}{r}\right)2\epsilon.$$

Now we assume that as $\epsilon \to 0$ the constant $C \to \infty$ in such a way that the product $2\epsilon C = \mu$ has a finite limit. If this holds for all points of the considered surface so as to have a function μ of the points on the surface, we are led to the expression

$$\Phi = - \int_{(A)} \mu(x',y',z') \frac{\partial}{\partial n}\left(\frac{1}{r}\right)dA'. \tag{2}$$

This value of Φ is a solution of Laplace's equation for any integrable function $\mu(x',y',z')$. We are speaking in this case of the potential of a <u>double layer</u>; μ is called the <u>density</u> of the double layer. The integral is defined only for points

(x,y,z) not on the surface.

We are first interested in the case of constant density where μ is the same for all points of the surface A'. Write $\mu = \frac{C}{4\pi}$ and the integral becomes

$$\Phi(x,y,z) = -\frac{C}{4\pi}\int_{(A')}\frac{\partial}{\partial n}\left(\frac{1}{r}\right)dA'$$
$$r = [(x-x')^2 + (y-y')^2 + (z-z')^2]^{\frac{1}{2}}. \tag{2a}$$

Let us analyze this distribution.

Since $\frac{\partial r}{\partial x'} = -\cos(x,r)$ we have

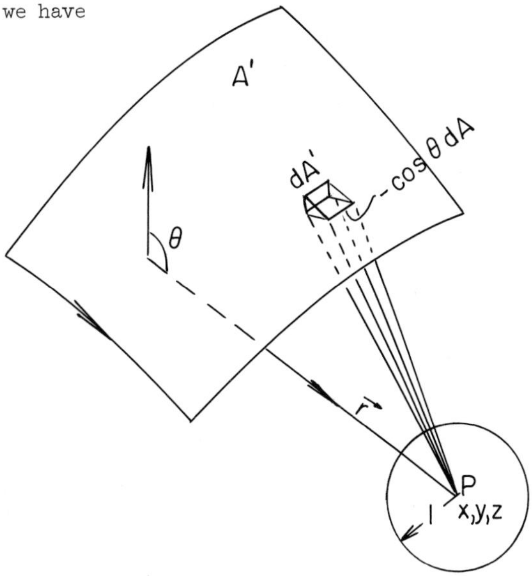

$$\frac{\partial}{\partial n}\left(\frac{1}{r}\right) = -\frac{1}{r^2}\frac{\partial r}{\partial n} = \frac{1}{r^2}\cos\theta$$

where (as shown in the figure) θ is the angle between the positive normal and the radius vector $\vec{r}$ from (x',y',z') to (x,y,z). Since we take the derivative $\frac{\partial r}{\partial n}$ at the point (x',y',z') we have for it $-\cos\theta$. Now

$$\Phi = -\frac{C}{4\pi}\int_{(A')}\frac{\cos\theta}{r^2}dA'.$$

This integral has a simple geometrical meaning which we consider with the help of the figure. The quantity $-\cos\theta\, dA'$ is the projection of dA' normal to the line from dA' to P while $-\frac{\cos\theta}{r^2}dA'$ is the central projection of dA' on a sphere of unit radius with center at P. Hence

$$-\frac{\cos\theta}{r^2}dA' = d\Theta$$

is the solid angle subtended at P by dA' and

$$-\int_{(A')} \frac{\cos\theta}{r^2}\, dA' = \int_{(A')} d\Theta = \Theta$$

is the solid angle subtended by the whole surface A' at P. We took here the convention that a solid angle of an element dA' is positive when viewed from the negative side of the surface to the positive. As a result the potential

$$\Phi(P) = \frac{C}{4\pi}\,\Theta \tag{3}$$

depends not so much on the surface A' as upon its boundary. We can distort the surface without changing the solid angle which it subtends at P, in so far as the surface does not cut across the point P.

For any closed curve in space which does not pass through any points of the surface A' the circulation is zero, according to the fact that the vortex is zero for all points of a surface subtended by this curve (Stoke's theorem). Thus

$$\oint \vec{q}\cdot d\vec{\ell} = \oint \operatorname{grad}\Phi\cdot d\vec{\ell} = \oint \frac{\partial\Phi}{\partial\ell}\, d\ell = 0.$$

However, this does not hold for a closed curve which cuts the surface of the double layer. If we attempted, in this case, to stretch a surface across the curve, this surface would cut the surface A', therefore include points where Φ is not determined, so that Stoke's theorem is not applicable.

If we examine the values of Φ, which equal except for the factor $\frac{C}{4\pi}$ the solid angle Θ, along the curve K, we find that Θ approaches $+2\pi$ as we approach the negative side of the surface (point B), and -2π as we approach the positive side (point A). As our circuit goes from A to B, we have

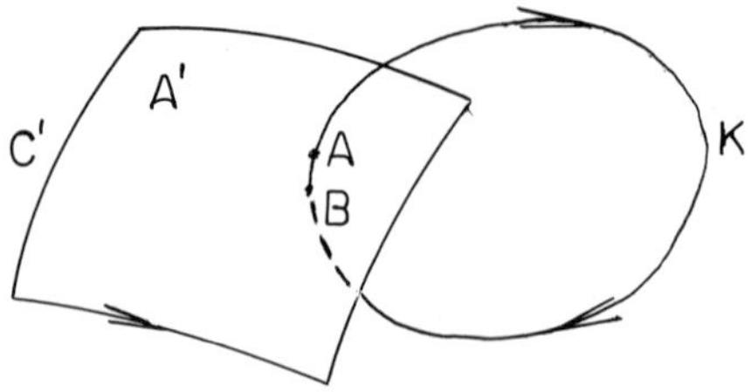

$$\oint d\Theta = \lim \int_A^B d\Theta = \lim [\Theta(B) - \Theta(A)] = 4\pi$$

The circulation along K is then

$$\oint \vec{q}\cdot d\vec{\ell} = \oint \frac{\partial\Phi}{\partial\ell}\, d\ell = \oint d\Phi = \frac{C}{4\pi} \oint d\Theta = \frac{C}{4\pi} \cdot 4\pi = C.$$

Thus C is the circulation for any closed curve K which cuts the surface A' or surrounds the boundary C' of A'. We can take for K very small circuits surrounding C' at one point. This shows that the only singular points of the velocity field are lying on C' and none are inner points of A'. In analogy with the expression used in the two dimensional case we call the curve C' a vortex line of vorticity C.

The fact that Φ or the solid angle Θ has a jump at the points of A' does not contradict this statement that the velocities given by Φ do not depend on the shape of A'. We may compare this to the angular "jump" of 2π which we get in the two dimensional case as we make a complete circuit around the origin. If we start with zero at the x-axis and come back to this axis the Φ value is increased by 2π. However, we could have selected an arbitrary curve through the origin as starting position and considered angles modula 2π from that starting point. Then we would have only changed the Φ-value for certain regions by a constant and the derivatives will not be affected.

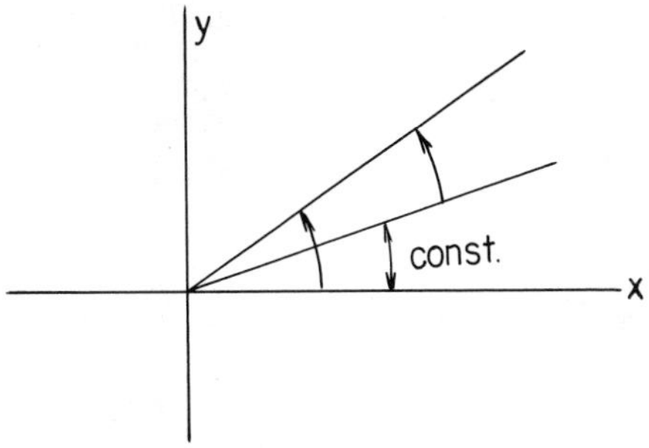

Analogously, different surfaces A' stretched across the closed circuit C' correspond to values of Φ differing in certain regions only by a constant, and the velocity distribution will be unaffected.

We now investigate the velocity distribution connected with the considered potential Φ of a double layer with constant $\mu = \frac{C}{4\pi}$. As it depends, as we saw, only on the boundary C' (and not on A') these velocities are often said to be "induced" by the single closed curve C' with circulation C or by the <u>vortex line C'</u>. For the potential function Φ we have from (2a)

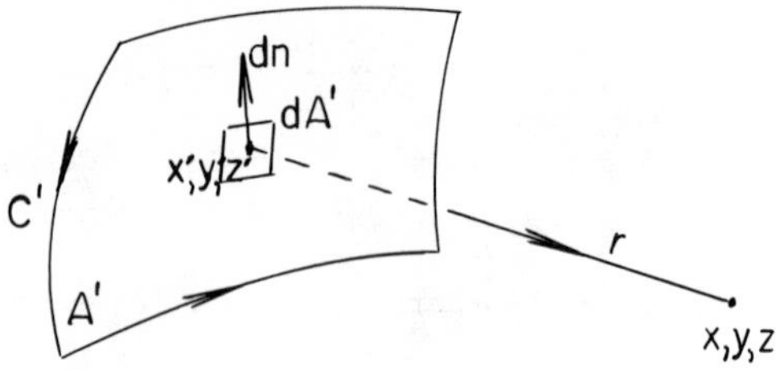

$$\Phi = -\frac{C}{4\pi}\int_{A'}\frac{\partial}{\partial n}\left(\frac{1}{r}\right)dA' \quad (2a)$$

If we let dn have the direction cosines ℓ', m', n' at the point x', y', z', then the above expression becomes

$$\Phi = -\frac{C}{4\pi}\int_{A'}\left(\ell'\frac{\partial}{\partial x'} + m'\frac{\partial}{\partial y'} + n'\frac{\partial}{\partial z'}\right)\left(\frac{1}{r}\right)dA'.$$

Since $\frac{\partial r}{\partial x} = -\frac{\partial r}{\partial x'}$, we get for the x-component u of the velocity $\vec{q}$

$$u = \frac{\partial\Phi}{\partial x} = \frac{C}{4\pi}\int_{A'}\left(\ell'\frac{\partial^2}{\partial x'^2} + m'\frac{\partial^2}{\partial x'\partial y'} + n'\frac{\partial^2}{\partial x'\partial z'}\right)\left(\frac{1}{r}\right)dA' \quad (4)$$

This surface integral can be transformed into a line integral by the application of Stokes formula. As we know, Stokes theorem states that for any vector $\vec{V}$

$$\int_A \text{curl}\ \vec{V}\cdot d\vec{A} = \oint \vec{V}\cdot d\vec{\ell}.$$

In order for this formula to apply to (4), the following relations must hold:

$$\frac{\partial V_z}{\partial y'} - \frac{\partial V_y}{\partial z'} = \frac{\partial^2}{\partial x'^2}\left(\frac{1}{r}\right);\ \frac{\partial V_x}{\partial z'} - \frac{\partial V_z}{\partial x'} = \frac{\partial^2}{\partial x'\partial y'}\left(\frac{1}{r}\right);\ \frac{\partial V_y}{\partial x'} - \frac{\partial V_x}{\partial y'} = \frac{\partial^2}{\partial x'\partial z'}\left(\frac{1}{r}\right).$$

It can easily be verified that

$$V_x = 0,\ V_y = \frac{\partial}{\partial z'}\left(\frac{1}{r}\right),\ V_z = -\frac{\partial}{\partial y'}\left(\frac{1}{r}\right)$$

is a solution of the last system of equations. In fact the last two equations hold by inspection, while the first equation follows from the fact that

$$\Delta\left(\frac{1}{r}\right) = \left(\frac{\partial^2}{\partial x'^2} + \frac{\partial^2}{\partial y'^2} + \frac{\partial^2}{\partial z'^2}\right)\left(\frac{1}{r}\right) = 0.$$

Hence we get the x-component of $\vec{q}$ in the form

$$u = \frac{C}{4\pi}\oint_{C'} \vec{V}\cdot d\vec{\ell} = \frac{C}{4\pi}\oint_{C'}\left(dy'\frac{\partial}{\partial z'} - dz'\frac{\partial}{\partial y'}\right)\left(\frac{1}{r}\right).$$

Making use of the fact that $-\frac{\partial r}{\partial z'} = \frac{\partial r}{\partial z} = \cos(r,z) = \frac{z - z'}{r}$, we get

$$u = \frac{C}{4\pi}\oint_{C'}\frac{1}{r^3}[(z-z')dy' - (y-y')dz'],$$

and by cyclic transformation

$$v = \frac{C}{4\pi}\oint_{C'}\frac{1}{r^3}[(x-x')dz' - (z-z')dx']$$

$$w = \frac{C}{4\pi}\oint_{C'}\frac{1}{r^3}[(y-y')dx' - (x-x')dy'].$$

(5)

Writing these results in vector form, we have

$$\vec{q} = \frac{C}{4\pi}\oint_{C'}\frac{1}{r^3}(d\vec{\ell'} \times \vec{r}). \qquad (5a)$$

This last equation gives the "induced velocity $\vec{q}$" at the point (x,y,z) due to the closed circuit C' with circulation C. If we let α be the angle $\vec{r}$ makes with $d\vec{\ell}'$, then (5a) states that the velocity at (x,y,z) due to the element $d\ell'$ is directed so as to be normal to the plane formed by $d\vec{\ell}'$ and $\vec{r}$, and in amount is proportional to $\frac{r \sin \alpha}{r^3} = \frac{\sin \alpha}{r^2}$. This is known as the rule of Biot-Savart.

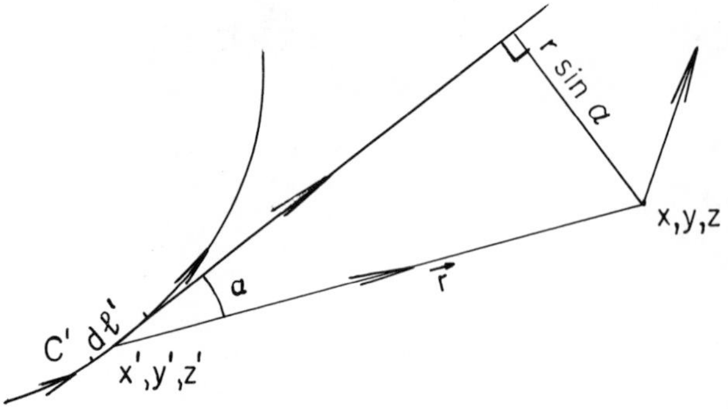

Even though it is clear that we can only consider vortex lines as closed circuits in three dimensions, it will be useful to discuss the value of the integral (5a) for a straight line segment. More precisely we can consider the straight segment as part of a closed circuit, or part of an infinite straight line which can be taken as the limiting case of a closed circuit.

For convenience let us take the segment on the z-axis between the points (1) and (2) (see figure), and let C be the circulation associated with the circuit. We also take, in particular, the point $(x,0,0)$ and consider the velocity at this point induced by the straight segment. Referring to equations (5) it is clear that $u = 0$, $w = 0$, and

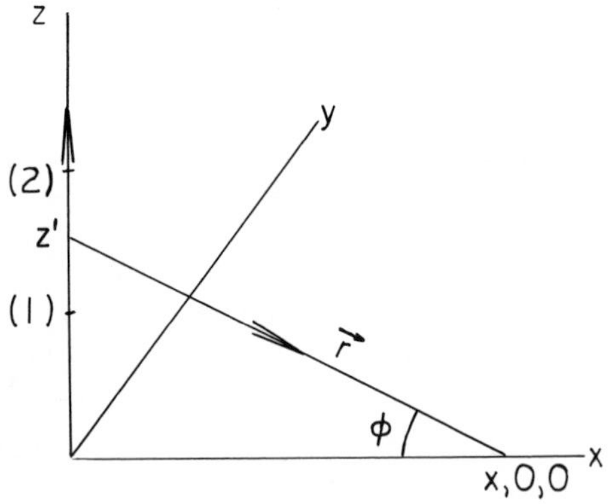

$$v = \frac{C}{4\pi} x \int_{(1)}^{(2)} \frac{1}{r^3} dz' .$$

Clearly, $r^2 = x^2 + z'^2$, and denoting by ϕ the angle indicated in the figure we have $z' = x \tan \phi$, $dz' = x \frac{d\phi}{\cos^2 \phi}$. Also $r^2 = x^2 \cdot \frac{1}{\cos^2 \phi}$, and we get

$$v = \frac{C}{4\pi} x \int_{(1)}^{(2)} \frac{\cos^3 \phi}{x^3} \cdot \frac{x d\phi}{\cos^2 \phi} = \frac{C}{4\pi} \frac{1}{x} \int_{(1)}^{(2)} \cos \phi \, d\phi = \frac{C}{4\pi} \frac{\sin \phi_2 - \sin \phi_1}{x} .$$

For the general case of a line segment anywhere in space, and an arbitrary point P: (x,y,z), we know that the induced velocity $\vec{q}$ will have the direction normal to the plane formed by the triangle consisting of the points (1), (2) and P. The amount q will be given by

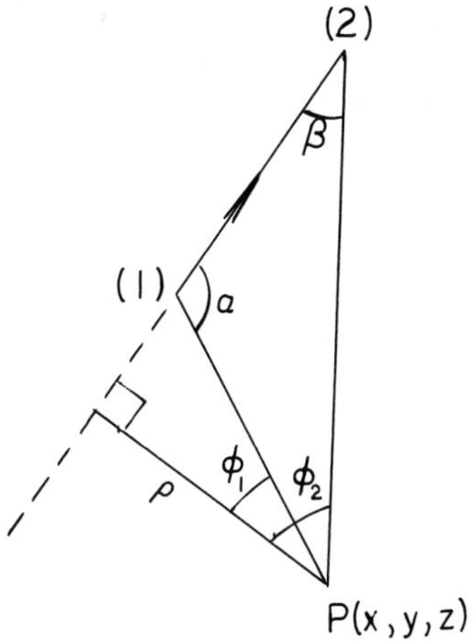

$$q = \frac{C}{4\pi} \cdot \frac{1}{\rho} (\sin \phi_2 - \sin \phi_1),$$

where ρ is the absolute value of the distance from P to the line containing the segment (1) - (2). If we introduce the angles α and β as shown in the figure, then $\sin \phi_2 = \cos \beta$ and $\sin \phi_1 = -\cos \alpha$. Thus the result for q becomes

$$q = \frac{C}{4\pi} \frac{1}{\rho} (\cos \alpha + \cos \beta). \tag{6}$$

Now, if instead of the line segment (1) - (2), we allow the points (1), (2) to approach infinity along this line, we see that the only quantities affected in (6) are the angles α and β. Thus for the case of an infinite straight line we have $\alpha = \beta = 0$ and

$$q = \frac{C}{2\pi} \cdot \frac{1}{\rho}. \tag{6a}$$

We remark that in the case of plane motion given by $w = \frac{C}{2\pi i} \log z$ we obtained the same result for q induced by the vortex line through the origin normal to the plane. A semi-ray of vorticity C would give, at the point P lying in the normal plane through the initial point, exactly half the value (6a) since $\cos \alpha = 0$, $\cos \beta = 1$.

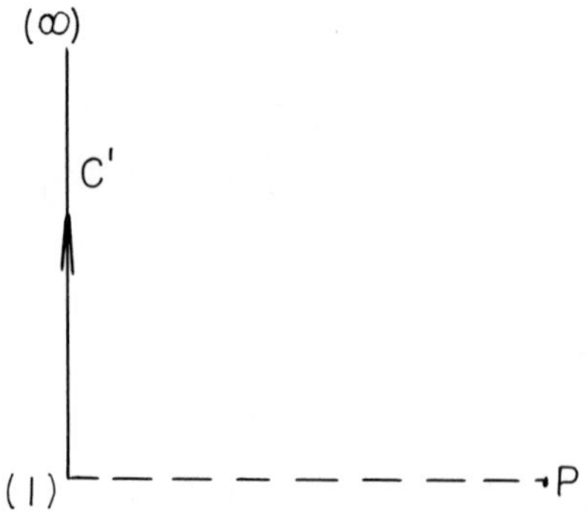

An extension of the results for the velocity induced by a vortex line is obtained by considering several closed curves $C'_1, C'_2, \ldots C'_n$ with circulations $C_1, C_2, \ldots C_n$ respectively and the superposition of the motions corresponding to each of them. For example say we have four curves C'_1, C'_2, C'_3, C'_4 across a surface A'. Denote by A'_1 the part of A' between C'_1 and C'_2. Similarly for A'_2, A'_3, A'_4 (see figure). The velocity $\vec{q}$ at a point P: (x,y,z) will be the sum of the velocities induced by each of the curves C'_1, C'_2, C'_3, C'_4. Also the potential function Φ will be the sum of the Φ values for each of the four motions. Using formula (3) we get

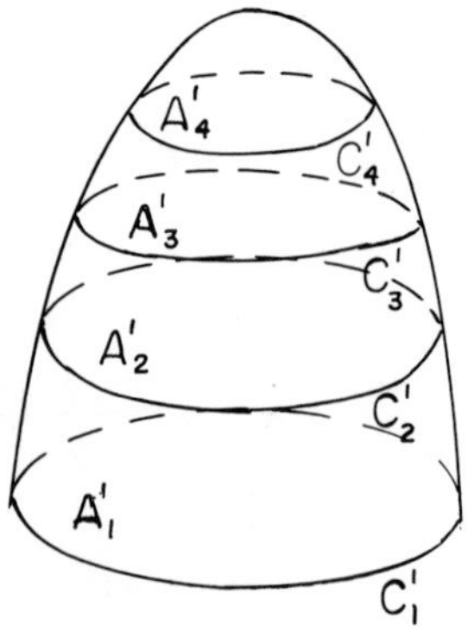

$$\Phi = \frac{C_1}{4\pi} \int_{(A'_1+\ldots+A'_4)} d\Theta + \frac{C_2}{4\pi} \int_{(A'_2+A'_3+A'_4)} d\Theta + \frac{C_3}{4\pi} \int_{(A'_3+A'_4)} d\Theta + \frac{C_4}{4\pi} \int_{(A'_4)} d\Theta,$$

since, for example, $A'_2 + A'_3 + A'_4$ is the surface extended across C'_2. We can break up this last expression as follows:

$$\Phi = \frac{1}{4\pi}\left[C_1 \int_{(A_1')} d\Theta + (C_1+C_2)\int_{(A_2')} d\Theta + (C_1+C_2+C_3)\int_{(A_3')} d\Theta + (C_1+\cdots+C_4)\int_{(A_4')} d\Phi\right]$$

$$= \frac{1}{4\pi}\sum_{i=1}^{4} \Gamma \int_{(A_i')} d\Theta \quad \text{where } \Gamma = \begin{cases} C_1 & \text{for } A_1' \\ C_1 + C_2 & \text{for } A_2' \\ C_1 + C_2 + C_3 & \text{for } A_3' \\ C_1 + \cdots + C_4 & \text{for } A_4' \end{cases} \tag{7}$$

We can immediately give Γ a physical significance. As was shown in Chapter II, Section 2, the circulation is additive. Thus the circulation around a curve K piercing the surface at one point P' in A_2' has the circulation $C_1 + C_2$. This means that the value of Γ at any point x',y',z' on the surface A' is the circulation of a closed curve piercing the surface at the single point x',y',z', and passing from the positive to the negative side of the surface.

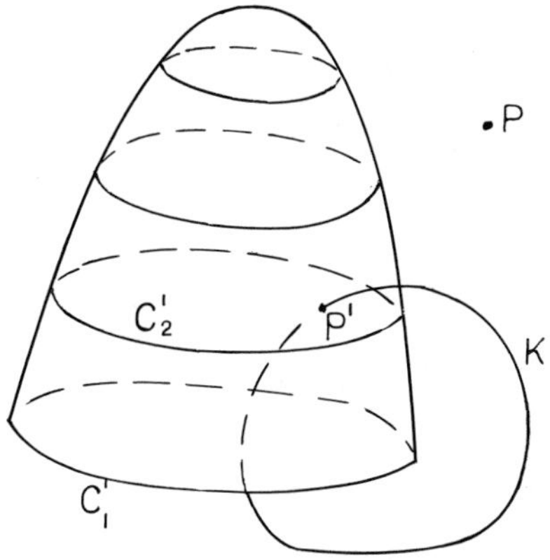

Now, instead of a finite number of vortex lines, we generalize to the case of a continuum of such lines so as to cover the surface A'. In this case we obtain Γ by summing up the circulations, which now change by infinitesimal amounts, as we move along the surface. Thus Γ is a point function on the surface, and the vortex lines become the lines $\Gamma = \text{const}$. Such a surface is called a vortex sheet. Here the potential Φ at any point $P: (x,y,z)$ outside A' is given by

$$\Phi(x,y,z) = \frac{1}{4\pi}\int_{(A')} \Gamma(x',y',z')d\Theta = -\frac{1}{4\pi}\int_{(A')} \Gamma(x',y',z')\,\frac{\cos\theta}{r^2}\,dA'. \tag{8}$$

This expression is essentially the potential function Φ for a double layer of non-constant density μ. If we let $\mu = \frac{\Gamma}{4\pi}$ in (2) we would obtain (8).

We observed that in the case of a single vortex line we had a "jump" in Φ of amount C when we made a circuit around the vortex line. For a vortex sheet, as we pass from the positive to the negative side of the surface we have a "jump" in Θ of amount 4π. This means that we have a jump in Φ of amount Γ. If we ascribe to the surface the arithmetic mean of the values of Φ on both sides of the surface, then the value on the positive side will be less by $\frac{\Gamma}{2}$ and the value on the negative side greater by $\frac{\Gamma}{2}$. Since we wish to investigate the velocity distribution we must consider grad Φ or the derivatives of Φ in various directions. Along a vortex line on the vortex sheet we have Γ = const. and the derivative of Φ in this direction will be the same for both sides of the surface. However, if we let dm be the direction normal to the vortex line in the tangent plane of the surface we will get different values for the component of $\vec{q}$ in this direction for the upper and lower sides. For the positive side we get $\frac{\partial\Phi_0}{\partial m} - \frac{1}{2}\frac{\partial\Gamma}{\partial m}$, while for the lower side we get $\frac{\partial\Phi_0}{\partial m} + \frac{1}{2}\frac{\partial\Gamma}{\partial m}$, and clearly the difference in $\vec{q}$ in this direction is $\frac{\partial\Gamma}{\partial m}$.

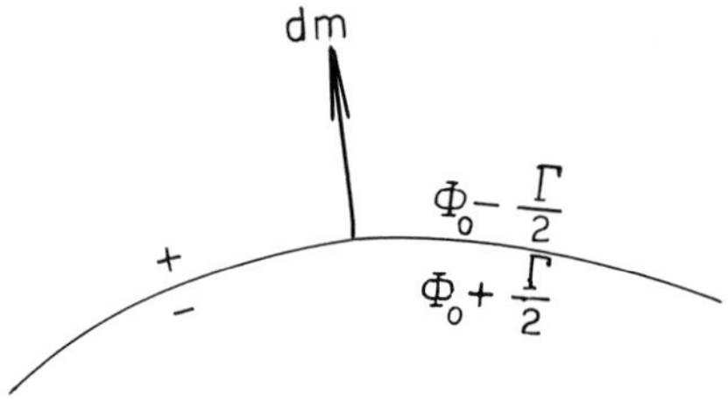

On the other hand, let us consider an irrotational motion and in it one surface where the velocities above and below this surface are not the same. That is we have a discontinuity in a direction in the tangential plane of the surface at each point. Since we assume a continuous pressure distribution the <u>amount</u> of the velocity on each side of the surface must be the same. Such a surface is equivalent to a vortex sheet. For, the difference in the velocities on the upper and lower sides give

$\frac{\partial \Gamma}{\partial m}$. The directions of Γ = const. are known as normal to the direction of $\frac{\partial \Gamma}{\partial m}$. Hence Γ is known (except for an additive constant).

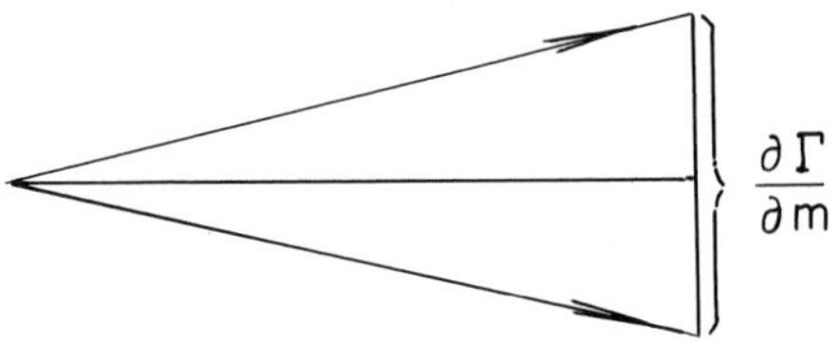

To fix the ideas we consider an example of a vortex sheet in plane motion. Let the sheet consist of a cylinder normal to the xy-plane, the plane of motion. The vortex lines of the sheet will be infinite straight lines normal to the xy-plane. In this case we can see clearly how a vortex sheet and a surface with discontinuous velocity are the same. The vortex sheet will intersect the plane in a curve (see figure) and the points on the curve will be vortices. If we take small circuits around the individual points, we see that the components of velocity of two neighboring circuits tend to cancel each other in a direction normal to the curve. However, on the upper side (see figure) the velocities add to give a flow in one direction while on the lower side they also add to give a flow in the opposite direction. We note that the direction of this flow is normal to the generators of the cylinder, the vortex lines.

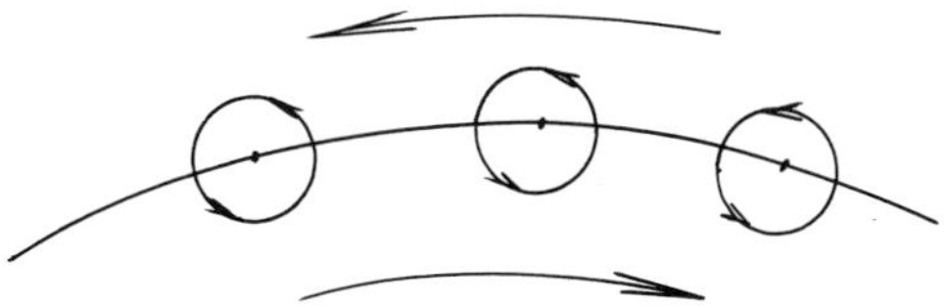

We now consider the case of a body (infinite cylinder) in two-dimensional flow. Let V be the velocity at infinity. We are going to prove the following

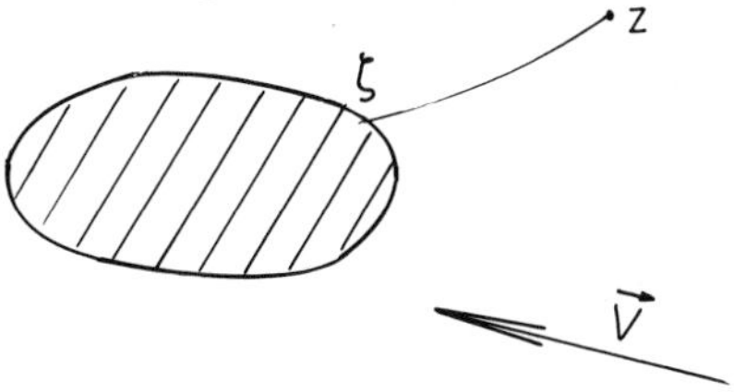

<u>Theorem: The flow of the fluid around the body can be represented as the sum of the uniform flow of velocity V and the flow induced by a vortex sheet which coincides</u>

with the surface of the body and has a certain vortex distribution Γ.

If we let ζ be a point on the vortex sheet then the vortex element $d\Gamma(\zeta)$ at the point ζ induces a velocity field whose complex potential is

$$dw = \frac{d\Gamma}{2\pi i} \log (z-\zeta). \tag{9}$$

The component of the (reflected) velocity at z due to this element dw is then given by

$$dw' = \frac{d\Gamma}{2\pi i} \frac{1}{z - \zeta} .$$

If we denote by w'_∞ the reflected velocity at infinity, the above theorem amounts to the following statement: The reflection velocity at any point z can be put in the form

$$w'(z) = w'_\infty + \frac{1}{2\pi i} \oint \frac{d\Gamma(\zeta)}{z - \zeta} \tag{10}$$

for a certain distribution $\Gamma(\zeta)$.

Proof. We use Cauchy's integral theorem which states that if $f(\zeta)$ is regular inside a closed curve C, then $\int_C f(\zeta)d\zeta = 0$. Let $|\zeta-z| = R$ be a circle about the point z, where the radius R is taken sufficiently large to include the body. Let $|\zeta-z| = \rho$ be another circle about the point z sufficiently small so that it does not include any part of the boundary (A') of the

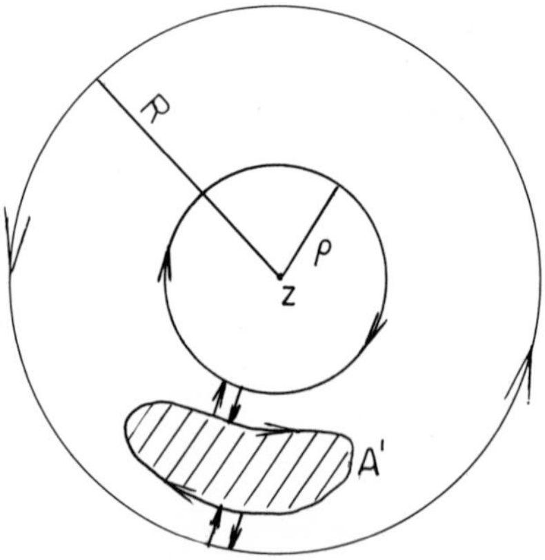

body. The function $w'(\zeta)$ is regular in any region not containing the body. The function $\frac{w'(\zeta)}{\zeta - z}$ will be regular in any region not containing the body and the point z. Thus we can apply Cauchy's theorem to the region shown in the figure, i.e., the domain inside the large circle R and outside the curve (A') and circle ρ. Then we obtain (as the boundary consists of the large circle taken positively and the other two curves negatively)

$$\oint_{|\zeta-z|=R} \frac{w'(\zeta)d\zeta}{\zeta - z} - \oint_{|\zeta-z|=\rho} \frac{w'(\zeta)d\zeta}{\zeta - z} - \oint_{(A')} \frac{w'(\zeta)d\zeta}{\zeta - z} = 0. \tag{11}$$

Setting $\zeta - z = Re^{\delta i}$, $d\zeta = Rie^{\delta i}d\delta$ in the first integral and allowing R to approach infinity, we get for the first term in (11)

$$i \int_0^{2\pi} w'(\zeta)d\delta = 2\pi i\, w'_\infty .$$

By the same method (or using Cauchy's integral formula) the second term in (11) becomes

$$\oint_{|\zeta-z|=\rho} \frac{w'(\zeta)d\zeta}{\zeta - z} = 2\pi i\, w'(z).$$

Hence from (11) we get

$$w'(z) = w'_\infty - \frac{1}{2\pi i} \oint_{(A')} \frac{w'(\zeta)}{\zeta - z}\, d\zeta. \tag{11a}$$

Comparing (11a) to (10) we see that we must identify $d\Gamma(\zeta)$ with $w'(\zeta)d\zeta = dw(\zeta)$ along the curve (A'). However, we know that this curve is a streamline (ψ = const.) and thus $dw = d\phi + i d\psi$ reduces to $dw = d\phi$ as $d\psi = 0$ along a streamline. Now $d\phi = \pm q d\ell$ depending upon how we define the sign of the element of length along (A'). Thus our theorem (10) is proved and, moreover, we see that the unknown Γ distribution is essentially given by the velocity values q along A'.

We remark that (10) gives a representation for the velocity at any point z, but does not actually give us a solution to the problem of finding this velocity since we do not know the value of $d\Gamma$ in the integral (or what would be the same thing, the velocities q along the boundary A'). However, we could find this velocity q by solving a certain integral equation, but this again would be equivalent to the problem of conformal mapping as we dealt with it in Chapter II.

The Thin Wing Theory which we discussed in Chapter II, Section 7 is often represented in a way which makes use of the vortex sheet. In computing the flow around the thin profile the assumption is made that we can consider the vortex sheet entering in the above theorem to have the chord of the air wing as its cross-section. Then one tries to find a distribution of Γ along this chord such that the velocity w' computed by (10) for the points of the wing profile has the direction given by the profile at these points. It is important to notice that this assumption introduces an error which, in general, will not approach zero as the airfoil thickness goes to zero.

If we used a real factor instead of the factor $\frac{d\Gamma}{2\pi i}$ in equation (9) we would have had, as we saw previously, a distribution of sources and sinks along A'. An argument similar to the one given in the proof of the above theorem holds here and we will get a representation for the flow around A' by a distribution of sources and sinks along A'. Such a method, using sources and sinks, is often applied in the case of flow with axial symmetry around a body of revolution. This method is correct only if we assume that the sources and sinks are distributed over the complete boundary of the body. If one assumes, as is often done, that the flow can be represented by admitting sources and sinks along the symmetry axis alone, it can be shown that for certain bodies it is impossible to get the correct value for $w'(z)$ in this way. However, this can be used as a method of approximation in some cases.

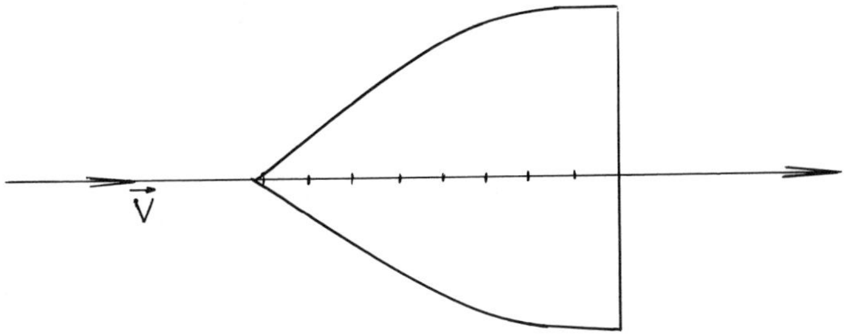

2. Horseshoe Vortex Lines and Sheets.

When we come to consider the airwing of finite span we shall have use for vortex lines shaped like a horseshoe with rectangular corners and two infinite sides. We take the coordinate axes (see figure) so that the finite side of the open rectangle coincides with the segment of the y-axis between $-\eta_1$ and $+\eta_1$, while the infinite sides lie along the lines $y = \pm\eta_1$ in the xy-plane, parallel and symmetric

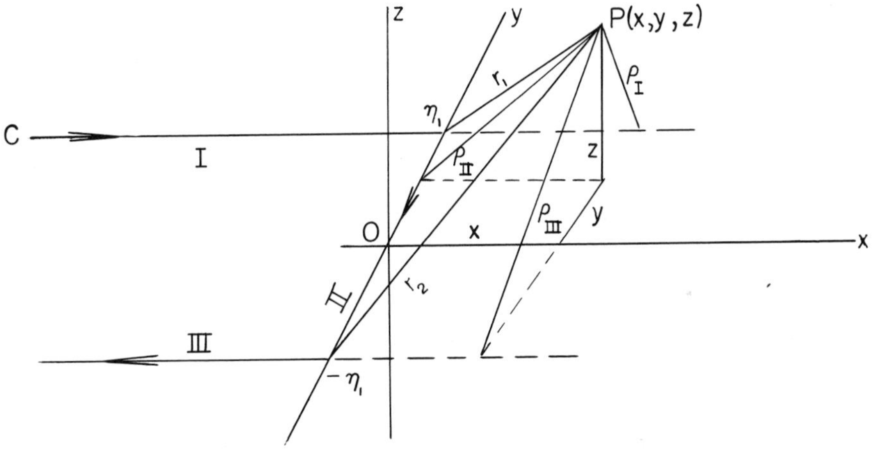

to the negative x-axis. Let C be the circulation of the vortex line along the horseshoe or open rectangle.

Our first task will be to find the velocity field induced by this vortex line; i.e., the velocity field which satisfies the differential equation $\Delta\Phi = 0$, which vanishes at infinity, which is regular at all points except those on the vortex line and which has the circulation C for any closed circuit around this line. In order to do this we break the line up into the three straight segments I, II, III as shown, and apply formula (6) to each of the three parts:

$$q = \frac{C}{4\pi}\frac{1}{\rho}(\cos\alpha + \cos\beta). \tag{6}$$

This expression gives the magnitude of $\vec{q}$. To find the components u, v, w we must use the result of Section 1 giving the direction of $\vec{q}$ as well, i.e., $\vec{q}$ is normal to the plane determined by the vortex line segment and the point $P(x,y,z)$ in the right handed sense. The components of the total velocity will be the sums of the components for the three parts.

We shall need the expressions for the distances of P from the lines I, II, III which we denote by $\rho_I, \rho_{II}, \rho_{III}$ respectively (see figure). They are given by

$$\rho_I^2 = (y-\eta_1)^2 + z^2 = r_1^2 - x^2$$

$$\rho_{II}^2 = x^2 + z^2$$

$$\rho_{III}^2 = (y+\eta_1)^2 + z^2 = r_2^2 - x^2$$

where

$$r_1^2 = x^2 + (y-\eta_1)^2 + z^2$$

$$r_2^2 = x^2 + (y+\eta_1)^2 + z^2$$

give the distances r_1 and r_2 of P from the ends of the finite line segment II.

For the line I, $\alpha = 0$ and $\cos\beta = -\frac{x}{r_1}$, so

$$q_I = \frac{C}{4\pi}\frac{1}{\rho_I}(1 - \frac{x}{r_1}).$$

The direction of $\vec{q}_I$ is normal to the x-axis, so if we take a plane cross section through P normal to the x-axis (see accompanying figure), we find the components of the velocity to be:

$u = 0$

$$v = \frac{C}{4\pi} \frac{1}{\rho_I} (1 - \frac{x}{r_1}) \cdot \frac{-z}{\rho_I} = - \frac{C}{4\pi} \frac{z}{r_1(r_1+x)}$$

$$w = \frac{C}{4\pi} \frac{1}{\rho_I} (1 - \frac{x}{r_1}) \cdot \frac{-(\eta_1 - y)}{\rho_I} = \frac{C}{4\pi} \frac{y - \eta_1}{r_1(r_1+x)} .$$

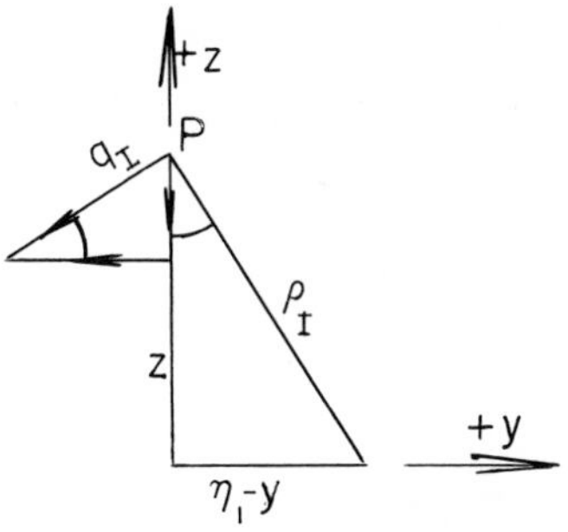

The singular points for v and w are those where $r_1 + x = 0$, i.e., all points on the segment I.

For the line II, $\cos \alpha = \frac{\eta_1 - y}{r_1}$ and $\cos \beta = \frac{\eta_1 + y}{r_2}$ and we obtain in the same way (figure is a cross-section normal to the y-axis):

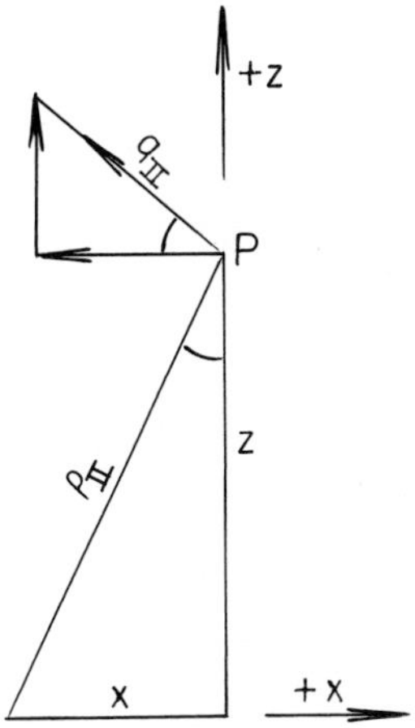

$$u = \frac{C}{4\pi} \frac{1}{\rho_{II}} (\frac{\eta_1 - y}{r_1} + \frac{\eta_1 + y}{r_2}) \cdot \frac{-z}{\rho_{II}}$$

$v = 0$

$$w = \frac{C}{4\pi} \frac{1}{\rho_{II}} (\frac{\eta_1 - y}{r_1} + \frac{\eta_1 + y}{r_2}) \cdot \frac{x}{\rho_{II}} .$$

These components are infinite only on the segment II; at other points of the y-axis they become indeterminate.

The result for line III is similar to that for line I. We need only replace C by $-C$, ρ_I by ρ_{III}, r_1 by r_2 and η_1 by $-\eta_1$ in the expressions above, to get for line III

$$u = 0$$

$$v = \frac{C}{4\pi} \frac{z}{r_2(r_2+x)}$$

$$w = - \frac{C}{4\pi} \frac{y + \eta_1}{r_2(r_2+x)} .$$

These components become infinite on the segment III where $r_2 + x = 0$.

We could have found these velocity components by first finding the potential function $\Phi(x,y,z)$ and then taking the derivatives of Φ. Let us now find an expression for Φ. We use the defining formula (3)

$$\Phi = \frac{C}{4\pi} \int_{(A')} d\Theta = -\frac{C}{4\pi} \int_{(A')} \frac{\cos\theta}{r^2} dA' \tag{3}$$

where, for us, A' is the semi-infinite strip bounded by the vortex line. With ζ and η as coordinates in the plane of the strip (see figure) we have $dA' = d\zeta d\eta$ and $\cos\theta = \frac{-z}{r}$. Thus

$$\Phi = \frac{C}{4\pi} \int_{-\eta_1}^{\eta_1} \int_{-\infty}^{0} \frac{1}{r^2} \cdot \frac{z}{r} d\zeta d\eta = \frac{Cz}{4\pi} \int_{-\eta_1}^{\eta_1} d\eta \int_{-\infty}^{0} \frac{d\zeta}{r^3} \tag{12}$$

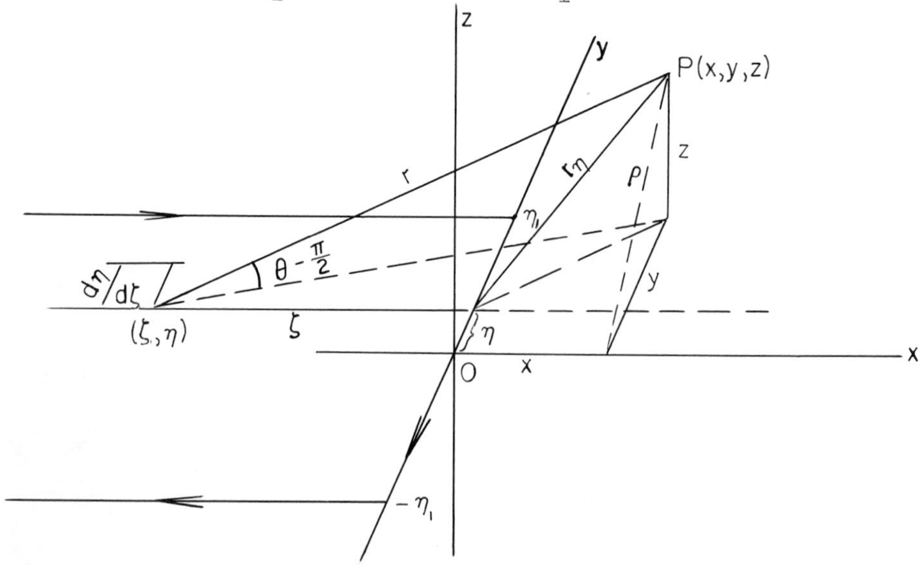

with $r^2 = (x-\zeta)^2 + (y-\eta)^2 + z^2$.

The integral to be taken with respect to ζ is the same as was used in Section 1, p. 132 for computing the velocity induced by a straight segment of a vortex line. We then had

$$\int_{\zeta_1}^{\zeta_2} \frac{d\zeta}{r^3} = \frac{1}{\rho^2} (\cos \alpha + \cos \beta)$$

and as now $\rho^2 = (y-\eta)^2 + z^2$ while $\alpha = 0$, $\cos \beta = - \frac{x}{r_\eta}$ we obtain

$$\int_{-\infty}^{0} \frac{d\zeta}{r^3} = \frac{1 - \frac{x}{r_\eta}}{(y-\eta)^2 + z^2} , \qquad r_\eta^2 = x^2 + (y-\eta)^2 + z^2 .$$

On the other hand

$$(y-\eta)^2 + z^2 = r_\eta^2 - x^2 ,$$

so the expression for Φ reduces to

$$\Phi(x,y,z) = \frac{Cz}{4\pi} \int_{\eta_1}^{\eta_1} \frac{d\eta}{r_\eta (r_\eta + x)} . \tag{13}$$

This integral can be evaluated in terms of elementary functions since r_η is the square root of a quadratic expression in η. We see that Φ is singular at any point $P(x,y,z)$ which lies on the strip. For then $x < 0$, $-\eta_1 \leq y \leq \eta_1$, $z = 0$ and therefore, $r_\eta + x$ vanishes at the point of the interval of integration where $\eta = y$.

We have now to study the more general case of a continuous distribution of horseshoe vortices on a semi-infinite strip of width $2\eta_1$ as shown in the next figure. The infinite sides of all horseshoes are parallel and symmetric to the x-axis while the finite sides coincide with each other and with the y-axis.

In order to find the corresponding value of Φ and of the velocity components we have to replace C by $d\Gamma(\eta)$ in the above expressions and then integrate over η from 0 to $+\eta_1$. The circulation Γ is a function defined at all points of the strip; it is constant along lines parallel to the x-axis and zero along the lines $y = \pm\eta_1$, $z = 0$. From the zero at the boundary $y = -\eta_1$, the function

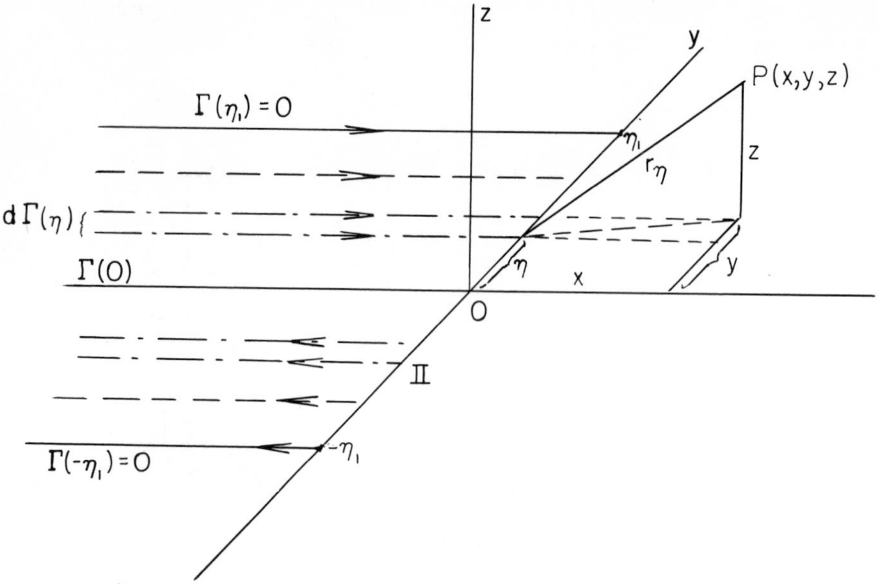

increases to a maximum value $\Gamma(0)$ at $\eta = 0$ and then decreases again to zero at $y = +\eta_1$. This function $\Gamma(\eta)$ is the circulation for any circuit cutting the strip once at a point of the line $y = \eta$, $(-\eta_1 \leq \eta \leq \eta_1)$. We may choose, for example, a circuit parallel to the yz-plane, then the circuit is pierced by those parts of certain vortex lines parallel to the x-axis which have as a whole the vorticity Γ. If we take a circuit parallel to the xz-plane, it surrounds at a certain point the segment II, which is common to all vortices, and Γ is then the circulation for each closed circuit in this plane. This segment II will later represent an airwing of the finite span and then the Γ for the circuit parallel to the xz-plane will enter into the lift formula.

We first recalculate $\Phi(x,y,z)$ for this vortex sheet using (8). The point x',y',z' in that formula becomes $\zeta,\eta,0$. Since Γ does not depend on ζ and $\cos\theta = \frac{-z}{r}$ we have as in equation (12)

$$\Phi(x,y,z) = \frac{z}{4\pi} \int_{-\eta_1}^{\eta_1} \Gamma(\eta)d\eta \int_{-\infty}^{0} \frac{d\zeta}{r^3} .$$

The integration with respect to ζ is carried out as above and we obtain

$$\Phi(x,y,z) = \frac{z}{4\pi} \int_{-\eta_1}^{\eta_1} \frac{\Gamma(\eta)d\eta}{r_\eta(r_\eta+x)}$$

with $r_\eta^2 = x^2 + (y-\eta)^2 + z^2$.

We are particularly interested in this function Φ when $x = 0$ and $x = -\infty$. We prove the following important statement:

The potential Φ in the plane $x = 0$ has half the value of Φ in the plane $x = -\infty$, and in both these planes Φ as a function of y,z satisfies the two dimensional Laplace equation $\frac{\partial^2\Phi}{\partial y^2} + \frac{\partial^2\Phi}{\partial z^2} = 0$. Expressed in symbols this says:

$$\Phi(0,y,z) = \frac{1}{2}\,\Phi(-\infty,y,z) = \phi(y,z) \tag{15}$$

where $\phi(y,z)$ denotes the potential function for a two dimensional flow in the yz-plane. For $x = 0$ we have $r_\eta^2 = (y-\eta)^2 + z^2$ and thus immediately get from (14)

$$\Phi(0,y,z) = \frac{z}{4\pi} \int_{-\eta_1}^{\eta_1} \frac{\Gamma(\eta)d\eta}{(y-\eta)^2 + z^2}\,.$$

Integrating this by parts we have, since $\Gamma(\eta_1) = \Gamma(-\eta_1) = 0$,

$$\Phi(0,y,z) = -\frac{1}{4\pi} \int_{-\eta_1}^{\eta_1} \text{arc tan}\,\frac{z}{y-\eta}\,d\Gamma(\eta).$$

It is easy to see that this function of y and z for any $\Gamma(\eta)$ is a two dimensional potential function. For, if we consider in the yz-plane a vortex distribution $-\frac{1}{2}\,\Gamma(\eta)$ along the y-axis from $-\eta_1$ to η_1, the corresponding complex potential will be

$$w = -\frac{1}{2}\frac{1}{2\pi i}\int_{-\eta_1}^{\eta_1} \log r_\eta d\Gamma(\eta)$$

where $r_\eta^2 = (y-\eta)^2 + z^2$. With $w = \Phi + i\psi$ and $\log r_\eta = \log |r_\eta| + i \tan^{-1} \frac{z}{y - \eta}$ we have

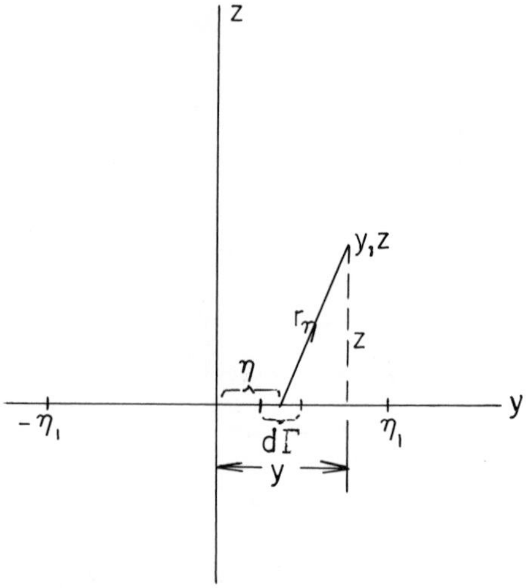

$$\Phi = \frac{-1}{4\pi}\int_{-\eta_1}^{\eta_1} \text{arc tan} \frac{z}{y - \eta} d\Gamma(\eta)$$

which is the same as the value of $\Phi(0,y,z)$. Thus we see that the expression $\Phi(0,y,z)$ is just the same as the two dimensional potential function $\Phi(y,z)$ which corresponds to the velocity field induced by a vortex distribution $-\frac{1}{2}\Gamma(\eta)$ from $-\eta_1$ to $+\eta_1$. If we denote by θ the angle in the plane $x = 0$ such that $\tan\theta = \frac{z}{y - \eta}$, then for the potential $\Phi(0,y,z)$ we have

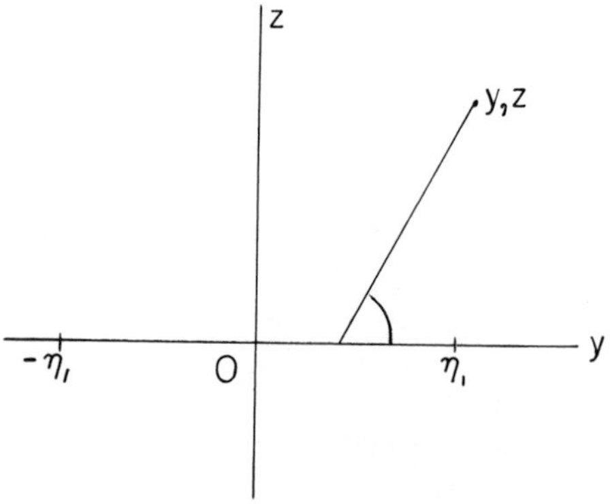

$$\Phi(0,y,z) = -\frac{1}{4\pi}\int_{-\eta_1}^{\eta_1} \theta(\eta)d\Gamma(\eta).$$

We now show that $\Phi(-\infty,y,z) = 2\Phi(0,y,z)$. By definition

$$r_\eta^2 = x^2 + (y-\eta)^2 + z^2 = x^2\left[1 + \frac{(y-\eta)^2 + z^2}{x^2}\right].$$

Since r_η is positive and we wish to examine this expression for large negative values of x, we must take $-x$ for $\sqrt{x^2}$ on the right side above. Doing this and expanding in a power series we get

$$r_\eta = -x\left[1 + \frac{1}{2}\frac{(y-\eta)^2 + z^2}{x^2} + \cdots\right]$$

$$r_\eta + x = -\frac{1}{2}\frac{(y+\eta)^2 + z^2}{x} + \cdots$$

$$r_\eta(r_\eta + x) = \frac{1}{2}[(y-\eta)^2 + z^2] + \cdots .$$

Then as $x \to -\infty$ we get in the limit

$$\Phi(-\infty,y,z) = \frac{z}{2\pi}\int_{-\eta_1}^{\eta_1} \frac{\Gamma(\eta)d\eta}{(y-\eta)^2 + z^2} = -\frac{1}{2\pi}\int_{-\eta_1}^{\eta_1} \theta(\eta)d\eta = 2\Phi(0,y,z). \quad \text{q.e.d.} \tag{16}$$

The result just obtained can be understood in the following way. We add to our original system of vortex lines a second system of horseshoe shaped vortex lines in opposite direction in the xy-plane, such that the finite segments coincide (see figure). Then the effect of the two finite segments will cancel, while the semi-infinite lines will add to give infinite straight vortex lines parallel to the x-axis. As we saw previously each such vortex line induces a two-dimensional (logarithmic) flow normal to the x-axis. The set of all horseshoe vortex lines of the first and second system combined would thus induce a two-dimensional motion in each plane x = const. with the potential $2\Phi(0,y,z)$. The above theorem proved, first, that half of this velocity field takes place in the plane $x = 0$ under the influence of the first system alone, and second, that the total velocity distribution corresponding to $2\Phi(0,y,z)$ exists in the plane $x = -\infty$ even under system (1) by itself. In fact, if we consider planes x = const. farther

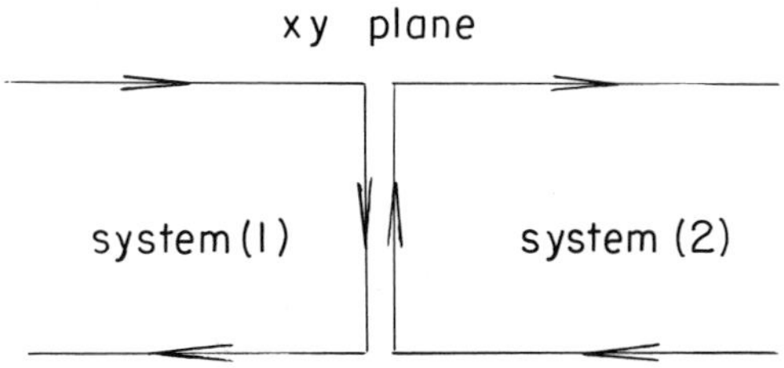

and farther along the negative x-axis the effect of the cutting off of the vortex lines at $x = 0$ diminishes. Thus for the plane $x = -\infty$ the flow is the same as that induced by straight vortex lines infinite in both directions, and for the potential we get the expression (16) holding for two dimensional motion.

To find the components of velocity for a vortex sheet we could use the components obtained for a horseshoe vortex line, replace the constant C by $-d\Gamma(\eta)$ (the minus sign since $d\Gamma$ is negative at the points where the vortex line has the $+x$ direction and is positive where its direction is negative), and integrate from 0 to η_1. Instead we may use equation (14) for Φ and consider its partial derivatives. For the x-component u of velocity we have

$$u = \frac{\partial\Phi}{\partial x} = \frac{\partial}{\partial x}\left[\frac{z}{4\pi}\int_{-\eta_1}^{\eta_1} \frac{\Gamma(\eta)d\eta}{r_\eta(r_\eta+x)}\right] = \frac{-z}{4\pi}\int_{-\eta_1}^{\eta_1} \frac{\Gamma(\eta)d\eta}{r_\eta^2(r_\eta+x)^2}\left[(r_\eta+x)\frac{x}{r_\eta} + r_\eta\left(\frac{x}{r_\eta}+1\right)\right]$$

$$= -\frac{z}{4\pi}\int_{-\eta_1}^{\eta_1} \frac{\Gamma(\eta)d\eta}{r_\eta^3} \ .$$

Making use of the fact that $\int \frac{d\eta}{r_\eta^3} = -\frac{1}{\rho_\eta^2}\frac{y-\eta}{r_\eta}$ where $\rho_\eta^2 = x^2 + z^2$, and integrating by parts, we get

$$u = -\frac{z}{4\pi}\int_{-\eta_1}^{\eta_1} \frac{y-\eta}{r_\eta(x^2+z^2)}\, d\Gamma(\eta). \qquad (17)$$

This form of u could have been obtained directly by integrating the expression for u above on page 143.

We do not examine the general form of the y-component of velocity, but consider it only for values near the vortex sheet. That is, we take $x < 0$, $-\eta_1 \le y \le \eta_1$ and $|z|$ small. The integral (14)

$$\Phi(x,y,z) = \frac{z}{4\pi}\int_{-\eta_1}^{\eta_1} \frac{\Gamma(\eta)d\eta}{r_\eta(r_\eta+x)} \qquad (14)$$

has an infinite integrand only at the one point on the vortex sheet where $y = \eta$. Hence for an $\epsilon > 0$ the integrals taken over the range $(-\eta_1, y-\epsilon)$ and $(y+\epsilon, \eta_1)$ will go to zero as $z \to 0$ since the integrand here has a positive upper bound and the factor z appears in front. Thus as $z \to 0$,

$$\Phi(x,y,z) \to \frac{z}{4\pi} \int_{y-\epsilon}^{y+\epsilon} \frac{\Gamma(\eta)d\eta}{r_\eta(r_\eta+x)} .$$

Since z is small and $y - \eta = \epsilon$ can be taken small we are entitled to use the expression we obtained for $r_\eta(r_\eta+x)$:

$$r_\eta(r_\eta+x) = \frac{1}{2}[(y-\eta)^2 + z^2] + \cdots .$$

Hence as z goes to zero .

$$\Phi(x,y,z) \to \frac{z}{4\pi} \cdot 2 \int_{y-\epsilon}^{y+\epsilon} \frac{\Gamma(\eta)d\eta}{(y-\eta)^2 + z^2} .$$

Since the distribution of $\Gamma(\eta)$ is continuous, upon integrating this last expression, we have

$$\Phi(x,y,z) \to \frac{z}{2\pi} \cdot \frac{1}{z}\Gamma(y)[\arc\tan\frac{\epsilon}{z} - \arc\tan(-\frac{\epsilon}{z})]$$

$$\to \frac{1}{2}\Gamma(y) \quad \text{as} \quad z \to +0$$

$$\to \frac{1}{2}\Gamma(y) \quad \text{as} \quad z \to -0.$$

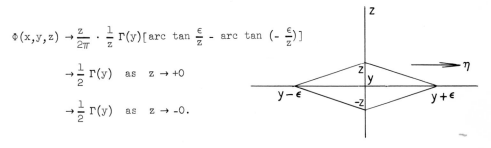

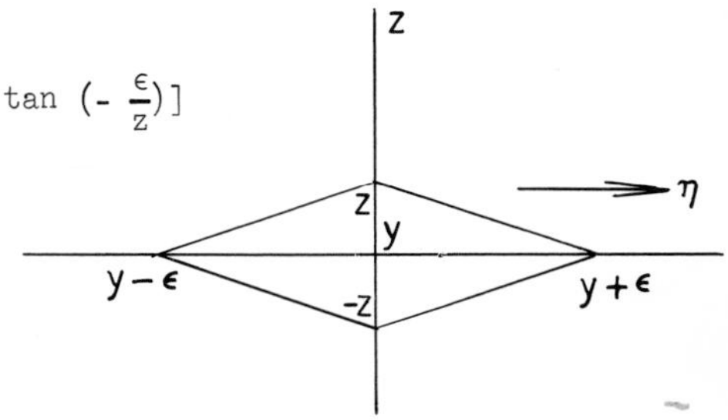

Then for the y component of velocity v, we have on the positive side of the vortex sheet

$$v = \frac{\partial\Phi}{\partial y} = \frac{1}{2}\frac{\partial\Gamma(y)}{\partial y} \qquad (18a)$$

and on the negative side

$$v = -\frac{1}{2}\frac{\partial\Gamma(y)}{\partial y}, \qquad (18b)$$

in accordance with our general statement in Section 1, p.136. If we define the potential Φ for the vortex sheet as the arithmetic mean of the potential as we approach both sides, we have $\Phi = 0$ for the vortex sheet.

For the z-component of velocity we consider the behavior of $\frac{\partial\Phi}{\partial z}$ in the plane $z = 0$. In this plane $r_\eta^2 = x^2 + (y-\eta)^2$ and

$$w = \frac{\partial\Phi}{\partial z}\bigg|_{z=0} = \frac{1}{4\pi}\int_{-\eta_1}^{-\eta_1}\frac{\Gamma(\eta)d\eta}{r_\eta(r_\eta+x)}.$$

Making use of the identity

$$\frac{d}{d\eta}\left[\frac{\eta - y}{x(r_\eta+x)}\right] = \frac{1}{x(r_\eta+x)} - \frac{(\eta-y)^2}{xr_\eta(r_\eta+x)^2} = \frac{1}{r_\eta(r_\eta+x)},$$

and integrating the above expression for w by parts we have

$$w(x,y,0) = \frac{1}{4\pi}\left[-\int_{-\eta_1}^{\eta_1}\frac{\eta - y}{x(r_\eta+x)}\,d\Gamma(\eta)\right].$$

Now applying the identity $\frac{\eta - y}{r_\eta + x} = \frac{r_\eta - x}{\eta - y}$ which holds for $z = 0$, we get

$$w(x,y,0) = -\frac{1}{4\pi}\cdot\frac{1}{x}\int_{-\eta_1}^{\eta_1}\frac{r_\eta - x}{\eta - y}\,d\Gamma(\eta).$$

It is clear that this last expression cannot be evaluated directly for $x = 0$. However, by making use of the fact (derived above) that the flow in the plane $x = 0$ is exactly $\frac{1}{2}$ of the flow in the plane $x = -\infty$, we can obtain a value for $w(0,y,0) = w_0$. For this purpose we write the expression for $w(x,y,0)$ in the form

$$w(x,y,0) = -\frac{1}{4\pi}\int_{-\eta_1}^{\eta_1}\frac{\frac{r_\eta}{x} - 1}{\eta - y}\,d\Gamma(\eta).$$

Since the expression $\frac{r_\eta}{x} \to -1$ as $x \to -\infty$, we have

$$w_\infty = w(-\infty,y,0) = +\frac{1}{2\pi}\int_{-\eta_1}^{\eta_1} \frac{d\Gamma(\eta)}{\eta - y} .$$

And for w_0 we get

$$w_0 = w(0,y,0) = \frac{1}{4\pi}\int_{-\eta_1}^{\eta_1} \frac{d\Gamma(\eta)}{\eta - y} . \tag{19}$$

Since the line $x = z = 0$ is a singular line of higher order (the vortex lines of the sheet all coinciding on this line) we do not have w_0 immediately as the value of $w(x,y,z)$ on this line. But it follows from our argument that the velocity $w(x,y,0) \to w_0$ as we approach this singular line.

If we consider the line $x = z = 0$ from $-\eta_1$ to η_1 to represent an airplane wing of span $2\eta_1$, then the expression (19) will be important for computing the lift force.

3. Lanchester-Prandtl Wing Theory.

In the theory of the airfoil in two dimensions we derived for the lift and the circulation the formulas

$$L = 4\pi a\rho V^2 \sin(\alpha+\beta)$$

$$\Gamma = 4\pi a V \sin(\alpha+\beta).$$

The second equation is the statement of the Joukowski condition that one stagnation point on the circle coincides with the map of the singular point on the profile. We can write for the lift

$$L = \rho\Gamma V.$$

Since $(\alpha+\beta)$ is generally small we set $\sin(\alpha+\beta) = \alpha + \beta = \alpha'$ whence α' is the effective angle of incidence. The constant $4\pi a$ may be written k for brevity.

We also remember that L is the lift force per unit span. Therefore, if we consider a wing in three dimensions (see figure) and denote by y the coordinate normal to the profile, we have to write

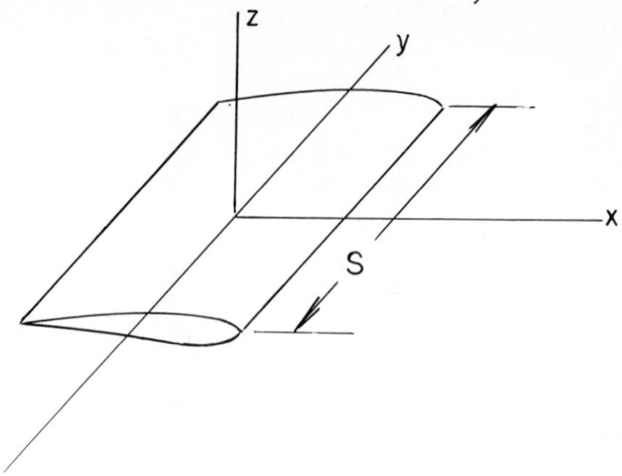

$$\frac{dL}{dy} = \Gamma \rho V, \ \Gamma = kV\alpha' \qquad (20)$$

now calling L the total lift over the whole span S.

Equation (20) would give the lift L for a given span S, if we were allowed to assume that the motion of the air is two dimensional, i.e., parallel to the xz-plane. If we apply this assumption to a wing of finite span and take for Γ the constant value we had in the two dimensional theory, we get too high a value for the lift force; the correct force is only about two thirds as much. On the other hand, it is impossible that Γ be constant along the span, since the lift per unit length $\frac{dL}{dy}$ cannot be as large at the ends of the span as at its center.

In the two dimensional theory the outside space of the cylindrical wing is doubly connected and Γ was the circulation for any closed curve about the airfoil. But the three dimensional space outside a finite wing is simply connected and any closed curve can be shrunk down to a point. Hence no curve has a circulation different from zero, if the flow is irrotational and continuous throughout the entire space. Thus we should have $\Gamma = 0$ and all the basis for computing a lift force seems to diappear. In fact, there is a classical theorem called the d'Alembert Paradox which states that a body, surrounded by an irrotational continuous flow with constant velocity at infinity, is not subjected to any force exerted by the fluid.

Now it is clear from experience that a lift force does exist. At the same time we also expect the essential features of two dimensional motion must hold at least for the center region of a wing of finite span. Thus we may assume we have a Γ in the center of a finite wing, which will diminish in value to zero as we move out along the wing to the ends. We saw in the foregoing sections that the two dimensional flow around the wing profile can be represented as being induced by vortex lines running along the surface of the wing normal to the profile. The sum of the vorticities of all these vortex lines is Γ. Now if we assume the circulation Γ diminishes as we go from the center to the ends of the span, the number of vortex lines must decrease. The lines cannot just disappear, so they must turn from the profile and extend backwards. (See figure). Then a circuit surrounding the wing at the central section 1-1 includes more vortex lines than the circuits surrounding the wing at 2-2 or 3-3.

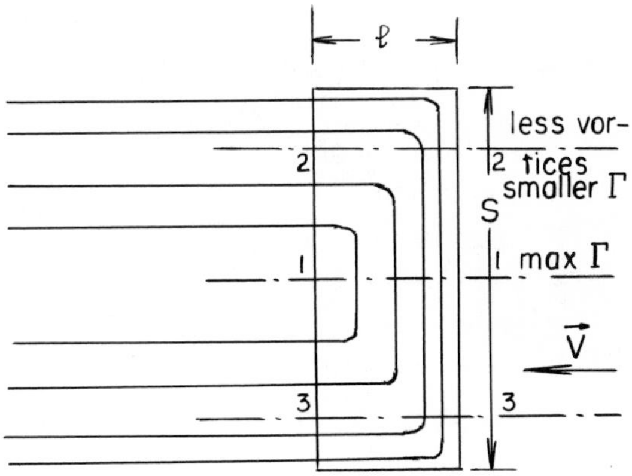

The turned portions of the vortex lines form a vortex sheet of almost the shape of the horseshoe vortex sheet considered in section 2. However, this vortex sheet differs in two respects from the simple horseshoe sheet. First, the finite segments do not coincide; they fill so to speak the total surface of the wing. Second, it is not possible that the infinite portions of the vortex lines form a plane horizontal sheet, if the velocity $\vec{V}$ is assumed to be horizontal, since the vortex sheet is a discontinuity surface for the velocity field and must at any point be tangent to the velocity vector itself. Now the velocity at a point of the vortex sheet is composed of $\vec{V}$ and the formerly computed $w(x,y,0)$, which is in general a small negative quantity. If we assume V is comparatively large, a vortex sheet with but a slight inclination will be possible. On the other hand, if we consider the motion as a whole in the infinite space, we may assume as a first approximation that the wing profile is reduced to one single point.

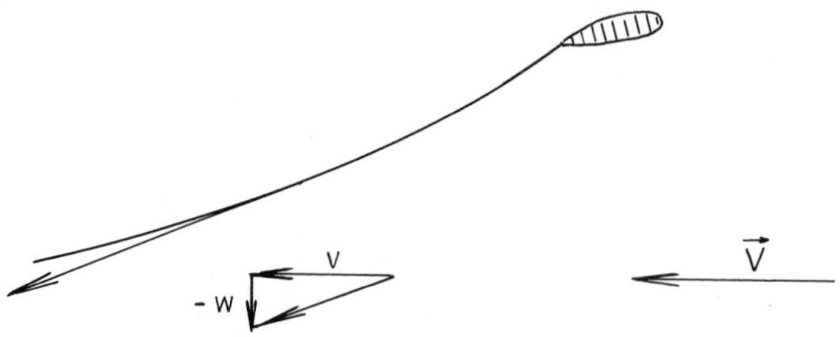

As a result, we consider the flow around a finite wing as the sum of the uniform flow with velocity $\vec{V}$ and the flow induced by a horseshoe vortex sheet with a certain distribution $\Gamma(y)$ of circulation. The d'Alembert statement is no longer applicable in this case, as we have a discontinuity surface stretching through the flow.

Thus far does the qualitative analysis of the problem go as it was essentially suggested by Lanchester as early as 1907. Prandtl supplemented it about 1917 by giving a quantitative analysis of $\Gamma(y)$, taking into account the two dimensional theory. The essential idea is this. We found at the end of Section 2 the velocity w_0, in the vertical direction, in the neighborhood of the line $x = z = 0$, $-\frac{S}{2} \leq y \leq \frac{S}{2}$ which now represents the wing:

$$w_0(y) = \frac{1}{4\pi} \int_{-S/2}^{S/2} \frac{d\Gamma(\eta)}{\eta - y} . \qquad (19a)$$

This integral, in general, is negative. For take the point $y = 0$, then $d\Gamma < 0$ when $\eta > 0$ and $d\Gamma > 0$ when $\eta < 0$ so the integrand is negative over the whole interval. Hence, the positive quantity $-w_0$ is called the <u>downwash velocity</u>.

(We have written (19a) as a Stieltjes integral, but if $\Gamma(y)$ has a derivative everywhere on the span we get an ordinary Riemann integral taking $d\Gamma(\eta) = \Gamma'(\eta)d\eta$. However, a limiting case occurs when we assume a single horseshoe vortex as was done by Lanchester. Then Γ is constant over the span and there $\Gamma' = 0$. But at the ends Γ becomes zero and $\Gamma'(\eta)$ is not defined, so (19a) must be taken strictly as a Stieltjes integral in such a case.)

As before, let $\vec{V}$ represent the velocity vector at infinity. The air particles approach the profile from the right with almost the velocity $\vec{V}$ in the case of infinite span. If we have finite span and, therefore, a trailing vortex sheet, the downwash velocity w_0 is superimposed. The effect is to alter the velocity to $\vec{V} + \vec{w}_0$ as shown in the figure. Since the ratio $\frac{w_0}{V}$ is small,

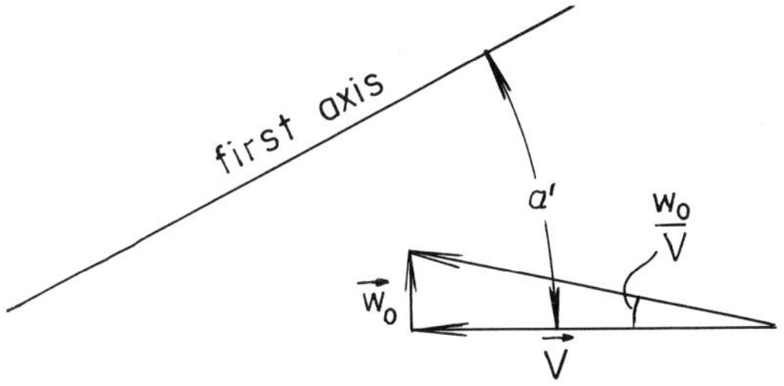

the direction of the new velocity makes the angle $\frac{w_0}{V}$ with $\vec{V}$ and the angle α' between $\vec{V}$ and the first axis is replaced by the new angle $\alpha' + \frac{w_0}{V}$. This is, in general, a decrease in the angle of attack, since $w_0 < 0$.

Then, as a first approximation to the lift force, Prandtl replaces the second term $\Gamma = kV\alpha'$ in (20) by

$$\Gamma = kV(\alpha' + \frac{w_0}{V}). \tag{21}$$

In general, the cross-section of an airwing may change as we move from the center toward either end. Hence k and α' are functions of y, because $k = 4\pi a$ and a is different for different profiles while α' is a variable because the direction of the first axis may vary from cross-section to cross-section. Substitute w_0 from (19a) and we get the first fundamental equation of the Prandtl theory

$$\Gamma(y) = k(y)V\left[\alpha'(y) - \frac{1}{V} \cdot \frac{1}{4\pi} \int_{-S/2}^{S/2} \frac{d\Gamma(\eta)}{y - \eta}\right] \tag{I}$$

This is an integro-differential equation for the determination of Γ. As $k(y)$ and $\alpha'(y)$ can be considered known functions (by solving the two dimensional problem for each cross-section), (I) has the form

$$\Gamma(y) = f(y) + \int_{-S/2}^{S/2} K(y,\eta)\Gamma'(\eta)d\eta$$

where $f(y)$ and $K(y,\eta)$ are given. If we had $\Gamma(\eta)$ instead of $\Gamma'(\eta)$ under the sign of integration, this would be a typical integral equation of second kind. As we have $\Gamma'(\eta)$ we call it an integrodifferential equation. Moreover, the kernel $K(y,\eta)$ has a singularity at $\eta = y$ so we have a singular integro-differential equation.

When Γ has been found, the integration of (20) gives the lift

$$L = \rho V \int_{-S/2}^{S/2} \Gamma(y)dy \tag{II}$$

which is the second fundamental equation.

According to the two dimensional theory, each element $d\vec{L}$ of $\vec{L}$ has to be supposed normal, not to the velocity at infinity $\vec{V}$, but to the effective velocity $\vec{V} + \vec{w}_0$. Thus $\vec{L}$ has a component parallel to $\vec{V}$. The component of the element $d\vec{L}$ normal to $\vec{V}$ is dL times the cosine of a small angle (see figure). This angle is approximately $\frac{w_0}{V}$ whose cosine can be taken as 1 and whose sine can be set equal to $\frac{w_0}{V}$. Then the component of $d\vec{L}$ normal to $\vec{V}$ is approximately dL while the component parallel to $\vec{V}$, the element of the drag force, dD, is

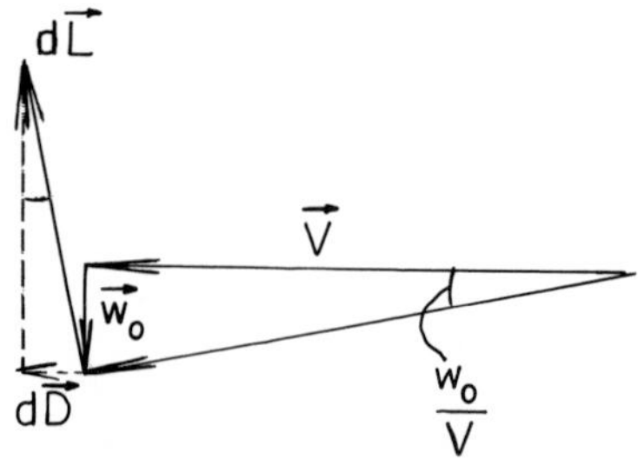

$$dD = -\frac{w_0}{V}\,dL = -\frac{w_0}{V}\,\rho\Gamma V dy$$

and thus the total drag D is

$$D = -\rho \int_{-S/2}^{S/2} w_0(y)\Gamma(y)dy. \tag{III}$$

This drag D is a force in the direction of $\vec{V}$ when $w_0 < 0$. The equation (III) is the third fundamental equation.

If we substitute in (III) the value of $w_0(y)$ given by (19a), D is expressed in terms of Γ. Thus, if we can solve (I) for Γ, we find L and D from (II) and (III).

We can give the expression for D a different form by using the two dimensional motion in the yz-plane as we did before. We then say that $\Phi(0,y,z)$ was equivalent to the motion induced by a vortex sheet on the y-axis extending from $y = -\frac{S}{2}$ to $\frac{S}{2}$ with a vorticity $\frac{1}{2}\Gamma(y)$. (See figure). If we surround the slit by the closed curve C_1, then $\phi(y,z)$ is regular outside C_1 and inside any circle C of large radius R. We apply the Gauss transformation to the integral of

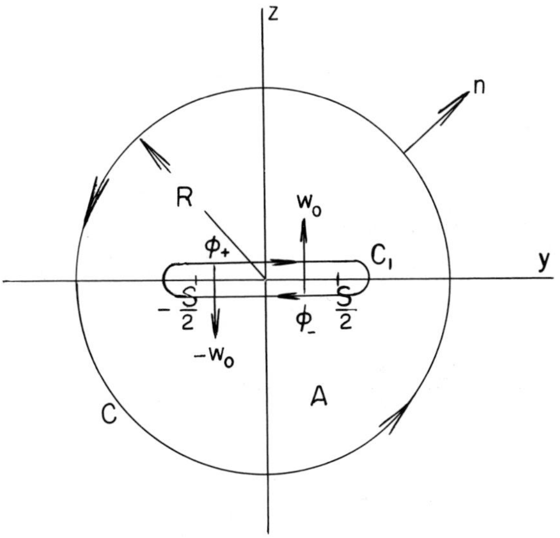

$$q_0^2 = \left(\frac{\partial\phi}{\partial y}\right)^2 + \left(\frac{\partial\phi}{\partial z}\right)^2$$

taken over this region A getting, since $\Delta\phi = 0$,

$$\iint_{(A)} q_0^2 \, dydz = \iint_{(A)} \left[\frac{\partial}{\partial y}\left(\phi \frac{\partial\phi}{\partial y}\right) + \frac{\partial}{\partial z}\left(\phi \frac{\partial\phi}{\partial z}\right)\right]dydz$$

$$= \oint_{(C+C_1)} \left[\phi \frac{\partial\phi}{\partial y} \cos(n,y) + \phi \frac{\partial\phi}{\partial z} \cos(n,z)\right]d\ell$$

$$= \oint_{(C+C_1)} \phi \frac{\partial\phi}{\partial n} \, d\ell = \oint_{(C)} \phi \frac{\partial\phi}{\partial n} \, d\ell + \oint_{(C_1)} \phi \frac{\partial\phi}{\partial n} \, d\ell .$$

This line integral is taken over the total boundary; $\vec{n}$ is the outward normal and

$d\ell$ is the element of length of the boundary.

Consider first the integral over C. There $\frac{\partial\phi}{\partial n} \sim \frac{1}{R}$ and $d\ell \sim R d\theta$. Now on a circle of large R the flow behaves like that induced by a single concentrated vortex line of vorticity

$$\frac{1}{2}\int_{-S/2}^{S/2} d\Gamma$$

and thus has the complex potential

$$\frac{1}{2\pi i} \cdot \frac{1}{2}\int_{-S/2}^{S/2} d\Gamma \cdot \log(y+zi).$$

As the integral of $d\Gamma$ is zero the potential $\phi(y,z)$ at infinity approaches zero. It follows that the integral of $\phi \frac{\partial\phi}{\partial n} d\ell$ over C vanishes. If we let the curve C_1 contract to the slit we have* $\phi = \phi_+$, $\frac{\partial\phi}{\partial n} = -\frac{\partial\phi}{\partial z} = -w_0$ and on the lower side $\phi = \phi_-$, $\frac{\partial\phi}{\partial n} = \frac{\partial\phi}{\partial z} = w_0$ since dn means the outward normal of the region A. Thus

$$\oint_{C_1} \phi \frac{\partial\phi}{\partial n} d\ell = -\int \phi_+ w_0 d\ell + \int \phi_- w_0 d\ell = -\int_{-S/2}^{S/2} (\phi_+ - \phi_-) w_0 dy.$$

But by Equation (3) the discontinuity in ϕ at a point of the vortex sheet is equal to the vorticity at this point, so

$$\phi_+ - \phi_- = \frac{1}{2}\Gamma(y).$$

Hence

$$\iint_{(A)} q_0^2 dydz = -\frac{1}{2}\int_{-S/2}^{S/2} \Gamma(y) w_0(y) dy$$

and we have

---

*on the upper side

$$D = 2\rho \iint q_0^2 \, dydz = \frac{\rho}{2} \iint q_\infty^2 \, dydz \qquad \text{(III')}$$

where $q_\infty = 2q_0$ is the velocity in the plane $z = -\infty$. Both integrals have to be evaluated over the whole plane exterior to the slit.

4. Airwing of Minimum Drag.

Before discussing the main problem, i.e., evaluating $\Gamma(y)$ from (I), and using this Γ for computing L and D, we consider a particular question which will lead to a simple and important result. We want to find a Γ distribution $\Gamma(y)$ so that D will be a minimum for a given L. This is a problem of the calculus of variations, for which we will give an outline of the solution.

According to well known rules, we make $D - \lambda L$, where λ is the Lagrange Multiplier, a minimum, i.e., we take the first variations of D and L and set $\delta D = \lambda \delta L$. It follows immediately from (II) that

$$\delta L = \rho V \int_{-S/2}^{S/2} \delta\Gamma(y)dy.$$

The calculation of δD is somewhat more lengthy. Using the form (III') for D, we have

$$\delta D = 2\rho\delta \iint \left[\left(\frac{\partial\phi}{\partial y}\right)^2 + \left(\frac{\partial\phi}{\partial z}\right)^2 \right] dydz.$$

Since $\phi(y,z)$ depends on Γ we then get

$$\delta D = 4\rho \iint \left[\frac{\partial\phi}{\partial y} \cdot \frac{\partial\delta\phi}{\partial y} + \frac{\partial\phi}{\partial z} \cdot \frac{\partial\delta\phi}{\partial z} \right] dydz.$$

by the well-known theorem for taking the variation of a derivative which allows us to set $\delta\partial$ equal to $\partial\delta$. Since $\Delta\phi = 0$ we can write this expression as

$$\delta D = 4\rho \iint \left[\frac{\partial}{\partial y} \left(\frac{\partial \phi}{\partial y} \delta\phi\right) + \frac{\partial}{\partial z} \left(\frac{\partial \phi}{\partial z} \delta\phi\right) \right] dydz.$$

Then by the Gauss transformation

$$\delta D = 4\rho \oint \delta\phi \left[\frac{\partial \phi}{\partial y} \cos(y,n) + \frac{\partial \phi}{\partial z} \cos(z,n) \right] d\ell$$

$$= 4\rho \oint \frac{\partial \phi}{\partial n} \delta\phi d\ell$$

$$= 4\rho \oint_{(\text{inf. circ.})} \frac{\partial \phi}{\partial n} \delta\phi d\ell + 4\rho \oint_{(\text{slit})} \frac{\partial \phi}{\partial n} \delta\phi d\ell.$$

The integral over the circle becomes zero as the radius approaches infinity according to the argument on pp. 159—160. On the slit we have, as before $\frac{\partial \phi}{\partial n} = \mp \frac{\partial \phi}{\partial z}$, $\delta\phi = \begin{cases} \delta\phi_+ \\ \delta\phi_- \end{cases}$, $\delta\phi_+ - \delta\phi_- = \frac{1}{2}\delta\Gamma$ and $d\ell = \pm dy$. Hence

$$\delta D = -2\rho \int_{-S/2}^{S/2} \frac{\partial \phi}{\partial z} \delta\Gamma dy = -2\rho \int_{-S/2}^{S/2} w_0(y)\delta\Gamma dy.$$

Substitute these values for δD and δL into our minimum principle and we have

$$\delta D - \lambda\delta L = -\rho \int_{-S/2}^{S/2} [2w_0 + V\lambda]\delta\Gamma(y)dy = 0.$$

This variation will be zero for all possible $\delta\Gamma$ if and only if the expression in brackets is zero at every point on the slit. This means that $w_0 = \frac{\partial \phi}{\partial z} = \text{const.}$ along the slit, or the "best airwing" is that for which the downwash velocity w_0 is a constant along the span.

In order to make use of this result we have to find the Γ distribution that leads to a constant value of w_0. According to (19a) we have to solve the integral equation of first kind

$$\int_{-S/2}^{S/2} \frac{d\Gamma(\eta)}{y - \eta} = \text{const.} \tag{22}$$

Instead of trying to solve this equation directly, we shall indicate a particular two dimensional flow with vortex sheet at $z = 0$, $-\frac{S}{2} \leq y \leq \frac{S}{2}$, and show that this flow satisfies the condition that $w_0 = \text{const.}$ along the sheet. Then the corresponding Γ distribution will be the solution of (22). Let $u = y + zi$ and c be a real constant, then we set

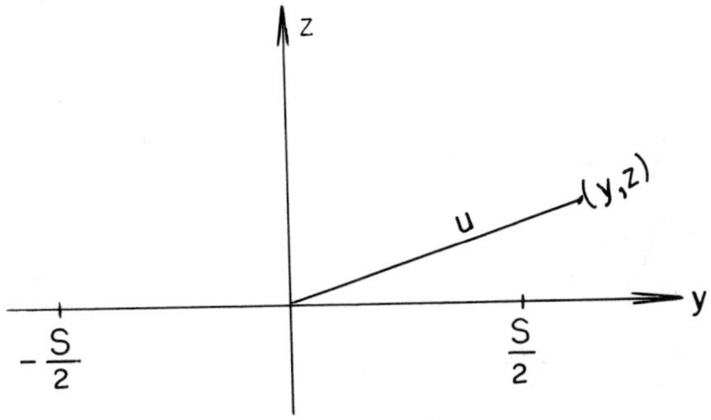

$$\phi + \psi i = ci\left[u - \sqrt{u^2 - \frac{S^2}{4}}\right].$$

Now u is real along the y-axis while the square root is real on the portion of the y-axis outside the slit and imaginary on the slit. Hence along the slit

$$\phi + \psi i = cyi + c\sqrt{\frac{S^2}{4} - y^2} = cyi + \text{real no.}$$

Hence $\psi = cy$ and so

$$w_0 = \frac{\partial\phi}{\partial z} = -\frac{\partial\psi}{\partial y} = -c$$

or the vertical velocity along the slit is constant, as we wished it to be.

We want now to find $\Gamma(y)$, which on the slit is the difference in the ϕ values or

$$\frac{1}{2}\Gamma = \phi_+ - \phi_- . \tag{23}$$

When we approach the slit from the positive or negative side ϕ approaches one of the values

$$\pm c\sqrt{\frac{S^2}{4} - y^2} ,$$

and thus the difference $\phi_+ - \phi_- = 2c\sqrt{\frac{S^2}{4} - y^2}$. To make this clearer we notice that on the slit

$$\phi^2 + \psi^2 = \frac{c^2S^2}{4} .$$

This is the equation of a circle in the $\phi\psi$ plane as shown in the figure. As $\psi = cy$ varies from $-\frac{cS}{2}$ to $\frac{cS}{2}$ we traverse the right half circle where

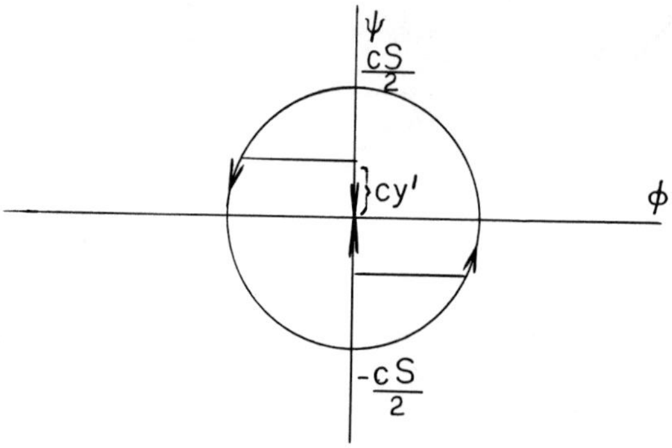

$\phi_+ = c\sqrt{\frac{S^2}{4} - y^2}$ and as ψ varies back again from $\frac{cS}{2}$ to $-\frac{cS}{2}$ we traverse the left half circle where $\phi_- = -c\sqrt{\frac{S^2}{4} - y^2}$. Substituting these values of ϕ_+ and ϕ_- into (23) we have

$$\Gamma(y) = 4c\sqrt{\frac{S^2}{4} - y^2} \tag{24}$$

as the definite solution of the minimum drag problem.

The grap of (24) is the upper half of the ellipse $\Gamma^2 + 16c^2y^2 = 4c^2S^2$

shown in the figure. Therefore, the distribution given by (24) is called <u>elliptic distribution</u> and as the lift per unit span is proportional to Γ, we say that in (24) we have an <u>elliptic distribution of lift</u>. Thus the airwing with minimum drag for given L is an airwing whose Γ distribution is a maximum at its center and drops to zero at the ends in the same way as the ordinates of an ellipse.

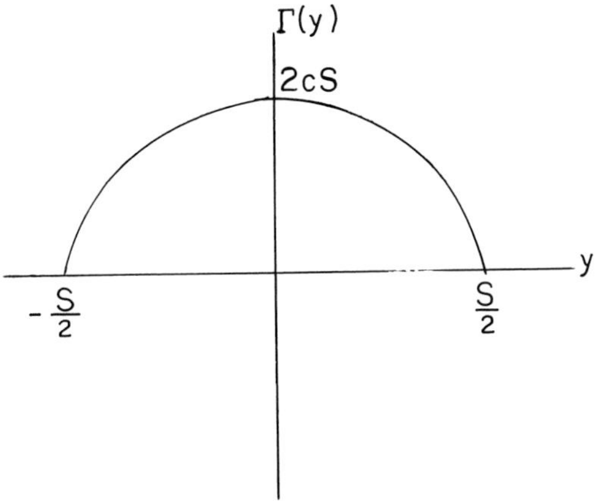

Substituting this elliptic distribution (24) of Γ into expression (II) for L we have,

$$L = \rho V \cdot 4c \int_{-S/2}^{S/2} \sqrt{\frac{S^2}{4} - y^2}\, dy = \frac{\pi}{2} \rho V c S^2. \tag{25}$$

The integral for D only differs from that for L in having V replaced by the constant $-w_0$ which is c here, so

$$D = \frac{\pi}{2} \rho c^2 S^2. \tag{26}$$

If we eliminate c between these equations we find that

$$D = \frac{L^2}{\pi \frac{\rho}{2} V^2 S^2} \quad \text{or} \quad L^2 = \pi \frac{\rho}{2} V^2 S^2 D \tag{27}$$

no matter what c is. If, for instance, c changes through a change in the angle of attack, this relation, which is seen to be parabolic, still holds between L and D.

Equation (27) was found for the airwing which has minimum drag for given lift. Hence we conclude that no matter what airwing we consider we always have

$$L^2 \leq \pi \frac{\rho}{2} V^2 S^2 D, \quad D \geq \frac{L^2}{\pi \frac{\rho}{2} V^2 S^2} . \tag{27'}$$

It is more customary to use the dimensionless coefficients

$$C_L = \frac{L}{\frac{\rho}{2} V^2 A}, \quad C_D = \frac{D}{\frac{\rho}{2} V^2 A}$$

where A is the area of the wing. In terms of these coefficients the above inequality becomes

$$C_D \geq \frac{C_L^2}{\pi} \cdot \frac{A}{S^2} .$$

The ratio $\frac{S^2}{A}$ is called the <u>aspect ratio</u> and denoted by Æ. For a rectangular wing $A = \ell S$ if ℓ is the chord and

$$\text{Æ} = \frac{S}{\ell} .$$

For any differently shaped wing the chord ℓ is a function of the coordinate y and

$$A = \int_{-S/2}^{S/2} \ell(y)dy = \bar{\ell} S$$

where $\bar{\ell}$ is the mean chord. Then $\text{Æ} = \frac{S}{\bar{\ell}}$. In any case we have

$$C_D \geq \frac{C_L^2}{\pi \text{Æ}} \tag{28}$$

as one of the most important results of the theory of Prandtl.

The equation $C_D = \frac{C_L^2}{\pi R}$ gives us a parabola (see figure) whose C_L shape depends on R (usually 6 for wind tunnel models). On the other hand, measurement of corresponding values of C_L and C_D at different angles of attack give points lying near but in all cases to the right of the parabola, in accordance with the theory. The horizontal distances between the theoretical parabola and the experimental curve can be considered a measure of the effectiveness of an airwing. These distances are due to the deviation of the actual Γ distribution from the elliptic one and to frictional influences. In no case can we ask for a wing which has a smaller drag than that corresponding to the parabola which belongs to its aspect ratio. For the infinite wing with $R = \infty$ the parabola becomes the vertical axis and we have the theoretical case of no drag.

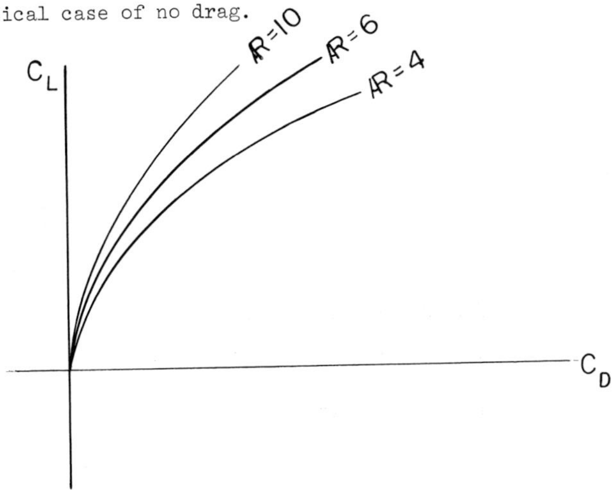

If the Γ distribution is not elliptic there will be a theoretical relation between C_D and C_L represented by a curve (curve (2) in next figure) lying between the parabola (1) and the curve obtained from experimental data (3). The horizontal distance between the parabola and the curve representing the Γ distribution gives for each angle α the drag due to the fact that the airwing is not the "best possible". The drag represented by the abscissas of curve (2) is called the <u>induced drag</u>. The horizontal distance between curve (2) and the experimental curve (3) then represents the friction drag. This last factor may be diminished by making the air wing smoother, etc., but the induced drag cannot be diminished except by changing the shape of the wing so that the Γ distribution is more nearly elliptic.

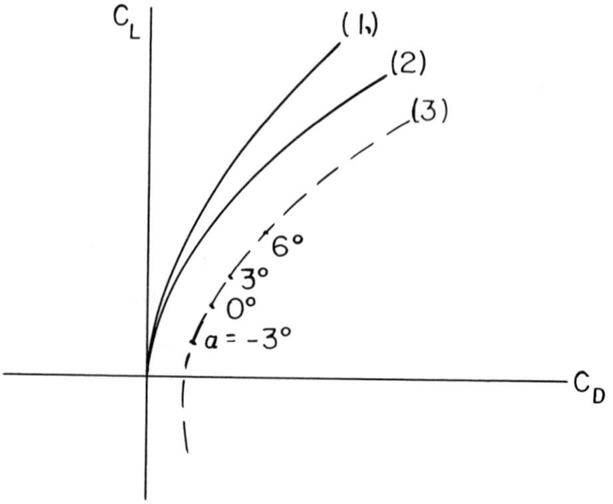

The air wing theory as developed here does not hold for all angles of attack α. Clearly when α increases sufficiently so that the difference in pressure between the upper and lower side of the wing is so great that a negative pressure would be needed on the upper side, the theory breaks down. Then the flow does not follow the outline of the profile and we get what is called a "separation". As a matter of fact, this separation occurs for angles of attack α which are a little smaller than those which would produce a negative pressure. This will be discussed later in the Boundary Layer Theory.

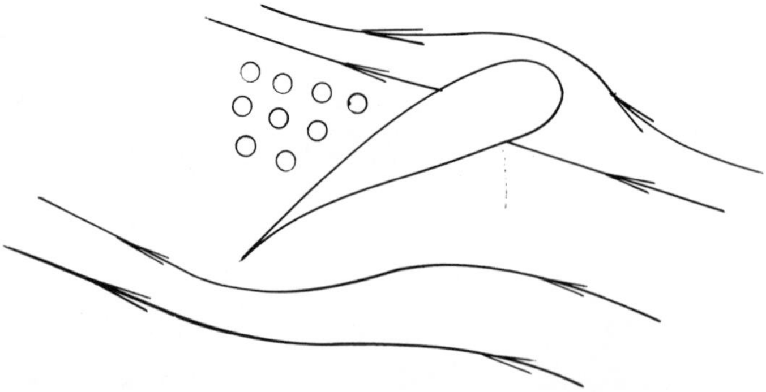

Until now we found that to get the "best airwing" we should have one such that the Γ distribution is elliptic. Now we wish to find out how such an airwing can be realized. From equation (I) we have

$$\Gamma(y) = k(y)V\left[\alpha'(y) + \frac{w_0}{V}\right].$$

In our case w_0 is equal to the constant $-c$. Thus

$$\Gamma(y) = k(y)V\left[\alpha'(y) - \frac{c}{V}\right]. \tag{29}$$

On the other hand, we saw that $\Gamma(y)$ is to be of the form

$$\Gamma(y) = 4c\sqrt{\frac{S^2}{4} - y^2}. \tag{29a}$$

In order to bring (29) and (29a) into agreement we can vary α' (which means varying β, the direction of the first axis) and k ($= 4\pi a$ which means varying the radius of the mapping circle of the profile). A simple way of finding an airwing satisfying (29) and (29a) is to keep α' constant, let the shape of the profile be constant, and vary only the length of the chord ℓ. In this case we have

$$k = 4\pi a = 4\pi \frac{a}{\ell} \cdot \ell, \quad \text{with} \quad \frac{a}{\ell} = \text{const.}$$

as this ratio depends only upon the shape of the profile. Thus k is proportional to the chord ℓ. If we let ℓ_0 be a fixed length and take the chord of the wing according to the law

$$\ell(y) = \ell_0\sqrt{1 - \frac{4y^2}{S^2}}, \tag{30}$$

where S is the span, then we see easily that (29) and (29a) are fulfilled. Upon substituting (30) in (29) and (29a), we have

$$\Gamma = 4c\sqrt{\frac{S^2}{4} - y^2} = 4\pi \frac{a}{\ell} \cdot \ell_0 \sqrt{1 - \frac{4y^2}{S^2}}\, V(\alpha' - \frac{c}{V}).$$

Or

$$2cS = 4\pi \frac{a}{\ell} \cdot \ell_0 V(\alpha' - \frac{c}{V}).$$

Once we take c fulfilling this condition our problem is solved.

For convenience let $\kappa = 4\pi \frac{a}{\ell}$ which is a constant (depending only upon the shape of the profile). Then solving this last equation for c,

$$c = \frac{\alpha' V}{1 + \frac{2S}{\kappa \ell_0}} .$$

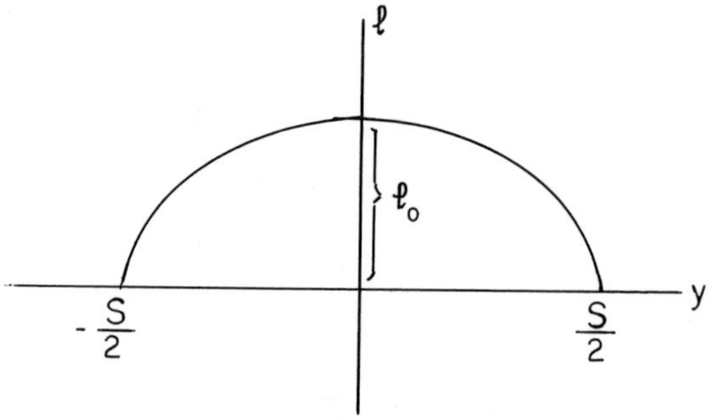

If we take the airwing to be half an ellipse or any other shape so long as the distribution $\ell(y)$ satisfies (30), we have

$$A = \int_{-S/2}^{S/2} \ell(y)dy = \frac{S\ell_0\pi}{4} .$$

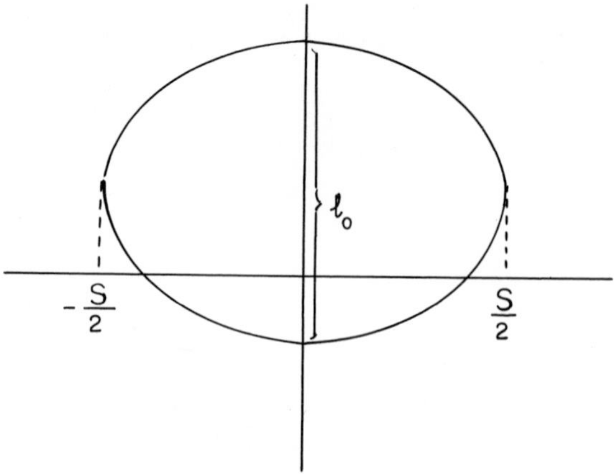

Since the aspect ratio AR is $\frac{S^2}{A}$, we get in this case $AR = \frac{S}{\ell_0} \cdot \frac{4}{\pi}$. Substituting in the above equation for c, we have

$$c = \frac{V\alpha'}{1 + \frac{\pi AR}{2\kappa}}$$

Then equation (25) for lift L becomes

$$L = \frac{1}{2}\rho V^2 s^2 \pi \frac{\alpha'}{1 + \frac{2s}{\kappa \ell_0}} ,$$

and the lift coefficient C_L takes the form

$$C_L = \frac{s^2}{A}\pi \frac{\alpha'}{1 + \frac{2s}{\kappa\ell_0}} = \frac{2\kappa\alpha'}{\frac{\kappa\ell_0}{2s}\left(1 + \frac{2s}{\kappa\ell_0}\right)} \text{ ...}$$

Now $\kappa = 4\pi \frac{a}{\ell}$, and if we assume $\frac{a}{\ell} \sim \frac{1}{4}$ or $\kappa \sim \pi$ we get

$$C_L = \frac{2\pi\alpha'}{1 + \frac{2}{AR}} . \qquad (31)$$

This equation is of great importance as it enables us to get a relation between the lift coefficients for the 2-dimensional case of wings of infinite span and wings of finite span. In the case of infinite span $\frac{1}{AR} = 0$ and we have

$$C_L^{\infty} = 2\pi\alpha' ,$$

a result we obtained previously (Chapter II, page 115). Putting this back in equation (30) we have

$$C_L = \frac{C_L^{\infty}}{1 + \frac{2}{AR}} . \qquad (32)$$

If we take a wind tunnel model with $AR = 6$ then $C_L^{\infty} = \frac{4}{3} C_L$ and we see that the values of C_L we get should be $\frac{3}{4}$ the value obtained in the two-dimensional theory.

We already mentioned that the real vortex sheet is not exactly a plane horseshoe sheet, but one that lies on a surface which has a slight inclination so as to be tangential to the velocity at each point. Besides the x and z components V and w_0 of the velocities which determine the inclination of the vortex sheet, there is still a v component as considered in (18a) and (18b). There we had $v = \pm \frac{1}{2} \frac{\partial \Gamma}{\partial y}$. In the case of elliptic Γ - distribution we have $\frac{\partial \Gamma}{\partial y} = 0$ at the center, $\frac{\partial \Gamma}{\partial y}$ negative as we go to positive values of y and positive as we go in the negative direction. Thus the particles have a flow around the wing as shown in the figure. This is connected with the fact that the pressure is smaller above the wing than below. We remark that, in general, the Γ distribution is quite similar to an elliptic one so that the general character of this circulating motion in the $\pm v$ direction holds for practically all cases.

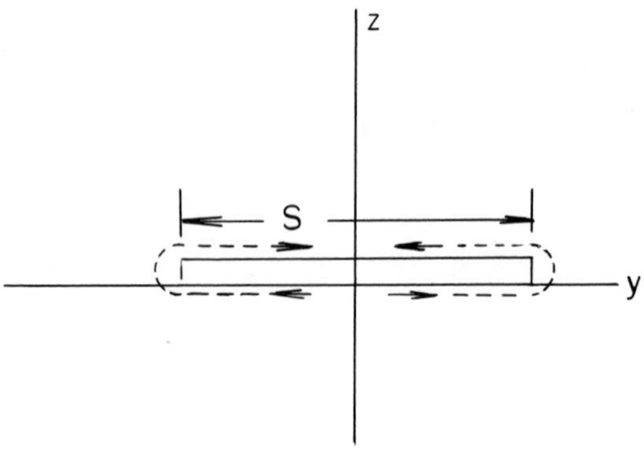

5. General Problems.

We now pass to the problem of solving the integrodifferential equation (I). This equation may be written

$$\Gamma(y) = k(y)V\left[\alpha'(y) + \frac{w_0}{V}\right].$$

Instead of introducing the integral (19) for w_0 we consider the plane $x = 0$ where, as we know, the velocity distribution v,w is determined by a potential function $\Phi(y,z)$ induced by the vortex distribution $\frac{\Gamma}{2}$. Then we have

$$w_0 = \frac{\partial \Phi}{\partial z} \quad \text{for} \quad \left\{ \begin{array}{l} z = 0 \\ -S/2 \leq y \leq S/2 \end{array} \right\}.$$

In order to find Φ we consider the conformal transformation

$$u = \frac{S}{4}\left(\zeta + \frac{1}{\zeta}\right), \tag{33}$$

where $u = y + zi$ and $\zeta = \xi + \eta i$. Clearly this transformation maps the slit $-\frac{S}{2} \le y \le \frac{S}{2}$, $z = 0$ into the circle $|\zeta| = 1$ such that the outside of the slit goes into the outside of the circle. From the transformation we have

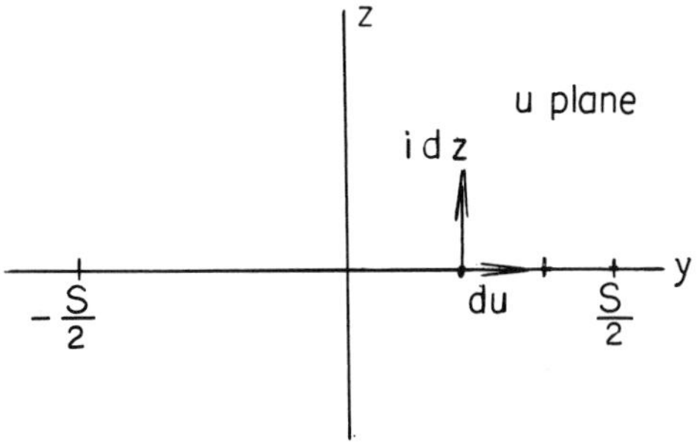

$$\frac{du}{d\zeta} = \frac{S}{4}\left(1 - \frac{1}{\zeta^2}\right) = \frac{S}{4\zeta}\left(\zeta - \frac{1}{\zeta}\right).$$

And for $\zeta = e^{i\theta}$, we get

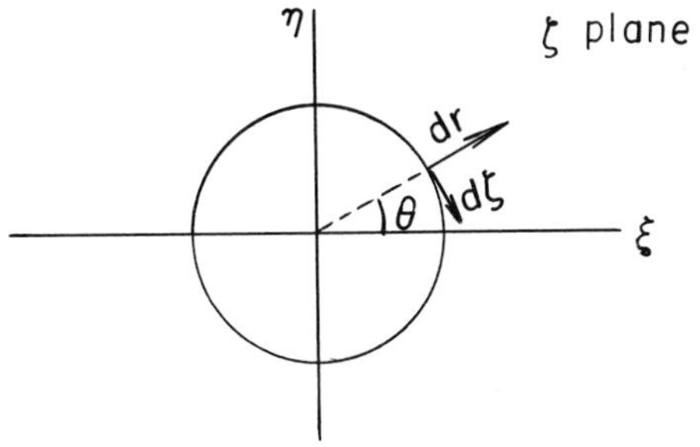

$$\frac{du}{d\zeta} = \frac{S}{4} e^{-i\theta} \cdot 2i \sin\theta,$$

or

$$d\zeta = \frac{2e^{i\theta} du}{i\, S \sin\theta}.$$

Thus to an element of length du along the slit (du = dy real) corresponds an element $d\zeta$ tangent to the circle at the point $e^{i\theta}$ (see figure). Now to an element $du = idz$ on the slit corresponds

$$d\zeta = \frac{2e^{i\theta} dz}{S \sin\theta},$$

and the element $d\zeta$ has the direction of the radius vector at the point $e^{i\theta}$. We denote its amount, as usual, by dr (see figure). Hence $\frac{dr}{dz} = \frac{2}{S \sin\theta}$, and we get

$$w_0 = \frac{\partial\Phi}{\partial z} = \frac{\partial\Phi}{\partial r} \cdot \frac{dr}{dz} = \frac{\partial\Phi}{\partial r} \cdot \frac{2}{S \sin\theta}$$

for points on the slit. Since the vorticity distribution on the slit is $\frac{1}{2}\Gamma$, we have for the difference in potential on the upper and lower sides of the slit

$$\Phi_+ - \Phi_- = \frac{1}{2}\Gamma.$$

We are allowed to assume that the potential on opposite sides of the slit is the same except for sign and thus we write

$$\Gamma = 4\Phi_+ = 4\Phi.$$

Hence under the transformation (33), equation (I) takes the form

$$4\Phi(\theta) = k(\frac{S}{2}\cos\theta)V\left[\alpha'(\frac{S}{2}\cos\theta) + \frac{2}{SV} \cdot \frac{1}{\sin\theta}\frac{\partial\Phi}{\partial r}\right] \tag{34}$$

for $r = 1$. This equation is of the form

$$\Phi(\theta) + A\frac{\partial\Phi(\theta)}{\partial n} = B \tag{34a}$$

where we put ∂n for ∂r, and A and B are given functions of θ. Equation (34a) gives a linear relation between the values of Φ and its normal derivative at the points on the unit circle. The problem of solving (I) is thus reduced to that of finding a function Φ which satisfies the Laplace equation $\Delta\Phi = 0$ outside the unit circle and satisfies (34a) on the unit circle. This is the so-called <u>third boundary problem of potential theory</u>. (The first problem is: given Φ on the boundary; the second, given $\frac{\partial\Phi}{\partial n}$ on the boundary; the third, given a linear relation between Φ and $\frac{\partial\Phi}{\partial n}$ on the boundary).

In many cases it is usual to transform the boundary value problem of a partial differential equation into the problem of an integral equation of second kind. Here we are going in the opposite direction, since we have on hand a form of solution for the boundary problem when the boundary consists of a circle.

We assume that exterior to the unit circle the flow has a complex potential $\phi + \psi i$ which can be expanded in inverse powers of ζ or

$$\phi + \psi i = \frac{1}{2} \text{SV} \sum_{n=0}^{\infty} C_n \zeta^{-n}, \quad C_n = A_n + iB_n \tag{35}$$

where A_n and B_n are real numbers. Take $\zeta = re^{\theta i}$ and then

$$\zeta^{-n} = \frac{1}{r^n} (\cos n\theta - i \sin n\theta).$$

As a result

$$\phi = \frac{1}{2} \text{SV} \sum_{n=0}^{\infty} \frac{1}{r^n} (A_n \cos n\theta + B_n \sin n\theta).$$

For $r = 1$ the series on the right-hand side must give the values of ϕ on the unit circle. It follows from our assumption that ϕ is an odd function on the circle or $\phi(-\theta) = -\phi(\theta)$. Therefore, the cosine terms must vanish and we have

$$\phi = \frac{1}{2} \text{SV} \sum_{n=1}^{\infty} \frac{B_n}{r_n} \sin n\theta. \tag{36}$$

Then $\frac{\partial\phi}{\partial r} = -\frac{1}{2} \text{SV} \sum_{n=1}^{\infty} \frac{nB_n \sin n\theta}{r^{n+1}}$, and the boundary condition (34) on the circle where $r = 1$ becomes

$$2S \sum_{1}^{\infty} B_n \sin n\theta = k\left[\alpha' - \frac{1}{\sin\theta} \sum_{1}^{\infty} nB_n \sin n\theta\right]. \tag{34b}$$

If we let $\mu = \frac{k}{2S}$ we have the equation for the infinite number of unknowns B_n:

$$\sum_{1}^{\infty} B_n \sin n\theta\left(1 + \frac{\mu n}{\sin \theta}\right) = \mu\alpha' . \tag{37}$$

This must be satisfied for each value of θ between 0 and 2π. We may consider (37) as a continuously infinite set of equations for a denumerably infinite number of unknowns B_n.

Assume that we have found the values of the B_n satisfying (37), then we show how L and D can be calculated. To do so we need $\Gamma(y)$ and $w_0(y)$ which are

$$\Gamma(y) = 4\phi\Big|_{r=1} = 2SV \sum_{1}^{\infty} B_n \sin n\theta \tag{38}$$

$$w_0(y) = \frac{2}{S \sin \theta} \frac{\partial\phi}{\partial r}\Bigg|_{r=1} = -V \sum_{1}^{\infty} nB_n \frac{\sin n\theta}{\sin \theta} \tag{39}$$

where $y = \frac{S}{2} \cos \theta$. Since $dy = -\frac{S}{2} \sin \theta d\theta$, we get

$$L = \rho V \int_{-S/2}^{S/2} \Gamma(y)dy = -\rho V^2 S^2 \int_{\pi}^{0} \sum_{1}^{\infty} B_n \sin n\theta \cdot \sin \theta d\theta .$$

Integrating the series term by term, every one gives zero except the one for which $n = 1$. Thus

$$L = \rho V^2 S^2 \int_{0}^{\pi} \sin^2\theta d\theta \cdot B_1 = \frac{\rho}{2} V^2 S^2 \pi B_1 . \tag{40}$$

We see that L depends only on the <u>first coefficient</u> B_1. In the same way

$$D = -\rho \int_{-S/2}^{S/2} \Gamma(y)w_0(y)dy = -\rho V^2 S^2 \int_{\pi}^{0} \sum_{1}^{\infty} B_m \sin m\theta \sum_{1}^{\infty} nB_n \sin n\theta d\theta .$$

The non-zero terms in this integration are only the ones in the product of the two series for which $m = n$ or

$$D = \frac{\rho}{2} V^2 s^2 \pi \sum_1^\infty n B_n^2 = \frac{\rho}{2} V^2 s^2 \pi B_1^2 (1+\delta) \tag{41}$$

where

$$\delta = \sum_2^\infty \frac{n B_n^2}{B_1^2} \tag{41'}$$

since $B_1 \neq 0$ for a non-zero lift. Thus D depends upon all the coefficients B_n.

These equations for L and D give another proof of the minimum drag problem, if we assume that Γ has a Fourier series development as in (38). Since δ is the sum of only positive numbers we see that $\delta \geq 0$. On the other hand given lift means a given B_1 according to (40). Hence we see by (41) that D is a minimum if $\delta = 0$. This is only possible if each term in the series for δ is zero or $B_2 = B_3 = \cdots = 0$. Then the above expression for Γ reduces to its first term

$$\Gamma(y) = 2sVB_1 \sin\theta = 4 \cdot VB_1 \cdot \frac{s}{2}\sqrt{1 - \frac{4y^2}{s^2}} .$$

The value of Γ is the same as the elliptic distribution (24) with $c = VB_1$.

In the general case $\delta > 0$ the dimensionless lift and drag coefficients are

$$C_L = \pi \mathcal{R} B_1, \quad C_D = \pi \mathcal{R} B_1^2 (1+\delta)$$

so

$$C_L^2 = \pi \mathcal{R} C_D \frac{1}{1+\delta} . \tag{42}$$

If $\delta = 0$ this equation becomes that of the parabola we had above. If the Γ distribution is not elliptic, equation (42) is no longer a parabola. The curve depends through δ on the B_n and in turn the B_n depend on α' and μ according to equation (37). Therefore, measurements made at different angles of attack α' give a curve other than a parabola. As the actual Γ distribution, in general, is not very different from an elliptic one we may assume that B_2, B_3, etc., and, therefore, δ are small numbers.

The question of finding the B_n from (37) still remains. Since the wing is given we know α' and μ as functions of y and hence of θ. So if we take a certain number, say p, of values of θ and set up equation (37) for $\theta = \theta_1, \theta_2, \ldots \theta_p$ (using the corresponding values of α' and μ and neglecting all B_n for $n > p$) we can solve this set of p linear equations for the first p coefficients $B_1, B_2 \ldots B_p$.

If α' is a constant, we take $\frac{B_n}{\alpha'}$ as the unknown and have only the parameter μ in our equations. When the B_n are found it is easy to calculate δ and find the correct drag formula. Glauert on p.146 of his "Aerofoil and Airscrew Theory" shows the work for the case of a rectangular wing, so that μ is constant, and for constant angle α'.

6. Formal Solution of the Integral Equation.

While we do not intend to study existence theorems for the integro-differential equation of Prandtl, we shall show how a direct solution similar in form to the solution of the partial differential equation of the last section can be found. We consider the problem in the form of a related pair of equations for $\Gamma(y)$ and $w_0(y)$

$$\Gamma(y) = k(y)V\left[\alpha'(y) + \frac{w_0(y)}{V}\right] \tag{43}$$

$$w_0(y) = -\frac{1}{4\pi}\int_{-S/2}^{S/2} \frac{\Gamma'(\eta)d\eta}{y - \eta} \tag{44}$$

where, as seen, we restrict ourselves to the case when Γ has a derivative. We do not use the two dimensional flow with potential Φ as above, but do make the following transformation of the independent variable

$$y = \frac{S}{2} \cos \theta, \qquad \eta = \frac{S}{2} \cos \psi.$$

We assume the solutions then have the form

$$\Gamma(y) = 2SV \sum_{n=1}^{\infty} B_n \sin n\theta \tag{45}$$

$$w_0(y) = -\frac{V}{\sin \theta} \sum_{n=1}^{\infty} nB_n \sin n\theta \tag{46}$$

and proceed to show that (45) and (46) do actually satisfy (43) and (44). We have from (45)

$$\Gamma'(y) = 2SV \sum_{n=1}^{\infty} nB_n \cos n\theta \cdot \frac{d\theta}{dy}.$$

But $dy = -\frac{S}{2} \sin \theta d\theta$ so $\frac{d\theta}{dy} = -\frac{2}{S \sin \theta}$ and

$$\Gamma'(y) = -\frac{4V}{\sin \theta} \sum_{n=1}^{\infty} nB_n \cos n\theta. \tag{47}$$

Substituting (45) and (46) into (43) we have

$$2S \sum_{n=1}^{\infty} B_n \sin n\theta = k\left[\alpha' - \frac{1}{\sin \theta} \sum_{n=1}^{\infty} nB_n \sin n\theta\right]$$

or

$$\sum_{n=1}^{\infty} B_n \sin n\theta\left(1 + \frac{k}{2S} \frac{n}{\sin \theta}\right) = \frac{k}{2S} \alpha'. \tag{48}$$

This is the same as equation (37) for the B_n in the preceding section. It is the condition placed upon the B_n in order that (45) and (46) be a solution of (43).

On the other hand, the same expressions for Γ and w_0 must satisfy equation (44). When (46) and (47) are substituted into (44) we get:

$$\frac{1}{\sin\theta}\sum_{n=1}^{\infty} nB_n \sin n\theta = -\frac{1}{\pi}\int_{-S/2}^{S/2}\sum_{1}^{\infty}\frac{nB_n \cos n\psi}{(y-\eta)\sin\psi}\,d\eta .$$

But $y - \eta = \frac{S}{2}(\cos\theta - \cos\psi)$, $d\eta = -\frac{S}{2}\sin\psi\,d\psi$ so the equation is

$$\sum_{n=1}^{\infty}\left\{ nB_n \frac{\sin n\theta}{\sin\theta} - \frac{1}{\pi}\int_{\pi}^{0}\frac{nB_n \cos n\psi}{\cos\theta - \cos\psi}\,d\psi \right\} = 0.$$

This must be true whatever the B_n are. Hence the coefficient of each B_n must vanish or

$$\sin n\theta = -\frac{1}{\pi}\int_{0}^{\pi}\frac{\cos n\psi \sin\theta}{\cos\theta - \cos\theta}\,d\psi . \qquad \text{(A)}$$

We started by assuming for Γ and w_0 the expressions (45) and (46) and we derived from this assumption the identity (A). Equation (A) is a purely trigonometric formula. If we had started by proving (A) we would have been able to derive (46) from (45), i.e., from the assumption that $\Gamma(y)$ has a Fourier development with respect to θ.

As a matter of fact, (A) is true and can be proven by elementary transformations (proof is on p. 93 of Glauert's book) as well as the following analogous formula

$$\cos n\theta = \frac{1}{\pi}\int_{0}^{\pi}\frac{\sin n\psi \sin\psi}{\cos\theta - \cos\psi}\,d\psi . \qquad \text{(B)}$$

Both the integrals (A) and (B) are improper, the integrand having an infinite value at the point on the interval of integration where $\psi = \theta$. They have a logarithmic infinity if the integration is carried out from 0 to θ or from θ to π. But they have a so-called Cauchy principle value which is defined by

$$\int_{0}^{\pi} = \lim_{\epsilon \to 0}\left[\int_{0}^{\theta-\epsilon} + \int_{\theta+\epsilon}^{\pi}\right].$$

As we saw, the assumption that $\Gamma(y)$ has a Fourier series expansion (45) leads, by use of formula (A), to equation (46). Under the same assumption the identity (B) can be used to obtain a formal solution of the equation (44) which reads

$$w_0(y) = -\frac{1}{4\pi}\int \frac{\Gamma'(\eta)d\eta}{y - \eta}. \tag{44}$$

If we apply the identity (B) to equation (47) which is the derivative of (45), we get

$$\Gamma'(y) = -\frac{4V}{\pi \sin \theta}\int_0^{\pi} \frac{\sum nB_n \sin n\psi \sin \psi}{\cos \theta - \cos \psi} d\psi.$$

Substituting in this last expression the value of $w_0(\eta)$ (from (46)):

$$w_0(\eta) = -\frac{V}{\sin \psi}\sum nB_n \sin n\psi$$

it becomes

$$\Gamma'(y) = \frac{4}{\pi}\frac{1}{\sin \theta}\int_0^{\pi} \frac{w_0(\eta) \sin^2\psi}{\cos \theta - \cos \psi} d\psi.$$

Since $\eta = \frac{S}{2}\cos \psi$, $y = \frac{S}{2}\cos \theta$ we can write this in the form

$$\Gamma'(y) = \frac{4}{\pi \sin \theta}\int_{-S/2}^{S/2} \frac{w_0(\eta) \sin \psi d\eta}{y - \eta}$$

$$= \frac{4}{\pi}\frac{1}{\sqrt{1 - \frac{4y^2}{S^2}}}\int_{-S/2}^{S/2} \frac{w_0(\eta)\sqrt{1 - \frac{4\eta^2}{S^2}}}{y - \eta} d\eta. \tag{49}$$

Equation (49) is the formal solution of (44) giving the Γ distribution when the w_0 distribution is known. In the case w_0 = const. which is the condition for minimum drag we could have used this solution to find the elliptic distribution of Γ. However, this method is not simpler than the one we used as the integral above is an improper one and difficult to evaluate.

The above solution is correct under the hypothesis that $\Gamma(y)$ has a Fourier expansion (and also under certain less restrictive hypotheses). The solution has an additional term if $\int \Gamma(y)dy \neq 0$, a fact which is excluded in our Fourier series for $\Gamma(y)$.

7. Application of the Theory to the Biplane.

We now wish to apply the results obtained for a single wing to the case of a biplane. When we consider two wings of infinite span in two-dimensional flow, the problem becomes a difficult one of conformal mapping. But it follows from other considerations that the mutual interference of the two wings in the case of infinite span is not important. On the other hand if we assume a finite span the mutual interference of the two vortex sheets takes on some importance.

As in the case of one wing, we consider each of the wings reduced to a line segment with a trailing horseshoe vortex sheet. We take the case where both wings are in the same vertical plane (<u>no stagger</u>), but allow them to have different spans. Let the span of the top wing (1) be S_1, that of the bottom wing (2) be S_2, and let the distance between them be a. Let L_1 and L_2 be the lift forces for wings (1) and (2). We may assume that they are connected

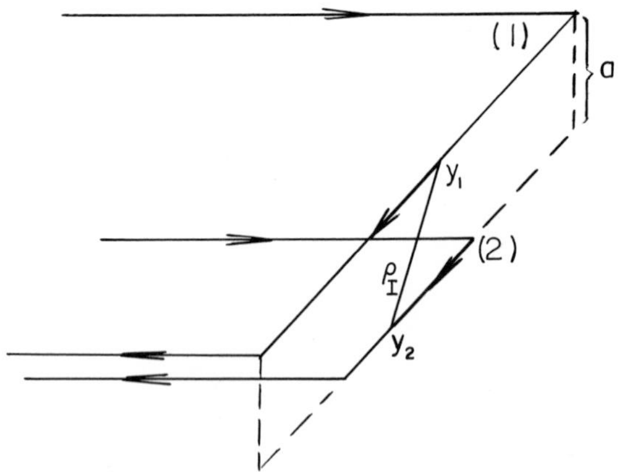

with their respective Γ-distribution $\Gamma_1(y_1)$, $\Gamma_2(y_2)$ by $L_1 = \rho V \int \Gamma_1(y_1)dy_1$, $L_2 = \rho V \int \Gamma_2(y_2)dy_2$. Then the total lift L is $L = L_1 + L_2$. The mutual effect of the vortex sheets enters into the formulas for the drag. The essential idea is that each vortex sheet induces at each point of the span a vertical velocity component and that this induced velocity at the points of the wings themselves affects the angle of attack. In the case of a single wing the change in angle was $\frac{w_0}{V}$. For the case of a biplane, instead of w_0, we have at the points of wing (1) the sum of two velocities w_{11} and w_{21} where w_{11} is induced by the first and w_{21} is induced by the second vortex sheet. In analogy to the formula for drag in the case of a single wing which was

$$D = -\rho \int_{-S/2}^{S/2} \Gamma(y) w_0(y) dy,$$

we get the drag D_1 exerted upon wing (1):

$$D_1 = -\rho \int_{-S_1/2}^{S_1/2} \Gamma_1(y_1)[w_{11}(y_1) + w_{21}(y_1)]dy_1.$$

In a similar way the drag D_2 for wing (2) is

$$D_2 = -\rho \int_{-S_2/2}^{S_2/2} \Gamma_2(y_2)[w_{22}(y_2) + w_{12}(y_2)]dy_2,$$

where w_{12} is the velocity induced by the first vortex sheet at a point of the second wing. In this way the total drag $D = D_1 + D_2$ can be decomposed into four parts

$$D = D_{11} + D_{12} + D_{21} + D_{22}$$

where

where

$$D_{11} = -\rho \int_{-S_1/2}^{S_1/2} \Gamma_1(y_1) w_{11} dy_1; \quad D_{21} = -\rho \int_{-S_1/2}^{S_1/2} \Gamma_1(y_1) w_{21}(y_1) dy_1$$

$$D_{12} = -\rho \int_{-S_2/2}^{S_2/2} \Gamma_2(y_2) w_{12}(y_2) dy_2; \quad D_{22} = -\rho \int_{-S_2/2}^{S_2/2} \Gamma_2(y_2) w_{22}(y_2) dy_2.$$

The quantities D_{11} and D_{22} are precisely those for the single wing obtained in section 3, p. 158, hence the problem is to find the quantities D_{12} and D_{21}. For this purpose, we must find the values of the quantities w_{21} and w_{12}. As w_{21} is the velocity induced by the total vortex sheet (2) at a point of wing (1) we can consider it as an integral of elements dw_{21} each of which is induced by one element $d\Gamma_2$ of the sheet (2). Let ρ_I be the distance from y_1 to y_2 (see figure) and β be the angle the line through (y_1, y_2) makes with the vertical. Then by formula (6):

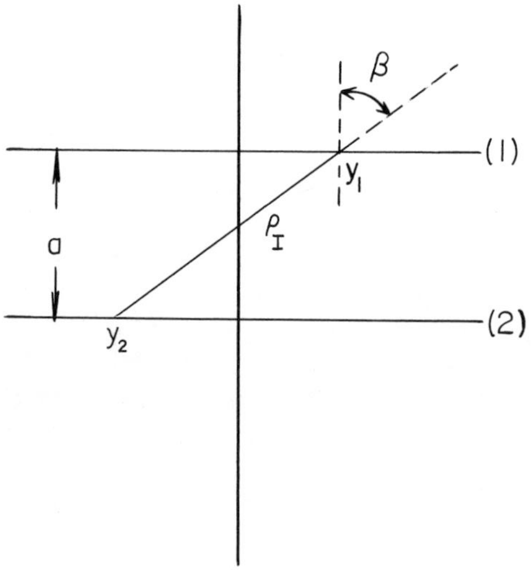

$$q = \frac{1}{4\pi} \frac{1}{\rho} (\cos \alpha + \cos \beta) = \frac{C}{4\pi} \frac{1}{\rho_I},$$

and since w_{21} is the z-component of $\vec{q}$ we have

$$dw_{21} = -\frac{1}{4\pi} d\Gamma_2 \frac{1}{\rho_I} \sin \beta.$$

(The negative sign since $d\Gamma_2$ is positive when vortex line has negative x-direction and negative when vortex line has positive x-direction). Thus

$$w_{21} = -\frac{1}{4\pi}\int_{-S_2/2}^{S_2/2} d\Gamma_2(y_2)\,\frac{\sin\beta}{\rho_I}\,.$$

Integrating by parts, we get (since $\Gamma(-\frac{S_2}{2}) = \Gamma(\frac{S_2}{2}) = 0$)

$$w_{21} = \frac{1}{4\pi}\int_{-S_2/2}^{S_2/2} \Gamma_2(y_2)\,\frac{d}{dy_2}\left(\frac{\sin\beta}{\rho_I}\right)dy_2\,.$$

Since $\rho_I^2 = a^2 + (y_1-y_2)^2$, $\sin\beta = \dfrac{y_1 - y_2}{\rho_I}$, we have

$$\frac{d}{dy_2}\left(\frac{\sin\beta}{\rho_I}\right) = -\frac{1}{\rho_I^2} + \frac{2(y_1-y_2)^2}{\rho_I^4} = \frac{-1 + 2\sin^2\beta}{\rho_I^2} = \frac{-\cos 2\beta}{\rho_I^2}\,.$$

Hence

$$w_{21} = -\frac{1}{4\pi}\int_{-S_2/2}^{S_2/2} \Gamma_2(y_2)\,\frac{\cos 2\beta}{\rho_I^2}\,dy_2,$$

and the formula for drag D_{21} becomes

$$D_{21} = +\rho\int_{-S_1/2}^{S_1/2} \Gamma_1(y_1)\left[\int_{-S_2/2}^{S_2/2} \Gamma_2(y_2)\,\frac{1}{4\pi}\,\frac{\cos 2\beta}{\rho_I^2}\,dy_2\right]dy_1\,. \qquad (50)$$

In computing D_{12} we simply have to replace β by $\pi + \beta$, and as $\cos 2\beta$ does not change in this case we get $D_{12} = D_{21}$.

We now suppose that we have an elliptic distribution in both wings. This means

$$\Gamma_1(y_1) = \text{const.}\sqrt{\frac{S_1^2}{4} - y_1^2}\,,$$

and a similar expression for $\Gamma_2(y_2)$. Then since we have

$$L_1 = \rho V \int \Gamma_1(y_1) dy_1,$$

we can find the constant in the above expression for $\Gamma_1(y_1)$, getting

$$\Gamma_1(y_1) = \frac{8L_1}{\pi\rho V S_1^2} \sqrt{\frac{S_1^2}{4} - y_1^2} ,$$

and a similar expression for $\Gamma_2(y_2)$. For convenience let $\eta_1 = \frac{2y_1}{S_1}$, $\eta_2 = \frac{2y_2}{S_2}$.

Then

$$\Gamma_1(y_1) = \frac{4L_1}{\rho\pi V S_1} \sqrt{1-\eta_1^2} , \quad \Gamma_2(y_2) = \frac{4L_2}{\rho\pi V S_2} \sqrt{1-\eta_2^2} . \tag{51}$$

Hence we can write

$$\frac{\rho_I^2}{S_1 S_2} = \frac{a^2}{S_1 S_2} + \frac{1}{4} (\eta_1^2 \frac{S_1}{S_2} - 2\eta_1\eta_2 + \eta_2^2 \frac{S_2}{S_1}).$$

We note that the quantity $\frac{\rho_I^2}{S_1 S_2}$ depends only upon the parameters $\frac{S_2}{S_1}$ and $\frac{a}{S_1}$ which are constants concerning the geometry of the wings only. Now substituting the above values for $\Gamma_1(y_1)$, $\Gamma_2(y_2)$ in equation (50), we get

$$D_{12} = D_{21} = \frac{L_1 L_2}{\rho\pi^3 V^2 S_1 S_2} \int_{-1}^{+1} \int_{-1}^{+1} \frac{S_1 S_2}{\rho_I^2} \cos 2\beta \sqrt{(1-\eta_1^2)(1-\eta_2^2)} d\eta_1 d\eta_2 . \tag{52}$$

We see that the expression under the integral sign depends only upon the geometric ratios $\frac{S_2}{S_1}$ and $\frac{a}{S_1}$. And if we let

$$\sigma = \frac{1}{2\pi^2} \int_{-1}^{1}\int_{-1}^{1} \frac{S_1 S_2}{\rho_I^2} \cos 2\beta \sqrt{(1-\eta_1^2)(1-\eta_2^2)}\, d\eta_1 d\eta_2 \, , \tag{53}$$

we have $D = D_{11} + D_{12} + D_{21} + D_{22}$

$$= \frac{1}{\pi \frac{\rho}{2} V^2} \left[\frac{L_1^2}{S_1^2} + 2 \frac{L_1 L_2}{S_1 S_2} \sigma + \frac{L_2^2}{S_2^2} \right] . \tag{54}$$

We note that this is a quadratic form in L_1 and L_2, which means a generalization of the parabolic relation we obtained for the case of an elliptic distribution for one wing.

In order to get an idea of the value of $\sigma(\frac{S_2}{S_1}, \frac{a}{S_1})$ given by (53) we consider the limiting case $a \to 0$. Then we have a monoplane with a non-elliptic distribution since the two distributions over each span add up so as to give $\Gamma(y) = c_1\sqrt{\frac{S_1^2}{4} - y^2} + c_2\sqrt{\frac{S_2^2}{4} - y^2}$. But with only one sheet instead of two the velocity induced at the points of (2) by the Γ distribution on (1) is constant if $S_2 < S_1$ or

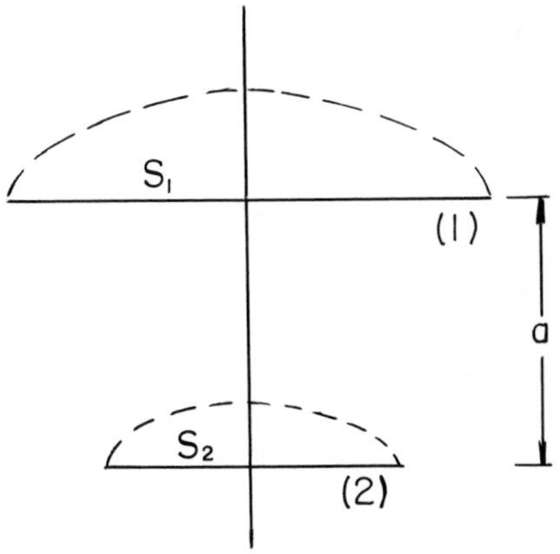

$$w_{12} = -c_1 .$$

Hence the drag of wing (2) due to (1) is

$$D_{12} = -\rho \int_{-S_2/2}^{S_2/2} \Gamma_2(y_2) w_{12}(y_2) dy_2 = \rho c_1 \int_{-S_2/2}^{S_2/2} \Gamma_2(y_2) dy_2 .$$

We know, however, c_1 in terms of L_1 since by (25)

$$L_1 = c_1 \frac{\rho}{2} V s_1^2 \pi$$

and we also have

$$L_2 = \rho V \int_{-s_2/2}^{s_2/2} \Gamma(y_2) dy_2 .$$

So we can write D_{12} in the following form

$$D_{12} = \frac{c_1}{V} L_2 = \frac{L_1 L_2}{\frac{\rho}{2} V^2 s_1^2 \pi} .$$

On the other hand, according to (54),

$$D_{12} = D_{21} = \frac{L_1 L_2 \sigma}{\pi \frac{\rho}{2} V^2 s_1 s_2} .$$

Hence for $a = 0$

$$\sigma = \frac{s_2}{s_1} \quad \text{if} \quad s_2 < s_1. \tag{55}$$

We might ask why there is lack of symmetry in this value for σ. The answer is that we cannot apply the same argument to D_{21} since w_{21} is constant only in the interval $-\frac{s_2}{2} \leq y \leq \frac{s_2}{2}$ and not over the whole interval $-\frac{s_1}{2} \leq y \leq \frac{s_1}{2}$. On the other hand, we learned by general considerations that $D_{12} = D_{21}$.

It is clear that σ will diminish as a increases, for as the distance between the two wings increases the interference must weaken When $a \to \infty$ we must have $\sigma = 0$. It really can be shown that σ diminishes monotonically as a increases, so $\frac{s_2}{s_1}$ is the maximum value for σ, and we always have

$$\sigma \leq \frac{s_2}{s_1} \leq 1 \leq \frac{s_1}{s_2} \quad \text{if} \quad s_2 < s_1. \tag{55'}$$

It is customary to plot σ against the ratio of a to the arithmetic mean of the two spans. The curves have then the general shape shown in the figure.

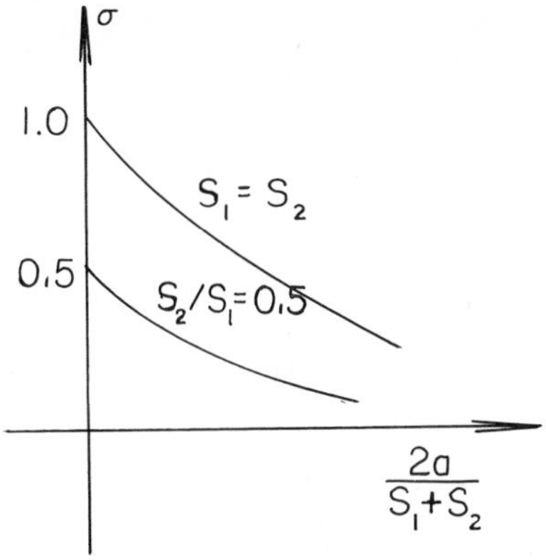

We may raise the problem of finding the value of σ for which the total drag is a minimum for a given total lift $L_1 + L_2$. The expression to be minimized is, using (54),

$$\frac{1}{2}\left[\frac{L_1^2}{S_1^2} + 2\,\frac{L_1 L_2}{S_1 S_2}\,\sigma + \frac{L_2^2}{S_2^2}\right] + \lambda(L_1 + L_2)$$

where λ is the Lagrange multiplier. Taking derivatives with respect to L_1 and L_2 we get

$$\frac{L_1}{S_1^2} + \sigma\,\frac{L_2}{S_1 S_2} + \lambda = 0$$

$$\frac{L_2}{S_2^2} + \sigma\,\frac{L_1}{S_1 S_2} + \lambda = 0.$$

We subtract in order to eliminate λ and get

$$\frac{L_1}{L_2} = \frac{S_1/S_2 - \sigma}{S_2/S_1 - \sigma} \tag{56}$$

or the value of σ which gives minimum drag is

$$\sigma = \frac{1}{L_1 - L_2}\left(\frac{S_2}{S_1}\,L_1 - \frac{S_1}{S_2}\,L_2\right).$$

Then, on substituting (56) into (54) and reducing,

$$D_{min} = \frac{(L_1+L_2)^2}{\pi \frac{\rho}{2} V^2 S_1^2} \cdot \frac{1 - \sigma^2}{1 - 2\sigma \frac{S_2}{S_1} + (\frac{S_2}{S_1})^2} . \qquad (57)$$

When $a = 0$ and $\sigma = \frac{S_2}{S_1}$ the second factor in this product is unity and we have

$$D_{min} = \frac{(L_1+L_2)^2}{\pi \frac{\rho}{2} V^2 S_1^2} ,$$

which is the minimum drag for a monoplane of span S_1 and given lift $L_1 + L_2$ with elliptic distribution according to (27). Thus the second factor of the product in (57) is a measure of the mutual interference when we have two wings with no stagger at distance a apart. We have assumed throughout that each of the wings has an elliptic distribution of Γ. As we know, in all cases the Γ distribution is not very different from an elliptic one. Therefore, the relations derived may be used as approximations in all cases.

We conclude our study of the biplane by considering the effect of stagger between the two wings. As we shall see there will be no change in our value for $D_{12} + D_{21}$. This was known empirically before the theory had been developed. The result, that stagger does not change the relation between drag and lift, was proven by Munk in 1919 and is called the theorem of Munk.

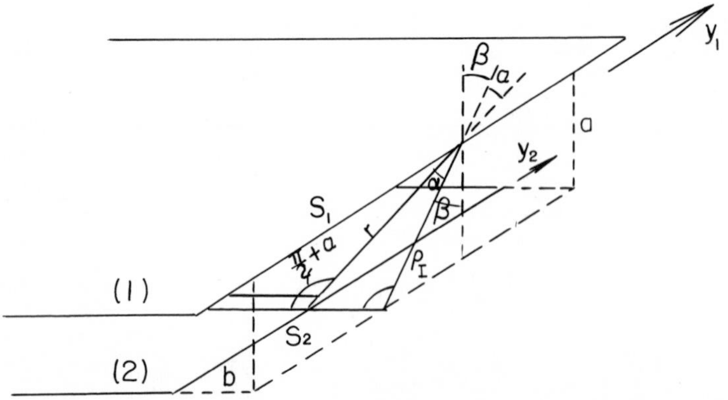

We calculate again the value of w_{21} which is more complicated in this case. We have (see above figure) besides the quantities a, β, ρ_I used before another constant b giving the amount of stagger, the distance r between elements of the two wings and the angle α which r makes with ρ_I. The relations between these quantities are seen to be

$$\sin \beta = \frac{y_1 - y_2}{\rho_I}, \qquad \sin \alpha = \frac{b}{r}$$

$$r^2 = b^2 + \rho_I^2, \quad \rho_I^2 = (y_1 - y_2)^2 + a^2.$$

The element of velocity dw_{21} at y_1 induced by the second vortex sheet at a point of the first wing now consists of two parts, the first being due to the semi-infinite strip from $-b$ to $-\infty$ of strength $d\Gamma_2$ and the second is due to the line element dy_2 along S_2 of strength Γ_2. Before, with no stagger, this second part would have given a velocity in the x-direction and thus not contributed to the z-component of the velocity. The first can be computed as in the previous case by replacing the factor 1 (for $\cos \alpha + \cos \beta$) by $1 - \sin \alpha$. For the second term we use simply the rule of Biot-Savart (5a). The result is

$$dw_{21} = -\frac{1}{4\pi} \frac{d\Gamma_2}{\rho_I} (1 - \sin \alpha) \cdot \sin \beta + \frac{\Gamma_2}{4\pi} \cdot \frac{r dy_2}{r^3} \sin \alpha.$$

Using partial integration on the first term we get

$$w_{21} = \frac{1}{4\pi} \int_{-S_2/2}^{S_2/2} \Gamma_2(y_2) \left[\frac{d}{dy_2} (1 - \sin \alpha) \frac{\sin \beta}{\rho_I} + \frac{b}{r^3} \right] dy_2$$

since $\Gamma_2(\frac{S_2}{2}) = \Gamma_2(-\frac{S_2}{2}) = 0$. Now on p. 185 we found that

$$\frac{d}{dy_2} \left(\frac{\sin \beta}{\rho_I}\right) = -\frac{\cos 2\beta}{\rho_I^2}$$

so

$$\frac{d}{dy_2}(1 - \sin\alpha)\frac{\sin\beta}{\rho_I} = -(1 - \sin\alpha)\frac{\cos 2\beta}{\rho_I^2} - \frac{\sin\beta}{\rho_I}\frac{d}{dy_2}\left(\frac{b}{r}\right)$$

$$= -(1 - \sin\alpha)\frac{\cos 2\beta}{\rho_I^2} - \frac{\sin^2\beta\,\sin\alpha}{r^2}\,.$$

Substituting the resulting value of w_{21} into the formula

$$D_{21} = -\rho\int_{-S_1/2}^{S_1/2}\Gamma_1(y_1)w_{21}(y_1)dy_1$$

we have

$$D_{21} = \frac{\rho}{4\pi}\int_{-S_1/2}^{S_1/2}\int_{-S_2/2}^{S_2/2}\Gamma_1(y_1)\Gamma_2(y_2)\left[(1 - \sin\alpha)\frac{\cos 2\beta}{\rho_I^2} - \frac{\cos^2\beta\,\sin\alpha}{r^2}\right]dy_2dy_1.$$

On the other hand, we easily get D_{12} in the same way only replacing α and β by $\alpha + \pi$ and $\beta + \pi$ respectively. Then

$$D_{12} = \frac{\rho}{4\pi}\int_{-S_1/2}^{S_1/2}\int_{-S_2/2}^{S_2/2}\Gamma_1(y_1)\Gamma_2(y_2)\left[(1 + \sin\alpha)\frac{\cos 2\beta}{\rho_I^2} + \frac{\cos^2\beta\,\sin\alpha}{r^2}\right]dy_2dy_1$$

and finally, as the additional terms cancel out,

$$D_{12} + D_{21} = \frac{\rho}{4\pi}\cdot 2\int_{-S_1/2}^{S_1/2}\int_{-S_2/2}^{S_2/2}\Gamma_1(y_2)\Gamma_2(y_2)\frac{\cos 2\beta}{\rho_I^2}dy_2dy_1$$

which is the same value we got from (50) in the case of no stagger.

Problem 11. Study the potential and velocity distribution in space due to

(a) one circular vortex line of radius R,

(b) a circular vortex sheet of radius R with a Γ distribution such that $\Gamma(0) = \max$ and $\Gamma(R) = 0$.

Use coordinates as shown in figure with vortex lines lying in the xy-plane.

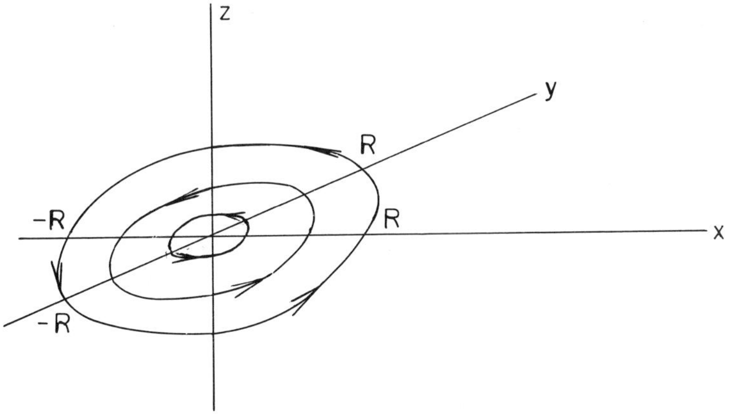

Problem 12. Develop the theory of lift and drag for an wirwing with triangular Γ distribution as shown in the figure. Use either a Fourier development for Γ or else obtain results in closed form by going through the same procedure used for the elliptic distribution. Do the same for a biplane with triangular distribution on each wing.

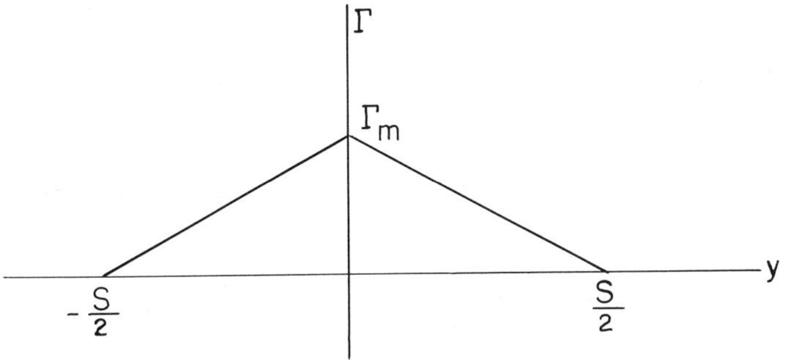

CHAPTER IV

THEORY OF VISCOUS FLUIDS

Viscous fluids as compared with ideal fluids can be characterized by two properties: (1) when a viscous fluid flows along a wall it will adhere to the wall, i.e., the layer of fluid in immediate contact with the wall has no velocity relative to it, (see Goldstein, vol. II, Appendix); (2) whenever the fluid particles are distorted, shearing stresses arise.

Before formulating these properties precisely and deriving the equations of motion from them, we consider the effect of the viscosity in two important and typical types of flow.

References:

S. Goldstein: Modern Developments in Fluid Dynamics, 2 vols.

W. Durand, Ed.: Aerodynamic Theory, vol. III, section G by L. Prandtl, Mechanics of Viscous Fluids.

1. Couette and Poiseuille Flow

(a) Plane Couette Flow

Consider the case of a fluid enclosed between two infinite, parallel, plane walls which move with a constant relative velocity. (Each wall moves so that it remains in the same plane.) If the fluid were perfect the motion of the walls would have no effect on the fluid. However a viscous fluid must adhere to the walls.

Let one wall $y = 0$ be considered at rest and the other, $y = h$ have the constant velocity $(U,0,0)$. It is natural to assume that when a steady state is reached the velocity distribution from one wall to the other will be linear. Then the velocity of the fluid particles between the walls is given by

$$u = U\frac{y}{h}, \quad v = w = 0.$$

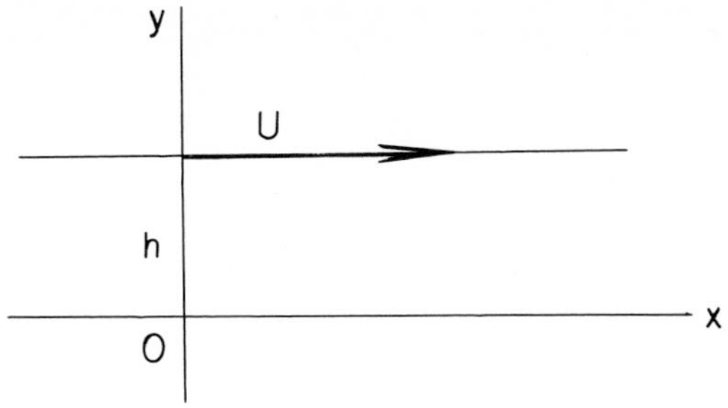

Consider a stream surface in the above flow $y = y_0$, y_0 a constant, and let us discuss the effect of the viscosity. The particles of fluid on one side of this surface, $y > y_0$, will exert a force in the x-direction upon the particles on the other side. This force will be proportional to the area upon which it acts, and the force per unit area is what we call the <u>shearing stress</u>, denoted by τ. The shearing stress τ will be proportional to the rate of change of the velocity across the surface $y = y_0$, i.e.,

$$\tau = \mu \frac{\partial u}{\partial y} = \mu \frac{U}{h}.$$

The coefficient of proportionality μ is called the "viscosity" and depends upon the nature of the fluid. We denote by D the drag on the upper wall per unit breadth, (i.e., for a strip $z = 0$ to $z = 1$). Since the shearing stress τ is the drag per unit area exerted on the walls, the drag D for a finite segment of wall of length ℓ $(x = a$ to $x = a + \ell)$ becomes

$$D = \mu U \frac{\ell}{h}.$$

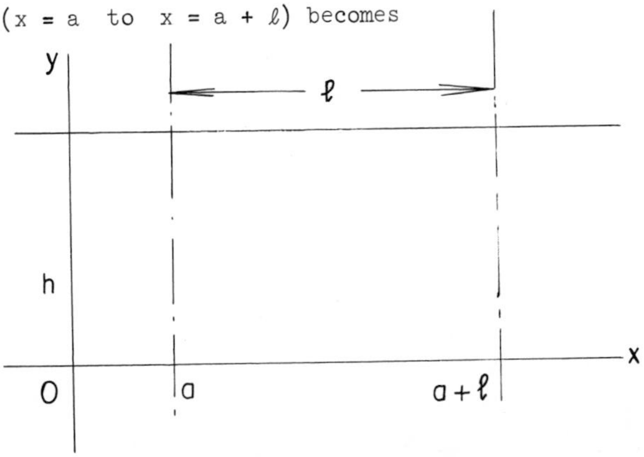

(b) Plane Poiseuille Flow

As a second example consider the flow that results when a viscous fluid is pressed between two fixed plane plates $y = h$ and $y = -h$. It is natural to assume that when a steady state is reached the velocity distribution will depend only upon the distance from the plates. That is

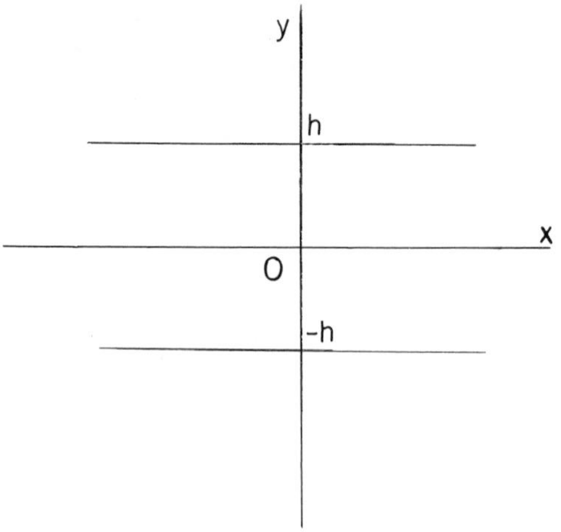

$$u = u(y), \quad v = w = 0. \qquad (1)$$

The condition that the fluid adhere to the walls is given by $u = 0$ for $y = \pm h$. Here again, due to the variation of u in the y-direction, there will be a shearing stress τ given by

$$\tau = \mu \frac{\partial u}{\partial y} = \mu u_y. \qquad (2)$$

The shearing stress τ at a point (x_0, y_0) is exerted by the part of the fluid $y > y_0$ against the fluid $y \leq y_0$. In accordance with the basic law of symmetry of stresses it is to be noted that the same shearing stress τ is exerted by the part of the fluid $x > x_0$ against the part $x \leq x_0$.

Since the fluid is being forced between the fixed plates, there will be, in addition to the shearing stress τ, a compressive normal stress, the pressure p. The plane stress system is thus given by

$$\begin{pmatrix} -p & \tau \\ \tau & -p \end{pmatrix}.$$

The condition that this system of stresses be in equilibrium is that the force per unit volume resulting from the variation of these stresses vanish. Consider a parallelopiped with one corner $P(x_0, y_0, z_0)$ and sides Δx, Δy, Δz.

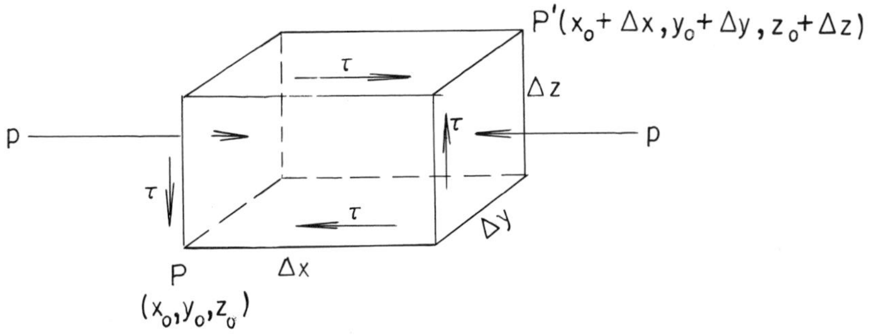

The x-component of the resultant stress forces against the boundary of the parallelopiped is given by

$$\left[\iint p \, dy \, dz\right]_{x_0}^{x_0+\Delta x} + \left[\iint \tau \, dx \, dz\right]_{y_0}^{y_0+\Delta y}.$$

This is clearly equal to

$$\iiint p_x dx \, dy \, dz + \iiint \tau_y \, dx \, dy \, dz.$$

Upon dividing by $\Delta x \Delta y \Delta z$ and letting the sides Δx, Δy, Δz approach zero simultaneously, the x-component of the resulting force per unit volume is

$$-p_x + \tau_y.$$

Similarly the y-component is found to be

$$-p_y + \tau_x.$$

From (1) and (2) it is obvious that the z-component vanishes. Thus we have the conditions for equilibrium (since by (2), $\tau = \tau(y)$):

$$\tau_y = p_x, \; p_y = 0. \qquad (3)$$

Since $p_y = 0$, p is a function of x alone, and since τ is a function of y alone, the first of the equations (3) implies that p_x and τ_y are constant. That $\tau_y (= \mu u_{yy})$ is constant implies that u is a quadratic in y. Taking into consideration the boundary conditions, $u(\pm h) = 0$, we have

$$u = u_0 \left(1 - \frac{y^2}{h^2}\right). \qquad (4)$$

Clearly the maximum velocity u_0 is reached midway between the walls. The shearing stress τ can now be found explicitly and is given by

$$\tau = -2\mu u_0 \frac{y}{h^2} .$$

The <u>flux per unit breadth</u>, i.e., the volume of fluid crossing the plane x = const. per unit time per unit breadth, (breadth means in z-direction), is

$$\int_{-h}^{h} u \, dy = \frac{4}{3} u_0 h.$$

Frequently it is more convenient to state results in terms of the average velocity $\bar{u}$. In the above case we have

$$\bar{u} = \frac{1}{2h} \int_{-h}^{h} u \, dy = \frac{2}{3} u_0.$$

As before, the drag D per unit breadth exerted on the upper wall for a segment of length ℓ can be found:

$$D = -\tau(h) \cdot \ell = 2\mu u_0 \frac{\ell}{h} = 3\mu \bar{u} \frac{\ell}{h} .$$

The above result that p_x = const. also enables us to find the <u>pressure drop</u>

Δp in the x-direction over a distance ℓ. Since p_x = const., p is a linear function of x. Then $\Delta p = -p_x \cdot \ell$, the negative sign since the pressure decreases when x increases, and we get

$$\Delta p = 2\mu u_0 \frac{\ell}{h^2} = 3\mu\bar{u}\frac{\ell}{h^2} .$$

Problem 13. Determine the velocity distribution, pressure drop and drag for the "proper Poiseuille flow", i.e., the flow of a viscous fluid in a cylindrical pipe with circular cross-section, the flow being assumed to have the direction of the axis of the cylinder and to be constant in this direction.

Answer:

$$u = 2\bar{u}\{1 - \frac{r^2}{r_0^2}\}$$

$$\Delta p = 8\mu\bar{u}\frac{\ell}{r_0^2}$$

$$D = 8\pi\mu\bar{u}\ell,$$

where r_0 is the radius of the circular cross-section.

2. Navier-Stokes Equation

Consider the flow of a viscous fluid given by the velocity field $\vec{q} = (u,v,w)$ depending upon the point given by the position vector $\vec{r} = (x,y,z)$. The stresses introduced into the field at any point by the viscosity can be characterized by the following five properties.

(1) The system of stresses

$$(\tau) = \begin{pmatrix} \tau_{xx} & \tau_{xy} & \tau_{xz} \\ \tau_{yx} & \tau_{yy} & \tau_{yz} \\ \tau_{zx} & \tau_{zy} & \tau_{zz} \end{pmatrix}$$

forms a symmetrical tensor. The symmetry means that $\tau_{yx} = \tau_{xy}$, $\tau_{xz} = \tau_{zx}$ and $\tau_{yz} = \tau_{zy}$. The fact that it is a tensor implies that when the system of stresses (τ) is given with respect to one coordinate system the stresses with respect to another coordinate system are determined by a certain linear transformation which is obtained from equilibrium considerations.

(2) The viscous stresses (τ) depend linearly upon the first derivatives of the velocity $\vec{q}$. Clearly this implies that viscous stresses do not arise in a state of rest.

(3) The viscous stresses vanish for a rigid body motion. Since this type of motion is a translation plus a rotation, the condition refers essentially to a rotation.

(4) The viscous stresses vanish for a motion that represents a homothetic deformation. By such a deformation is meant one in which a configuration of particles in the fluid has a shape after the deformation similar to the original configuration. Since no homothetic deformation exists for incompressible fluids this condition has a significance only for compressible fluid.

(5) For a flow given by $u = u(y)$, $v = w = 0$, the stress system is $\tau_{xy} = \mu u_y$, all other components of the stress tensor being zero.

The system of first derivatives of $\vec{q}$ form a tensor denoted by $\nabla\vec{q}$:

$$\nabla\vec{q} = \begin{pmatrix} u_x & u_y & u_z \\ v_x & v_y & v_z \\ w_x & w_y & w_z \end{pmatrix} .$$

If $(\nabla\vec{q})^*$ denotes the transpose of $\nabla\vec{q}$, the tensor <u>Curl $\vec{q}$</u> can be defined as follows:

$$\text{Curl } \vec{q} = \nabla\vec{q} - (\nabla\vec{q})^* = \begin{pmatrix} 0 & u_y - v_x & u_z - w_x \\ v_x - u_y & 0 & v_z - w_y \\ w_x - u_z & w_y - v_z & 0 \end{pmatrix}$$

Let I represent the unit tensor

$$I = \begin{pmatrix} 1 & 0 & 0 \\ 0 & 1 & 0 \\ 0 & 0 & 1 \end{pmatrix},$$

and define the tensor $\underline{\text{Div } \vec{q}}$ as

$$\text{Div } \vec{q} = \text{div } \vec{q} \cdot I = \begin{pmatrix} u_x + v_y + w_z & 0 & 0 \\ 0 & u_x + v_y + w_z & 0 \\ 0 & 0 & u_x + v_y + w_z \end{pmatrix}.$$

The tensor given by

$$\nabla\vec{q} - \frac{1}{2}\,\text{Curl } \vec{q} - \frac{1}{3}\,\text{Div } \vec{q}$$

is symmetric and has the property that its trace vanishes. We denote it by Def $\vec{q}$, and written explicitly it is:

$$\text{Def } \vec{q} = \begin{pmatrix} \frac{2}{3}u_x - \frac{1}{3}v_y - \frac{1}{3}w_z & \frac{1}{2}(u_y + v_x) & \frac{1}{2}(u_z + w_x) \\ \frac{1}{2}(v_x + u_y) & \frac{2}{3}v_y - \frac{1}{3}u_x - \frac{1}{3}w_z & \frac{1}{2}(v_z + w_y) \\ \frac{1}{2}(u_z + w_x) & \frac{1}{2}(v_z + w_y) & \frac{2}{3}w_z - \frac{1}{3}u_x - \frac{1}{3}v_y \end{pmatrix}.$$

It can be shown that every tensor which is linear in the first derivatives of $\vec{q}$, that is, which satisfies condition (2), is a linear combination of the tensors Def $\vec{q}$, Curl $\vec{q}$ and Div $\vec{q}$, i.e., is of the form

$$a \text{ Def } \vec{q} + b \text{ Curl } \vec{q} + c \text{ Div } \vec{q}. \qquad (5)$$

From condition (3) it follows that for a rigid body motion $b = 0$. To see this consider the rotation about the z-axis given by

$$u = -\omega y, \; v = \omega x, \; w = 0.$$

Clearly $\text{Def } \vec{q} = 0$ and $\text{Div } \vec{q} = 0$. Since $\text{Curl } \vec{q} \neq 0$, condition (3) states that the expression (5) must vanish or $b = 0$. The same result can be obtained by use of the symmetry property in condition (1). $\text{Def } \vec{q}$ and $\text{Div } \vec{q}$ are symmetric tensors. Since the complete expression (5) must be symmetric and $\text{Curl } \vec{q}$ is anti-symmetric we must have $b = 0$. For a flow representing a homothetic deformation

$$u = kx, \; v = ky, \; w = kz$$

it follows from condition (4) that $c = 0$. This is clear since $\text{Def } \vec{q} = 0$, $\text{Curl } \vec{q} = 0$ and $\text{Div } q \neq 0$. Hence in order to have the stress system vanish c must be zero. Condition (5) gives immediately $a = 2\mu$. Hence the viscous stresses are given by

$$2\mu \text{ Def } \vec{q}.$$

Since it was seen that $\text{Def } \vec{q} = 0$ for rigid body motion and homothetic deformations, the above expression gives a non-vanishing stress system only when there is an actual alteration of shape.

In addition to this system of viscous stresses there is another stress system consisting of a hydraulic pressure, p, so that the complete stress system is given by

$$(\tau) = -pI + 2\mu \text{ Def } \vec{q}.$$

To avoid unnecessary generality we assume from now on that the fluid is incompressible, and that consequently the continuity equation is

$$\operatorname{div} \vec{q} = 0.$$

When the stresses vary across the fluid, forces per unit volume arise. To calculate these forces consider the stress system on a parallelopiped which is an element of volume. The x-component of this force per unit volume is given by*

$$\tau_{xx/x} + \tau_{xy/y} + \tau_{xz/z}.$$

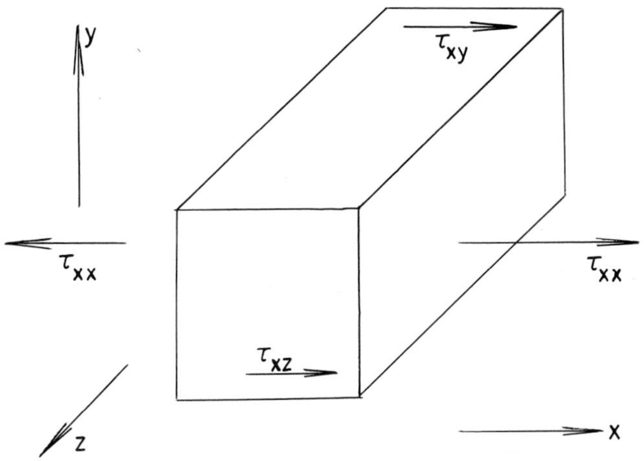

The contribution of the pressure is $-p_x$, and the tensor $2\mu \operatorname{Def} \vec{q}$ yields $\mu\Delta u$, where $\operatorname{div} \vec{q} = 0$ has been used. The usual notation

*A remark may be made about the notation. Subscripts will be used to represent both components and partial differentiation. In most cases the context will make the meaning clear. For example the vortex vector $\vec{\omega}$ has components $\omega_x, \omega_y, \omega_z$. However, $\vec{\omega}_x$ denotes the vector $(\frac{\partial\omega_x}{\partial x}, \frac{\partial\omega_y}{\partial x}, \frac{\partial\omega_z}{\partial x})$. If for example we wish to take the partial derivative of the component ω_z with respect to y, the notation $\omega_{z/y}$ will be used.

$$\Delta = \frac{\partial^2}{\partial x^2} + \frac{\partial^2}{\partial y^2} + \frac{\partial^2}{\partial z^2}$$

is employed. Thus the x-component becomes

$$\tau_{xx/x} + \tau_{xy/y} + \tau_{xz/z} = -p_x + \mu\Delta u.$$

Working out the corresponding expressions for the y- and z-components the force per unit volume becomes

$$-\text{grad } p + \mu\Delta\vec{q}.$$

The condition that the mass per unit volume, ρ, multiplied by the acceleration, $\frac{d\vec{q}}{dt}$, equals the force per unit volume then becomes

$$\rho \frac{d\vec{q}}{dt} = -\text{grad } p + \mu\Delta\vec{q}. \tag{6}$$

This equation is the <u>Navier-Stokes Equation</u>. In particular, if the flow is <u>steady</u>, the partial derivatives with respect to time vanish and using the fact that

$$\frac{d\vec{q}}{dt} = \frac{\partial\vec{q}}{\partial t} + u\vec{q}_x + v\vec{q}_y + w\vec{q}_z$$

equation (6) takes the form

$$\rho(u\vec{q}_x + v\vec{q}_y + w\vec{q}_z) = -\text{grad } p + \mu\Delta\vec{q}. \tag{7}$$

(It may be noted that the gravity g has been omitted throughout.)

The pressure p can be eliminated from (7) and the continuity equation $\text{div } \vec{q} = 0$. This can be done by introducing the vortex vector

$$\vec{\omega} = \text{curl } \vec{q} = (w_y - v_z,\ u_z - w_x,\ v_x - u_y),$$

the Bernoulli function

$$H = \frac{p}{g\rho} + \frac{q^2}{2g},$$

and Equation (14) page 24 which states

$$g \text{ grad } H = \vec{q} \times \vec{\omega}.$$

Then Equation (7) can be written in the form

$$-\rho(\vec{q} \times \vec{\omega}) = \text{grad } g\ H + \mu\Delta\vec{q}.$$

Application of the operation "curl" to (7) and using $\text{div } \vec{q} = 0$ yield

$$u\vec{\omega}_x + v\vec{\omega}_y + w\vec{\omega}_z - (\vec{q}_x\omega_x + \vec{q}_y\omega_y + \vec{q}_z\omega_z) = \nu\Delta\vec{\omega}, \tag{8}$$

where $\nu = \frac{\mu}{\rho}$ is the kinematic viscosity. Equation (8) is obviously satisfied for irrotational flow, since every term vanishes when $\vec{\omega} = 0$. In order to satisfy equation (7) it is sufficient to satisfy equation (8) by a flow $\vec{q}$; for then there always exists a pressure distribution $p = p(\vec{r})$ which satisfies (7). In fact, if a vector has the property that its curl vanishes then it is the gradient of some function. All that we have to do is identify this function with the pressure. Thus the pressure will be determined up to an additive constant, and this means that only pressure differences will be determined by the flow.

It is useful to specialize to the case of two-dimensional flow. Here $w = 0$ and $\vec{q}_z = 0$ as well as $\omega_x = \omega_y = 0$. In this case the continuity equation can be satisfied by the introduction of a stream function $\psi(x,y)$ from which the velocity is derived as follows:

$$u = \psi_y,\ v = -\psi_x.$$

We then have $\omega_z = -(\psi_{xx} + \psi_{yy}) = -\Delta\psi$. In this case equation (8) takes the form

$$\psi_y \Delta\psi_x - \psi_x \Delta\psi_y = \nu \Delta\Delta\psi. \tag{8a}$$

This differential equation is of the fourth order in ψ, the leading term involving the "biharmonic" operator $\Delta\Delta$. However the left side of this equation, involving terms of the third order, is not linear, and therefore the solution of such an equation is difficult.

3. Problems

The Navier-Stokes equation is to be solved under certain boundary conditions. In this section we consider different types of problems which arise from different boundary conditions.

The problem of finding a solution of the differential equations is called the problem of the first kind if for a certain domain D in the x,y,z-space the velocity is prescribed on the boundary, (in addition to μ and ρ being given). A part of the boundary may consist of a "wall" in which case the velocity is that of the wall and, in particular, if the wall is at rest the velocity is zero. Also the domain D may consist of a pipe of finite length. In this case one may give the velocity at the entrance (e.g., so as to be constant and in the direction of the axis) and at the exit. The problem of the flow around an immersed body with given velocity at infinity is of the first kind.

That the problem of the first kind is mathematically consistent with the differential equations is clearer in the case of two-dimensional flow. In this case the differential equation for the stream function ψ was found to be of the fourth order:

$$\psi_x \Delta\psi_y - \psi_y \Delta\psi_x = -\nu \Delta\Delta\psi.$$

To prescribe the velocity at the boundary means the first derivatives of ψ are given or what is equivalent (except for an arbitrary additive constant) to prescribe the value of ψ and its normal derivative at the boundary. It is well-known that for the differential equation

$$\Delta\Delta\psi = f$$

these are possible boundary data and from experience one expects that the terms of highest order in (8a) determine the possible boundary data.

In cases, however, where the domain extends to infinity it is not obvious what should be prescribed at infinity; e.g., in the case of a half-infinite pipe, or of immersed bodies.

Problems of a different kind would arise when the pressure is prescribed, e.g., instead of the velocity, the pressure at both ends of a pipe is given.

In the case of non-steady flow questions of an entirely different nature appear. Here one may be interested in how a motion begins. Then aside from the boundary data the flow is described at an initial time. Another type of problem is that of the final state of the motion approached after a long period of time

Problem 14. Same as problem 13 for a pipe of elliptical cross-section.

Answer:

$$u = 2\bar{u}\{1 - \frac{y^2}{b^2} - \frac{z^2}{c^2}\}$$

$$\Delta p = 4\mu\bar{u}\{\frac{1}{b^2} + \frac{1}{c^2}\}\ell$$

$$D = 4\pi\mu\bar{u}\{\frac{1}{b^2} + \frac{1}{c^2}\}bc\ell,$$

where b and c are the semi-axes.

Problem 15. (a) Determine the flow between two co-axial circular cylinders with radii r_0, r_1 $(r_0 < r_1)$ moving with velocities U_0, U_1 in the direction of their axis. The pressure is assumed to be the same at both ends of the cylinders.

(b) Determine the drag exerted on the cylinders.

(c) What limit will the flow approach when $r_0 \to 0$, while r_1, U_0, U_1 remain fixed?

Answer: (a) $u = (U_1 \log \frac{r}{r_0} + U_0 \log \frac{r_1}{r})/\log \frac{r_1}{r_0}$.

(b) The drag per unit area, $T = \mu \dfrac{U_1 - U_0}{r_0 \log \frac{r_1}{r_0}}$ for $r = r_0$

$$= \mu \frac{U_1 - U_0}{r_1 \log \frac{r_1}{r_0}} \quad \text{for} \quad r = r_1.$$

The drag over length ℓ, $D = \dfrac{2\pi\mu(U_1 - U_0)\ell}{\log \frac{r_1}{r_0}}$ for $r = r_0$

$$D = \frac{-2\pi\mu(U_1 - U_0)\ell}{\log \frac{r_1}{r_0}} \quad \text{for} \quad r = r_1.$$

(c) $u = U_1$; $D = 0$

Problem 16. (a) Determine the proper Couette flow, i.e., the flow between two co-axial cylinders with radii r_0, r_1 $(r_0 < r_1)$ which rotate with the angular velocities ω_0, ω_1. The pressure is assumed to be the same at both ends of the cylinder.

(b) Determine the pressure distribution and drag on the cylinders.

(c) What limit will the flow approach when $r_0 \to 0$ while r_1, ω_0, ω_1 remain fixed?

Answer: (a) $u = -\omega y$, $v = \omega x$, $w = 0$ where $\omega = A + Br^{-2}$ and

$$A = (r_1^2\omega_1 - r_0^2\omega_0)/(r_1^2 - r_0^2), \quad B = (\omega_0 - \omega_1)/(r_0^{-2} - r_1^{-2}).$$

(b) $\rho^{-1}p = \frac{1}{2}A^2r^2 - \frac{1}{2}B^2r^{-2} + 2\,AB \log r.$

$$\rho^{-1}(p_1-p_0) = \frac{1}{2}\left(r_1^2\omega_1-r_0^2\omega_0\right)^2/(r_1^2-r_0^2).$$

$$+\frac{1}{2}(\omega_0-\omega_1)^2/(r_0^{-2}-r_1^{-2}) + 2\text{ AB}\log\frac{r_1}{r_0}.$$

$$\tau_0 = -2Br_0^{-2}\mu;\ \tau_1 = 2Br_1^{-2}\mu;\ D_0 = 4\pi Br_0^{-1}\ell\mu;$$

$$D_1 = -4\pi Br_1^{-1}\ell\mu$$

(c) $p \to \frac{1}{2}\rho\omega_1^2r^2,\ \tau_1 \to 0,\ D_1 \to 0,\ \tau_0 \to -2(\omega_0-\omega_1)\mu,\ D_0 \to 0$

as $r_0 \to 0$.

4. Similarity

In the case of a perfect fluid, a solution to a particular problem yields immediately the solution to any "kinematically similar problem". When the domain undergoes a homothetic deformation and the velocities are all multiplied by a constant the solution is obtained by the same operation. However, in the case of a viscous fluid this does in general not hold.

Let a flow be given by $\vec{q} = \vec{q}_0(\vec{r})$. Then a flow given by $\vec{q} = \beta\vec{q}_0(\alpha\vec{r})$ where α and β are constants is called a "kinematically similar flow". It is obtained when all distances are stretched by a constant factor α and all velocities multiplied by a constant β.

Consider a problem of the first kind represented symbolically by P_0; a "kinematically similar problem" P_1 is obtained when the domain is stretched by a constant factor α and when to boundary points of the new domain correspond velocities of the original boundary multiplied by a constant factor β. It may be left open whether or not P_1 refers to the same fluid, i.e., ρ and μ are the same.

Clearly when the flow $\vec{q}_0$ solves the problem P_0, the kinematically similar flow satisfies the boundary conditions for problem P_1. However the kinematically similar flow may not solve the problem P_1, i.e., satisfy the Navier-Stokes equation. In fact, this flow will solve the problem with the same kinematic viscosity only under rather special circumstances. Otherwise it will be a solution for P_1 for a fluid with a different value of ν. Consider for example the special similarity transformation given by $\alpha = 1$, $\beta = -1$ which means that only the direction of the flow is reversed; in general the reversed flow is not the solution of the "reversed" problem obtained by this similarity transformation.

To investigate these questions it is customary to introduce a "significant length" L and a "significant velocity" U for problem P_0 and flow $\vec{q}_0$. For the similar problem P_1 these significant quantities will be αL and βU. For example, in flow through a pipe the significant quantities may be the diameter of the pipe and the average velocity. Quantities that are unchanged under the transformation taking problem P_0 into P_1 will be called "invariants". Examples of these are $\vec{r}/L$ and $\vec{q}/U$.

Now we write the Navier-Stokes equation in the form

$$\frac{\frac{d\vec{q}}{dt}}{U^2L^{-1}} = -\frac{\text{grad } p}{\rho U^2 L^{-1}} + R^{-1}\frac{\Delta\vec{q}}{U L^{-2}} \tag{9}$$

where the dimensionless quantity

$$R = \frac{UL}{\nu}$$

is the Reynolds number of the problem. It is evident that the terms $\frac{d\vec{q}}{dt}/U^2L^{-1}$ and $\frac{\Delta\vec{q}}{U L^{-2}}$ are invariant.

Consider the case where the Reynolds number R for problem P_1 is the same as for problem P_0. (Then the problem P_1 is called "dynamically similar" to problem P_0). In this case either the problem P_1 refers to a different fluid with different μ and ν or the transformation from P_0 to P_1 was such that UL is unchanged. Then the kinematically similar flow $\vec{q}_1 = \beta\vec{q}_0(\alpha\vec{r})$ is a solution of P_1; the pressure distribution for P_1 is easily adjusted so that the Navier-Stokes equation (9) is satisfied. Let Δp be the pressure difference between any two points; then it is only necessary to assign for P_1 pressure differences between corresponding points so that

$$\frac{\Delta p}{\rho U^2}$$

is unchanged.

The importance of the Reynolds number may be seen in the following general considerations. Let P_0 be a given problem and consider all problems kinematically similar to P_0 with all possible Reynolds numbers. We assume that each such problem has a unique solution $\vec{q}$. Then it follows that each of the quantities

$$\frac{\vec{q}}{U}, \frac{\Delta p}{\rho U^2}, \frac{\tau}{\rho U^2}, \frac{D}{A\rho U^2},$$

where Δp is any pressure difference, τ any shearing stress and D is the drag on a surface A, depends only on the Reynolds number R.

This result that the dimensionless quantities arising from the flow depend only on the Reynolds number R for all kinematically similar problems has frequently been derived in an apparently simpler way. The derivation is performed by so-called "dimensionless analysis". Since it is easily seen that the only dimensionless quantity that can be formed from the significant data of the problem is just the Reynolds number, all other dimensionless quantities entering the problem must depend only upon R. However this argument was also applied in cases where the Navier-Stokes equation are considered to be no longer valid, e.g., in turbulent flow. But it is clear that this argument, though much simpler makes the tacit physical hypothesis that the viscosity μ and density ρ are still the only physical constants that determine the flow.

It is clear that there are cases where the solution of the Navier-Stokes equation does not depend on the Reynolds number and consequently each kinematically similar flow is a solution for the corresponding kinematically similar problem.

Case I. If the flow satisfies the equation

$$\Delta\vec{q} = 0 \quad \text{or} \quad \operatorname{curl} \vec{\omega} = 0$$

the term in (9) involving the Reynolds number vanishes. For two-dimensional flow we have $\omega_x = \omega_y = 0$ and the above condition becomes $\omega_z = \text{const.}$ In particular, this holds for irrotational flow as then $\vec{\omega} = 0$. The above conditions, $\Delta\vec{q} = 0$, is satisfied in plane Couette flow (cf. Sec. 1 (a)), proper Couette flow (cf. Problem 16) and the flow between two co-axial cylinders each moving with a constant velocity in the direction of the axis (cf. Problem 15). In all these cases the stress ratios

$$\frac{\tau}{\mu U L^{-1}} = \frac{\tau}{\rho U^2 R^{-1}} \quad \text{and} \quad \frac{\Delta p}{\rho U^2}$$

are independent of R, where τ is any significant shearing stress and Δp is a pressure difference.

Case II. A second class of flows which are independent of the Reynolds number consists of those with vanishing acceleration:

$$\frac{d\vec{q}}{dt} = 0. \tag{10}$$

In this case (9) reduces to

$$\frac{\Delta\vec{q}}{UL^{-2}} = \frac{\operatorname{grad} p}{R^{-1}\rho U^2 L^{-1}} .$$

Thus the pressure depends upon R but the flow does not. By adjusting the pressure properly the kinematically similar flow can always be made the solution of the kinematically similar problem for every Reynolds number. Illustrations of flows satisfying condition (10) are the Poiseuille flow (cf. Sec. 1 (b)), plane Couette flow (cf. Sec. 1 (a)) and the flow between two co-axial cylinders each moving in the direction of the axis (cf. Problem 15). Here the ratios

$$\frac{\Delta p}{\rho U^2 R^{-1}} \quad \text{and} \quad \frac{\tau}{\rho U^2 R^{-1}}$$

are independent of R. In particular, for the plane Poiseuille flow we may set $L = 2h$, $U = \bar{u}$, hence $R = \frac{2h\bar{u}}{\nu}$. Then we have

$$\frac{\Delta p}{\rho U^2} = 6(\ell/h)R^{-1}, \quad \frac{T}{\rho U^2} = 6R^{-1},$$

T being the drag per unit area on the wall, in accordance with the preceding statement.

It is very likely that slight generalizations of Couette and Poiseuille flows cover all flows falling under cases I and II; and that cases I and II cover all cases of flows satisfying the Navier-Stokes equations which are independent of the Reynolds number.

5. Small Reynolds Numbers

Case II above (i.e., vanishing acceleration) arises approximately when the Reynolds number is small. Small Reynolds numbers will occur when the viscosity is large or when either the significant length or significant velocity is small. A flow with small Reynolds number is sometimes called "slow" motion.

The Navier-Stokes equation can be written

$$R \frac{\frac{d\vec{q}}{dt}}{U^2 L^{-1}} = - \frac{\text{grad } p}{\rho U^2 L^{-1} R^{-1}} + \frac{\Delta \vec{q}}{UL^{-2}}. \tag{11}$$

If it is assumed that the terms involving the velocity in (11) are of the same order of magnitude, the fact that the Reynolds number is small allows us to neglect the acceleration term in taking a first approximation. Then (11) reduces to

$$\mu \Delta \vec{q} = \text{grad } p$$

or taking the curl

$$\Delta\vec{\omega} = 0$$

We make use of the above assumption in examining the "slow" flow around an infinite circular cylinder. This flow may be considered the two-dimensional flow around a circle of radius r_0 with center at the origin. Take the velocity at infinity as $(U,0)$. Also select the units so that $U = 1$, $r_0 = 1$ and $\mu = 1$. Then the equations reduce to

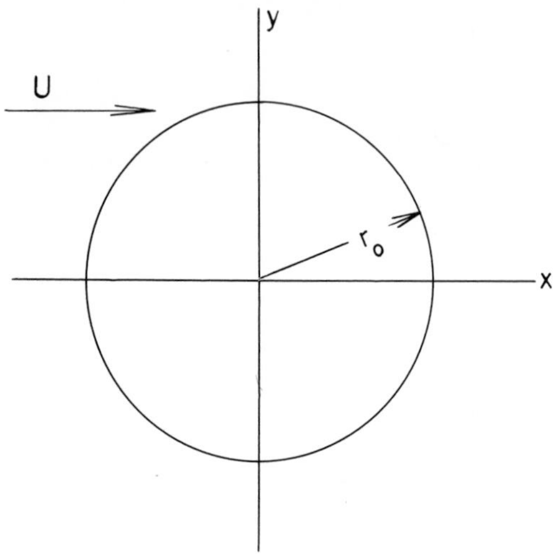

$$\text{div}\ \vec{q} = 0,\ \Delta\vec{q} = \text{grad}\ p;$$

they are to be solved subject to the boundary conditions

$$u = 1,\ v = 0 \quad \text{for} \quad r = \infty$$

$$u = v = 0 \quad \text{for} \quad r = 1.$$

If the stream function $\psi(x,y)$ is introduced we have

$$u = \psi_y,\ v = -\psi_x,\ \Delta\psi_y = p_x,\ -\Delta\psi_x = p_y;$$

by eliminating p the equation to be solved becomes the biharmonic equation

$$\Delta\Delta\psi = 0. \tag{12}$$

The boundary conditions are

$$\psi_x = 0,\ \psi_y = 1 \quad \text{for} \quad r = \infty,$$

$$\psi = \psi_x = \psi_y = 0 \quad \text{for} \quad r = 1.$$

To find a solution for equation (12), the complex variable $z = x + iy$ is introduced. It is known that every solution of (12) is of the form

$$\psi = \operatorname{Im} f(z,\bar{z})$$

with $f(z,\bar{z}) = \bar{z}g(z) + h(z)$, where $g(z)$ and $h(z)$ are analytic functions of z and $\bar{z} = x-iy$. Thus the problem reduces to finding the functions $g(z)$ and $h(z)$. For this purpose we note that $\Delta f(z,\bar{z}) = 4g'(z)$ and hence $\Delta\psi = \operatorname{Im}(4g'(z))$. By taking into account the boundary conditions at infinity we can state that $f(z,\bar{z})$ must behave like the function $(1+a)z + a\bar{z} + b$ at $z = \infty$, where a and b are real constants. This will make $\operatorname{Im} f(z,\bar{z})$ behave like y at $z = \infty$ and thus $\psi_y = 1$, $\psi_x = 0$. Since the imaginary part of $a(z+\bar{z})+b$ is zero, we may subtract $a(z+\bar{z})+b$ from f without changing ψ; hence we may assume $f(z,z) \sim z$, i.e., $g(z) \sim 0$, $h(z) \sim z$ for $z \to \infty$.

It should be noted that possibly the additional term $ic \log z$ is to be added to the above expression. However, the flow around the cylinder is expected naturally to be symmetric with respect to the x-axis. Since adding this term would destroy the symmetry we can correctly discard it.

Now as $f(z,\bar{z})$ is also equal to $\bar{z}g(z) + h(z)$ we deduce that for $z = \infty$ $g(z) \sim 0$ and $h(z) \sim z$. Invoking the boundary conditions for $|z| = 1$ we get $\operatorname{Im}\{\bar{z}g(z) + h(z)\} = 0$ for $|z| = 1$, or since $|z| = 1$ can be written $z = z^{-1}$, this expression is

$$\operatorname{Im}\{z^{-1}g(z) + h(z)\} = 0. \tag{13}$$

Taking the derivative in the x-direction this becomes

$$\psi_x = \operatorname{Im}\{f_z + f_{\bar{z}}\} = \operatorname{Im}\{\bar{z}g'(z) + g(z) + h'(z)\};$$

hence on the circle

$$\operatorname{Im}\{z^{-1}g'(z) + g(z) + h'(z)\} = 0. \qquad (14)$$

The fact that the function $k(z) = z^{-1}g(z) + h(z)$ is real-valued on the unit circle implies that it can be analytically continued to the interior of the unit circle (with the possible exception of the origin) by the principle of reflection*. Since $k(z) \sim z$ as $z \to \infty$, $k(z) \sim \frac{1}{z}$ as $z \to 0$. Now consider the function $k(z) - z - z^{-1}$. Clearly this function is regular in the extended plane, for $k(z) - z - z^{-1} \sim 0$ as $z \to \infty$ and $k(z) - z - z^{-1} \sim 0$ as $z \to 0$. Applying Liouville's theorem the function is a constant and this constant must be zero. Hence $k(z) = z + \frac{1}{z}$. The second condition (14) can be written

$$z^{-1}g'(z) + g(z) + h'(z) = k'(z) + (1+z^{-2})g(z)$$

which is real on the unit circle. Again using the principle of reflection and noting that $k'(z) + (1+z^{-2})g(z) \sim 1$ as $z \to \infty$ we see that this function is regular in the extended plane. Hence by Liouville's theorem this function is a constant, the constant being 1. Thus $k'(z) + (1+z^{-2})g(z) = 1$. Solving for $g(z)$ and substituting for $k'(z)$ its value $1 - z^{-2}$ we find

$$g(z) = \frac{z^{-1}}{z+z^{-1}}.$$

Now

---

*cf. Bieberbach: <u>Einführung in die konforme Abbildung</u>. Sammlung Göschen, p. 58 (or any text on complex variable).

$$f(z,\bar{z}) = \bar{z}g(z) + h(z) = k(z) + (\bar{z}-z^{-1})g(z) = z+z^{-1} + z^{-1}\frac{\bar{z}-z^{-1}}{z+z^{-1}} .$$

Therefore the solution for ψ becomes

$$\psi = \operatorname{Im}\{z + z^{-1} + z^{-1}\frac{\bar{z}-z^{-1}}{z+z^{-1}}\}.$$

However, when we attempt to check this result we find that the boundary conditions are not satisfied. By taking $z = iy$ we have

$$f(z,\bar{z}) = i(y-y^{-1}) + (-iy^{-1})\frac{-iy + iy^{-1}}{iy - iy^{-1}} = iy,$$

and $\psi = \operatorname{Im} f(z,\bar{z}) = y \to 1$ as $y \to 1$. Thus on the unit circle $\psi \neq 0$ which contradicts the boundary conditions. Since the preceding deductions were based on the assumption that a solution exists, we now conclude that this assumption was erroneous, i.e., the problem has no solution.

On the other hand, we may assume that the physical problem has a solution; therefore, we must further conclude that our original assumption was false. That assumption was that the acceleration is of the same order of magnitude as $\Delta\vec{q}$. As a matter of fact, it turns out that when the distance from the body is large the acceleration term becomes dominant.

The preceding result illustrates how careful one must be when the existence of a solution is assumed on "physical grounds". We are now tempted to assume that our problem also has no solution for the three-dimensional case. It turns out that this is not correct; the solution does exist and gives the correct approximate value for the resistance.

For the three-dimensional flow around a sphere the differential equations are

$$\mu\Delta\vec{q} = \operatorname{grad} p \quad \text{and} \quad \operatorname{div} \vec{q} = 0.$$

The boundary conditions are $\vec{q} = 0$ for $r = r_0$, r_0 the radius of the sphere (center at the origin), and $\vec{q} = (U,0,0)$ for $r = \infty$. If we introduce such units of time, space and mass that $r_0 = 1$, $U = 1$, $\mu = 1$, the equations become

$$\Delta\vec{q} = \text{grad } p, \quad \text{div } \vec{q} = 0,$$

and the boundary conditions become

$$\vec{q} = 0 \quad \text{for} \quad r = 1, \ \vec{q} = (1,0,0) \quad \text{for} \quad r = \infty.$$

In attempting to find a solution it appears that a reflection procedure is not easily available as for the two-dimensional case. However, we may proceed as follows: By applying the divergence operator to $\Delta\vec{q} = \text{grad } p$ and using the fact that $\text{div } \vec{q} = 0$ we get

$$\Delta p = 0.$$

Thus p is a harmonic function and can be developed with respect to spherical harmonics. We now make use of the symmetry property: $p = p(x,r)$, $r^2 = x^2 + y^2 + z^2$. The spherical harmonics enjoying this symmetry property are

$$r^{-1}, \quad \nabla_x r^{-1}, \quad \nabla_x^2 r^{-1}, \ldots,$$

where for brevity $\nabla_x = \frac{\partial}{\partial x}$. Hence p may be expanded in a series:

$$p = 2 \sum_{\mu} a_\mu \nabla_x^\mu r^{-1}.$$

We now introduce a function ϕ such that $p = \Delta\phi$. Then $\Delta\vec{q} = \text{grad } p = \text{grad } \Delta\phi$ and $\Delta(\vec{q} - \text{grad } \phi) = 0$. This implies $\vec{q} = \text{grad } \phi + \vec{q}_1$ where $\vec{q}_1$ is a harmonic function. If we observe that $\Delta r = 2r^{-1}$, then

$$\phi = \sum_{\mu=0} a_\mu \nabla_x^\mu r$$

is a function such that $\Delta\phi = p$, as application of the operator Δ term by term gives p. Since the function ϕ is only determined within an additional potential function, it is of the form

$$\phi = \sum_{\mu=0} a_\mu \nabla_x^\mu r + \sum_{\mu=0} b_\mu \nabla_x^\mu r^{-1}.$$

If, for brevity, the symbol ∇ is used instead of grad, we can write

$$\nabla\phi = \sum_{\mu=0} a_\mu \nabla\nabla_x^\mu r + \sum_{\mu=0} b_\mu \nabla\nabla_x^\mu r^{-1}. \tag{15}$$

Since the function $\vec{q}_1$ in the expression $\vec{q} = \text{grad}\ \phi + \vec{q}_1$ is harmonic, we may develop it with respect to

$$\vec{i},\ r^{-1}\vec{i},\ \nabla_x r^{-1}\vec{i},\ldots$$

where $\vec{i} = (1,0,0)$, and also with respect to

$$\nabla r^{-1},\ \nabla_x \nabla r^{-1},\ldots .$$

However, these latter terms can be considered as already included in the second sum in expression (15) for $\nabla\phi$. Then $\vec{q}$ can be expressed as follows

$$\vec{q} = \sum_{\mu=0} \{a_\mu \nabla\nabla_x^\mu r + b_\mu \nabla\nabla_x^\mu r^{-1} + c_\mu \nabla_x^\mu r^{-1}\vec{i}\} + \vec{i}.$$

The term $\vec{i}$ has been added to satisfy the condition that $\vec{q} = \vec{i}$ at $r = \infty$. To insure this condition all other terms must vanish at $r = \infty$. For this reason the term $a_0 \nabla r = a_0 \frac{\vec{r}}{r} = 0$, or $a_0 = 0$. Applying now the condition $\text{div}\ \vec{q} = 0$, and taking into consideration $\Delta\nabla_x^\mu r = \nabla_x^\mu \Delta r = \nabla_x^\mu 2r^{-1}$, we get

$$0 = \operatorname{div} \vec{q} = 2 \sum_{\mu=0} a_\mu \nabla_x^\mu r^{-1} + \sum_{\mu=0} c_\mu \nabla_x^{\mu+1} r^{-1} = 0,$$

whence $c_\mu = -2a_{\mu+1}$.

All that remains now is to insure that the boundary conditions for $r = 1$ are satisfied. For this purpose we first write down some elementary calculations which hold for $r = 1$:

$$\nabla_x r^{-1} = -\frac{x}{r^3} = -x, \quad \nabla r = \vec{r}; \quad \nabla\nabla_x r = \vec{i} - x\vec{r};$$

$$\nabla r^{-1} = -\frac{\vec{r}}{r^3} = -\vec{r}; \quad \nabla\nabla_x r^{-1} = 3x\vec{r} - \vec{i}.$$

Hence, for $r = 1$,

$$0 = \vec{q} = a_0 \nabla r + (-b_0\vec{r}) + c_0 r^{-1}\vec{i} + \vec{i} + (-a_1 x\vec{r}) - b_1\vec{i}$$

$$+ 3b_1 x\vec{r} - c_1\vec{i} + \dots .$$

Making use of the relation $c_\mu = -2a_{\mu+1}$ and the fact that $a_0 = 0$ we can write

$$0 = -b_0\vec{r} - (a_1 - 3b_1)x\vec{r} + \dots + (1 - a_1 - b_1)\vec{i} + \dots .$$

We see that this relation can easily be satisfied by setting $a_\mu = b_\mu = 0$ for $\mu \geq 2$ and $b_0 = 0$. Then $a_1 = 3b_1$, $1 = a_1 + b_1$ from which $b_1 = \frac{1}{4}$, $a_1 = \frac{3}{4}$, $c_0 = -\frac{3}{2}$. The final solution for $\vec{q}$ is

$$\vec{q} = \frac{1}{4}\{3\ \nabla\nabla_x r + \nabla\nabla_x r^{-1}\} - \frac{3}{2} r^{-1}\vec{i} + \vec{i}, \tag{16}$$

and the pressure p is given by

$$p = \frac{3}{2}\nabla_x r^{-1}.$$

Since the viscous stresses are given by $\nabla \vec{q} + (\nabla \vec{q})^*$, we obtain

$$(\tau) = \frac{1}{2} \nabla\nabla\{3 \nabla_x r + \nabla_x r^{-1}\} - \frac{3}{2}(\nabla r^{-1} \vec{i} + \vec{i}\nabla r^{-1})$$

For the vector of viscous stress against the sphere $r = 1$ where the normal vector is $\vec{r}$, we have

$$\vec{r}(\tau) = \frac{1}{2} \nabla_r \nabla\{3\nabla_x r + \nabla_x r^{-1}\} - \frac{3}{2}(\nabla_r r^{-1} \vec{i} + x \nabla r^{-1}).$$

The vector of the pressure against the sphere $r = 1$ is

$$-p\vec{r} = -\frac{3}{2} \vec{r} \nabla_x r^{-1} = \frac{3}{2} \vec{r}x.$$

After evaluating these expressions one finds surprising enough that the resultant is a constant vector in the x-direction: the resultant of viscous stresses and pressure against the sphere $r = 1$ is

$$3/2 \; \vec{i} \tag{17}$$

By integrating over the sphere, the drag $\vec{D}$ exerted on the sphere is found to be

$$\vec{D} = 6\pi\vec{i},$$

or in terms of general units

$$\vec{D} = 6\pi\mu \, U r_0 \vec{i}. \tag{18}$$

This is the famous <u>formula of Stokes</u>. The drag per unit area can be written

$$\frac{3}{2}\mu U r_0^{-1} = \frac{3}{2}\rho U^2 R^{-1}$$

with $R = U r_0 \nu^{-1}$ in accordance with the results of the similarity considerations.

A further conclusion we can draw is that the flow is symmetric in x in the sense that $u(-x,y,z) = u(x,y,z)$, $v(-x,y,z) = -v(x,y,z)$, $w(-x,y,z) = -w(x,y,z)$. This follows immediately from the explicit result (16) obtained for $\vec{q}$.

Problem 17. Carry out the evaluations which prove the result (17).

Problem 18. Discuss the stream function $\psi(x,r)$ and the stream surfaces $\psi(x,r) = \text{const.}$ for the "Stokes" flow. Also discuss the results obtained when observed from the sphere, i.e., when the fluid is at rest and the sphere moves with the velocity $(-U,0,0)$.

Now we investigate the validity of our assumption that the acceleration term can be neglected. Consider the acceleration term; for $y = z = 0$ and $|x|$ large it reduces to uu_x, while the viscosity term is $\nu\Delta u$. For large $|x|$ we have

$$u \sim 1 - \frac{3}{2}r^{-1}, \quad u_x \sim \frac{3}{2}r^{-2}, \quad \Delta u = \nabla_x p = -3r^{-3}.$$

Thus the ratio $\frac{uu_x}{\Delta u} \sim \frac{r}{2}$, or in general units $\frac{uu_x}{\nu\Delta u} \sim \frac{Ur}{2\nu} \to \infty$ as $r \to \infty$. Hence we conclude that the terms we neglected are larger than those considered. This indicates that for points which are at a large distance from the sphere the Stokes solution is not valid as an approximation. However, it does not invalidate the Stokes solution as being a correct approximation for points near the sphere. That this is so has been shown by Oseen in his improved treatment of the problem.

Oseen*, in his improvement, does not neglect the acceleration term altogether, but does replace the acceleration $u\vec{q}_x + v\vec{q}_y + w\vec{q}_z$ by

---

*See Aerodynamic Theory, W. F. Durand, ed., Vol. III, p. 75. Lamb, Hydrodynamics, 6th Edition, §342.

$$U\,\vec{q}_x.$$

This certainly represents an approximation to the flow at great distances from the sphere where $\vec{q} \sim U\,\vec{i}$, but not near the sphere where $\vec{q} = 0$. The differential equations become linear after this replacement; they are

$$\mu\Delta\vec{q} - \rho U\,\nabla_x\vec{q} = \nabla p, \quad \operatorname{div}\vec{q} = 0.$$

The boundary conditions are

$$\vec{q} = U\vec{i} \quad \text{for} \quad r = \infty; \quad \vec{q} = 0 \quad \text{at} \quad r = r_0.$$

As usual select units so that $r_0 = \mu = U = 1$, then $\rho = R$ where $R = \dfrac{Ur_0}{\nu}$. The solution of this set of equations can, after Lamb, be obtained as follows: Set

$$\vec{q} = \vec{q}_1 + \vec{q}_2$$

where $\Delta\vec{q}_1 = 0$, $R\,\nabla_x\vec{q}_1 = -\nabla p$ and $\Delta\vec{q}_2 = R\,\nabla_x\vec{q}_2$. We can select $\vec{q}_1$ satisfying these conditions by taking a potential function Φ and letting $\vec{q}_1 = \nabla\Phi$, $p = -R\,\nabla_x\Phi$. Then $\operatorname{div}\vec{q}_1 = 0$ and $R\,\nabla_x\vec{q}_1 = -\nabla p$. To find $\vec{q}_2$ satisfying $\Delta\vec{q}_2 = R\,\nabla_x\vec{q}_2$ and $\operatorname{div}\vec{q}_2 = 0$ we introduce the function X defined by

$$\Delta X - R\,\nabla_x X = 0, \tag{19}$$

and let

$$\vec{q}_2 = \nabla X - RX\vec{i}.$$

It is seen immediately that $\operatorname{div}\vec{q}_2 = 0$ and $\Delta\vec{q}_2 = R\,\nabla_x\vec{q}_2$ hold. Thus the problem reduces to solving (19). To do this set

$$\chi = e^{\frac{1}{2}Rx} h.$$

Then $\nabla_x \chi = e^{\frac{1}{2}Rx}\{\nabla_x h + \frac{1}{2}Rh\}$; $\Delta\chi = e^{\frac{1}{2}Rx}\{\Delta h + R\nabla_x h + \frac{1}{4}R^2 h\}$, and equation (19) gives

$$\Delta\chi - R\nabla_x \chi = e^{\frac{1}{2}Rx}\{\Delta h - \frac{1}{4}R^2 h\} = 0.$$

Now the equation

$$\Delta h - \frac{1}{4}R^2 h = 0$$

possesses solutions depending only upon r, i.e., $h = h(r)$. If we observe that $\Delta h(r) = h_{rr} + 2r^{-1}h_r = r^{-1}(rh)_{rr}$ then we only have to solve

$$(rh)_{rr} - \frac{R^2}{4}(rh) = 0.$$

The solutions of this clearly are $rh = e^{-\frac{1}{2}Rr}$ and $rh = e^{+\frac{1}{2}Rr}$. However, the solution $e^{+\frac{1}{2}Rr}$ is excluded since this would make $\vec{q}$ increase indefinitely as r became large. Thus the solution for h is

$$h = r^{-1}e^{-\frac{1}{2}Rr}.$$

Other solutions of $\Delta h - \frac{1}{4}R^2 h = 0$ having the necessary symmetry can be obtained by differentiation of this solution with respect to x. Without entering into a detailed discussion, from $h = Cr^{-1}e^{-\frac{1}{2}Rr}$, we see that $\chi = \frac{C}{r}e^{-\frac{1}{2}R(r-x)}$, and $\vec{q}_2$ can be found to be

$$\vec{q}_2 = -\frac{1}{2}RC\, r^{-1}e^{-\frac{1}{2}R(r-x)}\{r^{-1}\vec{r} + \vec{i}\} + Ce^{-\frac{1}{2}R(r-x)}\nabla r^{-1}.$$

Near the sphere the quantity $e^{-\frac{1}{2}R(r-x)}$ varies between 1 and e^{-R}. Hence in the neighborhood of the sphere we may say that on the average

$$\vec{q}_2 \sim -\tfrac{1}{2}RC_1 r^{-1}\{r^{-1}\vec{r} + \vec{i}\} + C_1 \nabla r^{-1}$$

where C_1 is between C and Ce^{-R}. From $\vec{q}_1 + \vec{q}_2 = 0$ at $r = 1$, it can then be shown that the contribution of $\vec{q}_1$ to $\vec{q}$ will essentially behave like

$$-\tfrac{1}{2}(2+R)C_1 \nabla r^{-1} + \tfrac{1}{2}RC_1 \vec{i}, \quad C_1 > 0.$$

The discussion of the flow is clearer when the fluid is considered as being at rest at infinity and the sphere as moving with the velocity $(-U,0,0)$. At great distances from the sphere the term $-\frac{1}{2}(2+R)C_1 \nabla r^{-1}$ in $\vec{q}_1$ represents a source, and this will be balanced by $\vec{q}_2$ only when $r - x$ is above a fixed positive quantity, that is, inside a paraboloid $r^2 \geq (x+C)^2$ which opens in the positive x-direction. Outside this paraboloid the flow behaves as from a source. The fluid appears to radiate from the body outside the parabola, while inside the parabola the fluid moves towards the body. (cf. Aerodynamic Theory, Vol. III, p. 79.) Observed from the sphere the motion is different: the fluid particles have little vorticity to the left of the sphere, gain vorticity and slow down as they approach the sphere, and then follow a rather parallel flow as they are drawn into the parabolic "wake". The non-symmetry of the flow is a remarkable feature. Nevertheless it turns out that in the neighborhood of the sphere the flow behaves like Stokes' flow and that it gives nearly the same value of the resistance, provided that the Reynolds number $R = \frac{Ur_0}{\nu}$ is small enough.

It may be remarked that for the two-dimensional case Oseen's method can be used. However, in finding the solution of the differential equations in that case, Bessel functions arise instead of the simple exponentials.

6. Unsteady Flow

We confine ourselves to two simple examples.

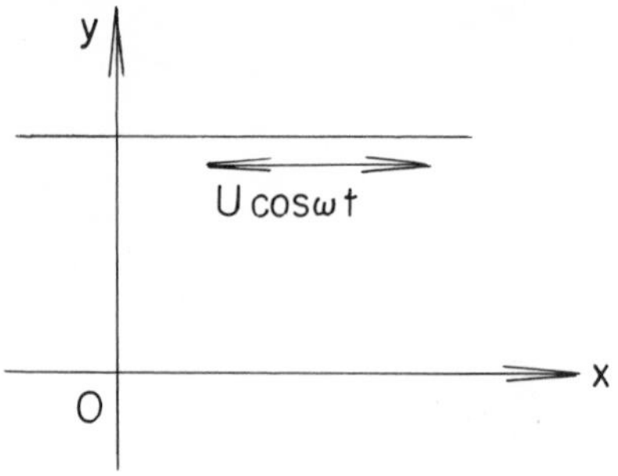

Problem 19. (a) Determine the velocity distribution of the oscillating plane Couette flow, i.e., the flow between two infinite, parallel, plane plates one of which is fixed and the other moving with a periodic motion $u = U \cos \omega t$, $v = w = 0$. Assume the pressure p is constant. [In the solution assume $u = \mathscr{R}\{f(y)e^{i\omega t}\}$].

(b) Find the energy put in per unit area over a period as a function of the period.

Problem 20. Discuss the two-dimensional vortex flow given by the stream function

$$\psi = \psi_0(x,y) = A \sin \pi \frac{x}{a} \sin \pi \frac{y}{b}.$$

Investigate the decay of the vortices due to viscosity.

Hint: Assume the stream function ψ is of the form

$$\psi = \psi_0(x,y) \cdot f(t),$$

where $f(t)$ is a function of the time alone.

7. Flow in Convergent and Divergent Channels

It is possible to determine exactly the flow of a viscous fluid between two inclined, plane plates, assuming that the particles move in straight lines through the intersection of the plates.* The flow will be two-dimensional and therefore can be described by a stream function $\psi(x,y)$. It is more convenient to use polar coordinates r, θ, and then the assumption about the flow is that ψ is a function of θ alone. We set $\psi'(\theta) = \nu g(\theta)$ and then the velocity is given by

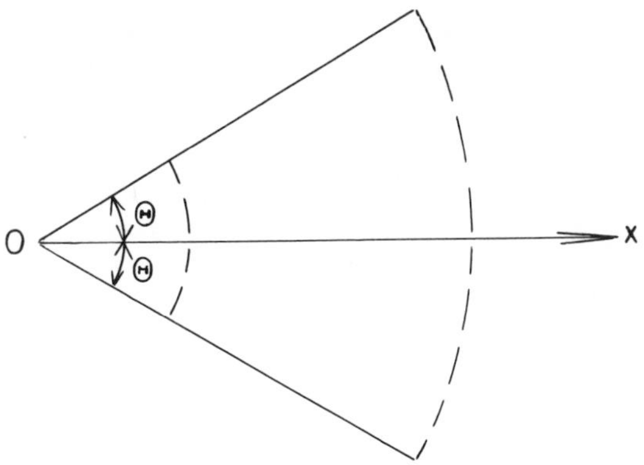

$$u = r^{-1}\nu g(\theta).$$

If the plates are given by $\theta = \pm\,\Theta$, the boundary conditions for the flow are

$$g(\theta) = 0 \quad \text{for} \quad \theta = \pm\,\Theta.$$

An element of flux dQ across the circle $r = \text{const.}$ is the product of the normal component of velocity multiplied by the area $rd\theta$. Hence the total flux Q is given by

$$Q = \nu\int_{-\Theta}^{\Theta} g(\theta)\,d\theta.$$

In addition to the boundary conditions, Q may also be prescribed. In the case of two-dimensional flow the Navier-Stokes differential equation was found (after eliminating the pressure) to be

$$\nu\Delta\Delta\psi = \psi_y\Delta\psi_x - \psi_x\Delta\psi_y. \tag{8a}$$

*For a more complete discussion, see A. Rosenblatt: Solutions Exactes des Equations du Movement des Liquides Visqueux.

Since $\Delta\psi(\theta)$ reduces to $r^{-2}\psi'' = \nu^2 r^{-2} g'(\theta)$, equation (8a) simplifies to

$$g'''(\theta) + 4g'(\theta) = -2gg'(\theta).$$

This can be integrated directly to give

$$g'' + 4g + g^2 = c.$$

Multiplying this last equation by the factor $2g'$ enables us to integrate again, obtaining

$$g'^2 + 4g^2 + \frac{2}{3}g^3 = 2cg + d.$$

Letting

$$H(g) = -\frac{2}{3}g^3 - 4g^2 + 2cg + d$$

we can write

$$g' = \pm (H(g))^{1/2} \quad \text{or} \quad \frac{d\theta}{dg} = \pm \frac{1}{(H(g))^{1/2}}.$$

In order that the boundary conditions, $g(\theta) = 0$ at the plates be satisfied, $(H(g))^{1/2}$ must be real when $g = 0$, or what is the same thing,

$$H(0) \geq 0.$$

Plotting the curve $H = H(g)$ with g as independent and H as dependent variable, we see that it must cross the H-axis above the origin. Since $H(g)$ is negative for large positive values of g, the condition $H(0) \geq 0$ can be satisfied only if $H(g) = 0$ has one or three positive roots. Since the sum of the roots is -6, the equation must have exactly one positive root. The function $H(g)$ can be written

$$H(g) = (a-g)K(g),$$

where $K(g) = \frac{2}{3}\{g^2 + (6+a)g + a_1\}$.

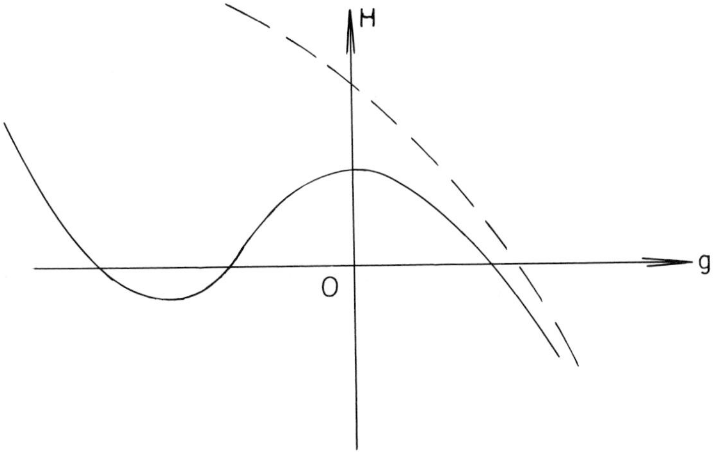

(We put $c = \frac{1}{3}(a^2 + 6a - a_1)$ and $d = \frac{2}{3}aa_1$). Now $K(g)$ has its minimum value for $g = -\frac{(6+a)}{2}$. Hence the condition that $K(g)$ be non-negative for g non-negative is equivalent to

$$a_1 \geq 0.$$

The curve $\frac{d\theta}{dg} = +\frac{1}{(H(g))^{1/2}}$ may be plotted as shown in the figure with two possible curves according as $H(g)$ has one (dotted curve in figure) or three (solid curve) real roots. From

$$\frac{d\theta}{dg} = \pm\frac{1}{(H(g))^{1/2}}$$

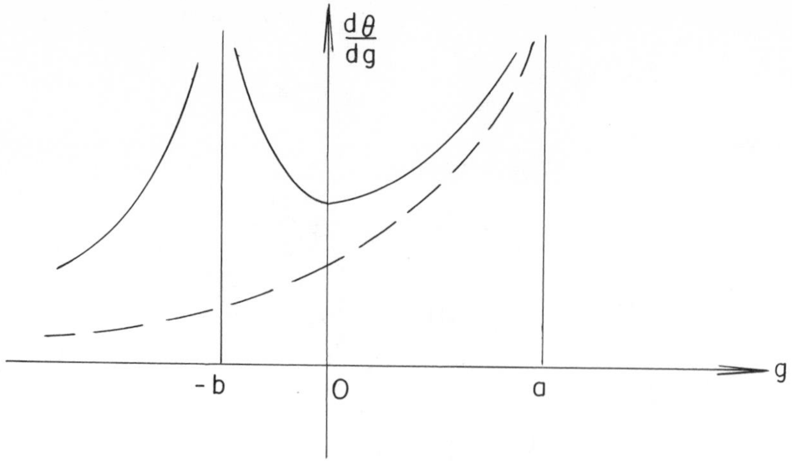

θ can be found as an elliptic integral

$$\theta = \int_g^a \frac{dg}{[(a-g)K(g)]^{1/2}} + \text{const.}$$

For the case where there are three real roots, a, -b, $-b_1$, (b,b_1 positive), g is a periodic function of θ, oscillating between a and -b, where -b is the larger of the negative roots. The period is given by $2(\Theta_0+\Theta_1)$:

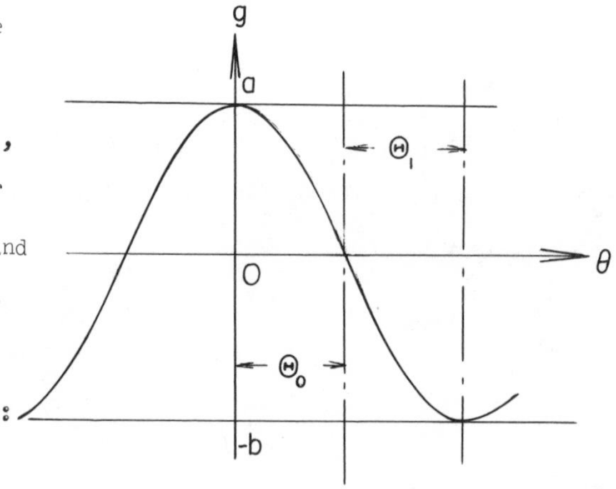

3 real roots

$$\Theta_0 = \int_0^a \frac{dg}{(H(g))^{1/2}} ;$$

$$\Theta_1 = \int_{-b}^0 \frac{dg}{(H(g))^{1/2}} .$$

The flux across the sectors swept out by the angle $2(\Theta_0+\Theta_1)$ is given by $2\nu(Z_0-Z_1)$ where

$$Z_0 = \int_0^a g(\theta)d\theta = \int_0^a \frac{d\theta}{dg} g(\theta)dg = \int_0^a \frac{gdg}{(H(g))^{1/2}} ; \quad Z_1 = -\int_{-b}^0 \frac{gdg}{(H(g))^{1/2}} .$$

In case there are no negative roots, we may simply consider b as approaching infinity in the preceding case, and g as a function of θ is shown in the accompanying figure.

The resulting flows are indicated in the diagrams below.

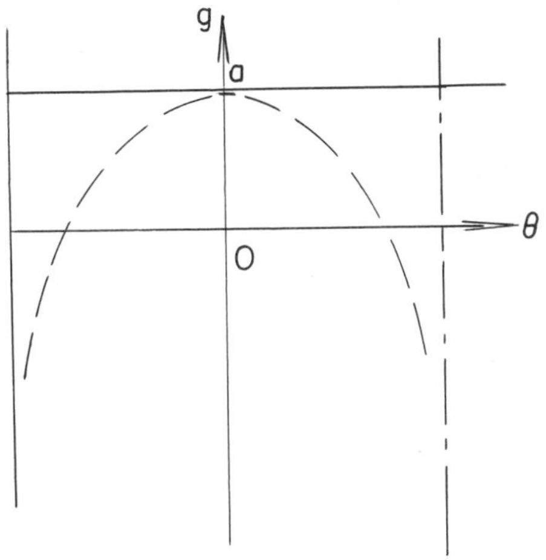

I real root

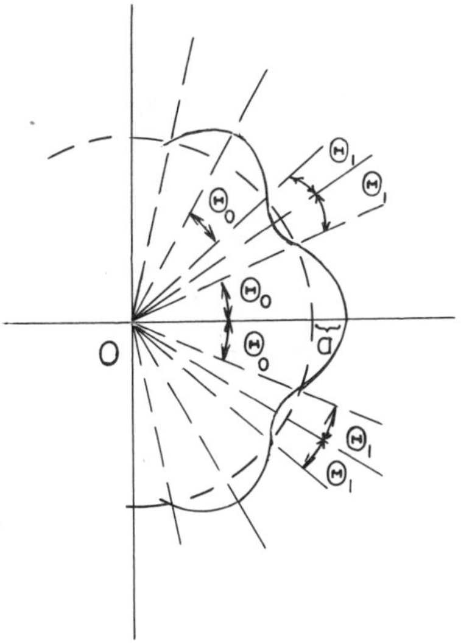

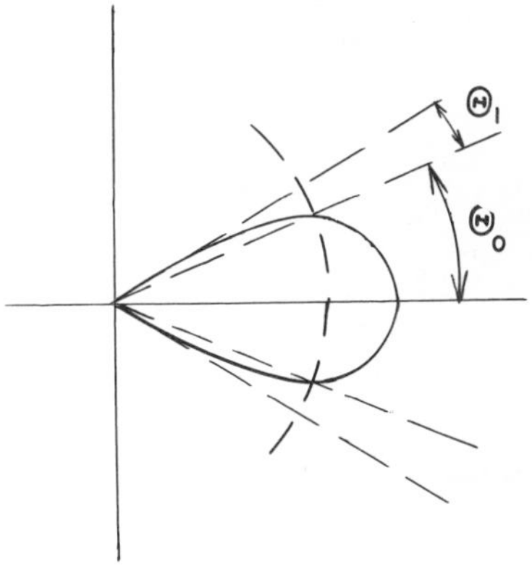

The solid curves show the flow as represented by g, and the points where these curves meet the intersection of the rays with the circular arc indicate the places where $g = 0$. The flow is to satisfy the boundary conditions, $g = 0$ for $\theta = \pm \Theta$ and also have a prescribed flux. We note that $\theta = \theta(g)$ is only determined up to an arbitrary constant. These angles depend upon $H(g)$ which in turn depend on a and a_1. Thus the values of a and a_1 are to be determined so that

$$\int_{-\Theta}^{\Theta} g(\theta)d\theta = \frac{Q}{\nu}$$

is satisfied.

First take the case of a purely divergent flow, i.e., $g(\theta) \geq 0$ everywhere, which implies $Q > 0$. In this case the angles $\theta = \pm \Theta_0$ must be identified with Θ, and Z_0 must be identified with Q/ν. That is, we have the two equations

$$\begin{aligned} \Theta_0(a, a_1) &= \Theta \\ Z_0(a, a_1) &= Q/\nu \end{aligned} \tag{20}$$

where the quantities on the right are known, and a, a_1 are to be determined. However these two transcendental equations may have no solution, many solutions, or even an infinite number of solutions. For example, if Θ or Q is too large there will be no solution; or if ν is too small there will also be no solution. We are interested in the maximum values that Θ and Q/ν can have for us to get a solution. To do this we minimize H. For a fixed a, since $a_1 \geq 0$, we know that $H(g)$ has its minimum for $a_1 = 0$. For $a_1 = 0$ we find, if we let $g = az$,

$$\Theta_0 = \Theta = a^{-\frac{1}{2}} \int_0^1 \frac{dz}{[\frac{2}{3}(1-z)z(z+1+6a^{-1})]^{1/2}}$$

$$Z_0 = \frac{Q}{\nu} = a^{\frac{1}{2}} \int_0^1 \frac{zdz}{[\frac{2}{3}(1-z)z(z+1+6a^{-1})]^{1/2}}$$

As $a \to \infty$ we see that $\Theta \to 0$, $Q/\nu \to \infty$, while the product $\Theta \cdot \frac{Q}{\nu}$ approaches a finite value; as $a \to 0$ we have $\Theta_0 \to \pi/2$ and $Z_0 \to 0$. The values of the pair (Θ_0, Z_0) attained for $a \geq 0$, $a_1 = 0$ determine a curve in the Θ_0-Z_0 plane which together with the lines $\Theta_0 = 0$, $Z_0 = 0$ bounds the region (shaded in the figure) of values which the pair (Θ_0, Z_0) can attain for $a \geq 0$, $a_1 \geq 0$. Only when $(\Theta, Q/\nu)$ falls in this region is a pure outward flow $(u > 0)$ possible. This shows that for small values of Θ a greater flux is possible than for large values of Θ; in addition it shows that for a given angle Θ the smaller the viscosity the smaller the flux that can be transported.

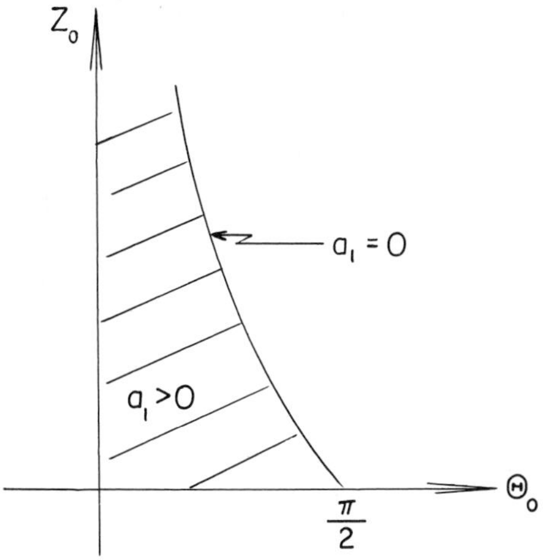

If the data Θ, Q/ν are not such that both conditions of (20) can be met, or what is the same, the point $(\Theta, Q/\nu)$ should be in the shaded region in the above figure, a flow through the channel is possible only if back-flow is admitted. One may select any pair of angles for which $g = 0$ (so that the boundary conditions are satisfied) including one or more regions of back-flow and then see if this selection of angles has for a resultant flux the prescribed amount Q/ν. It is likely that every value of Θ or Q/ν can be attained in this way, but probably not in a unique manner. Further, it may be conjectured that increasing values of Q/ν require selection of angles so that there are more and more sectors of back-flow. In particular, for given values of Θ and Q, if ν is allowed to approach zero the flow will not approach the source flow $\psi = Q \frac{\theta}{\Theta}$ for non-viscous fluids.

The above ideas concerning the non-uniqueness of the flow, the occurrence of back-flow and the non-convergence when $\nu \to 0$ are of the greatest importance in understanding the nature of the flow of viscous fluids. The presence of back-flow may be expected whenever the region to be traversed by the fluid is divergent. For example, the flow around an obstacle may be considered to be divergent on the rear side (see figure). In this case the flow will separate from the obstacle and regions of back-flow will occur.

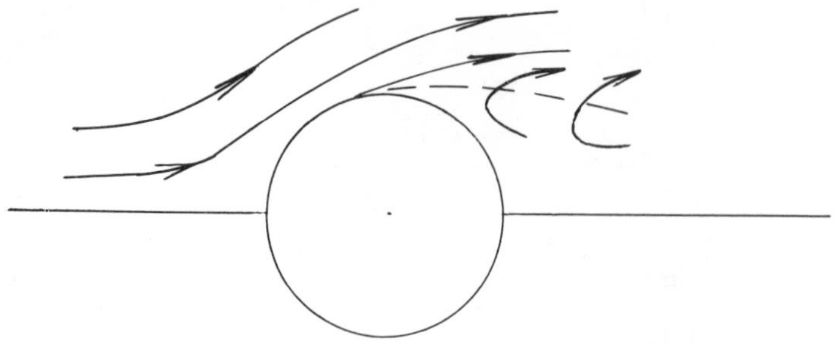

The situation for convergent flow is quite different from that for divergent flow. Here we want $Q < 0$, and hence we consider a flow with $g < 0$ throughout the sector. This can be done by identifying the angle Θ with Θ_1 where

$$\Theta_1 = \int_{-b}^{0} \frac{dg}{(H(g))^{1/2}}, \tag{21}$$

and $-Q/\nu$ with Z_1, where

$$Z_1 = -\int_{-b}^{0} \frac{gdg}{(H(g))^{1/2}}. \tag{22}$$

Since we are interested in a sector where there is pure back-flow, we know that we must consider the case for which $H(g)$ has three real roots (since the case of one real root yielded a purely divergent flow). In this case there are no limitations on the possible values of Θ and Q/ν since the integrals (21) and (22) can be made infinite by letting the two negative roots b, b_1 of $H(g)$ approach each other.

It is of interest to see what happens when the fluid is considered to have smaller and smaller viscosity, i.e., to approach a perfect fluid. Here $\nu \to 0$ and hence $-Q/\nu \to \infty$. In order that Θ remain finite while $-Q/\nu \to \infty$ one must at the same time let b approach infinity. The contributions of the integrands of the integrals (21) and (22) will then be small for all values of g except those near to $-b$. For the value $g = -b$, the integrands differ by the factor $g = -b$ and we can conclude that

$$\frac{Z_1}{\Theta_1 b} \to 1 \quad \text{as} \quad b \to \infty. \tag{23}$$

This is equivalent to $-\frac{Q}{\nu\Theta b} \to 1$ or $\nu b \to -\frac{Q}{\Theta}$. From a similar argument it follows that as $b \to \infty$, $\theta \to \Theta$ when $-g/b < 1$; hence g behaves like $-b$, i.e., $g/b \to -1$ for values of $|\theta| < \Theta$. Therefore $\nu g \to \frac{Q}{\Theta}$, and if the radial velocity is denoted by u, we have

$$\nu g = ur \to -\frac{Q}{\Theta} \quad \text{for} \quad |\theta| < \Theta. \tag{24}$$

Equation (24) then states that for fluids with small viscosity the flow interior to the walls corresponds to the simple case of flow towards a sink. On the other hand, at the wall the velocity is zero. Hence there is a transition from the value zero to $-\frac{Q}{r\Theta}$ over a region that shrinks to zero with ν. This transition region is known as the boundary layer.

In order to investigate the boundary layer a little more closely we introduce the new variable $z = -g/b$. Then $ru = g\nu = -b\nu z$ and from

$$\theta = -\int_{-b}^{g} \frac{dg}{(H(g))^{1/2}}$$

we obtain

$$\sqrt{b}(\Theta+\theta) = \sqrt{b}\int_g^0 \frac{dg}{(H(g))^{1/2}} = \int_0^{-\frac{ru}{b\nu}} \frac{dz}{[(b^{-1}a+z)(1-z)(b_1b^{-1}-z)]^{1/2}} .$$

On allowing νb to approach $-Q/\Theta$, $b^{-1}b_1 \to 1$, $b^{-1}a \to 2$ and we have

$$\lim \sigma = \int_0^{ru\frac{\Theta}{Q}} \frac{dz}{(1-z)(2+z)^{1/2}} \tag{25}$$

where σ is defined by

$$\sigma = (\frac{-Q}{\nu\Theta})^{1/2}(\Theta+\theta).$$

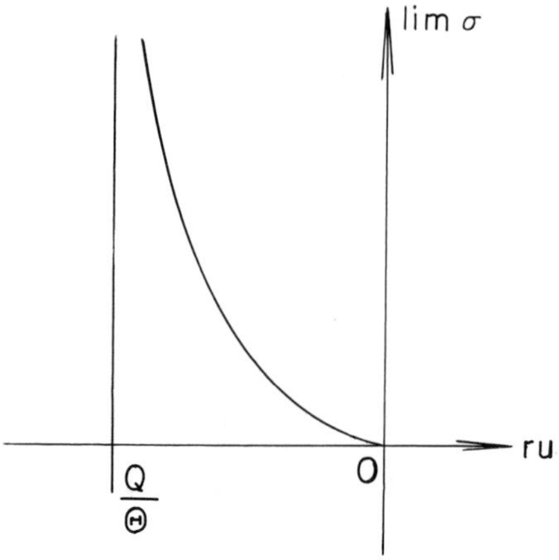

(See figure.) The integral on the right hand side of (25) approaches ∞ as $ru \to Q/\Theta$.

When ru is considered a function of $\sigma = (\Theta+\theta)(\frac{-Q}{\nu\Theta})^{1/2}$, (rather than of θ), ru approaches as $\nu \to 0$ a limit function obtained by inversion from (25). This limit function approaches the value Q/Θ as $\sigma \to \infty$ (see figure), it represents approximately the transition for small finite values of ν and sets in evidence that the transition region shrinks to zero like $\sqrt{\nu}$.

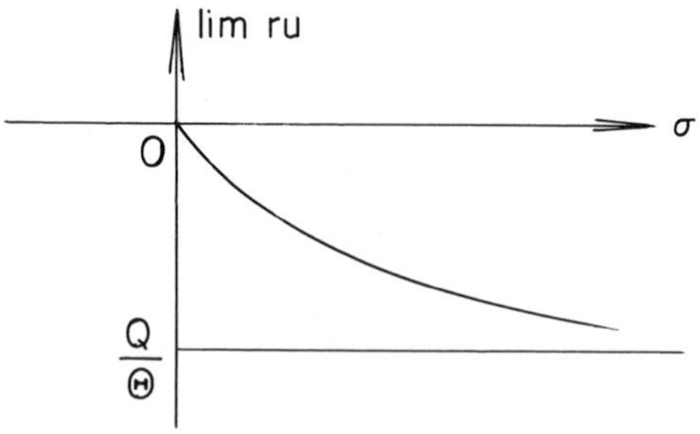

8. Flow Towards a Plane Plate

An important class of flows are flows against a plane plate. The flow at its beginning may also possess a radial motion. The results will be typical of behavior in the neighborhood of stagnation points. It so happens that such flows against a plane plate can be determined exactly, i.e., their determination can be reduced to the solution of an ordinary differential equation.

We assume the flow to be rotationally symmetric with respect to the z-axis, that is the plate lies in the xy plane at $z = 0$ and for all points of the fluid $z > 0$. We shall use cylindrical coordinates r, θ, z so that $x = r\cos\theta$, $y = r\sin\theta$ and denote the corresponding components of the velocity vector $\vec{q}$ by u, v, w, i.e., u is the component in the direction of r and v in the direction θ. We assume, however, that $v = 0$ (no rotation); then the x and y components of $\vec{q}$ are $u\cos\theta$ and $u\sin\theta$ respectively. The rotational symmetry tells us that u, w and the pressure p are functions only of r and z.

The Navier-Stokes equations then assume the form

$$\begin{aligned} uu_r + wu_z &= -\rho^{-1}p_r + \nu(\Delta u - r^{-2}u) \\ uw_r + ww_z &= -\rho^{-1}p_z + \nu\Delta w \end{aligned} \qquad (26)$$

where $\Delta = \nabla_r^2 + r^{-1}\nabla_r + \nabla_z^2$. The continuity equation is

$$u_r + r^{-1}u + w_z = 0. \qquad (26a)$$

Instead of trying to solve these partial differential equations for suitable boundary conditions, we simplify the problem by specifying a type of solution which will reduce the equations to ordinary differential equations in the single independent variable z. This is done by assuming

$$u = \nu r F(z) \qquad w = -2\nu H(z)$$

$$p = \rho\nu^2[-\tfrac{1}{2}kr^2 + 2P(z)], \quad k = \text{const.}$$

where F, H and P are functions to be determined. The first of the Navier-

Stokes equations (26) becomes

$$\nu^2 rF^2 - 2\nu^2 rHF' = k\nu^2 r + \nu^2 rF''$$

or

$$F''(z) + 2H(z)F'(z) - F^2(z) = -k$$

which is an equation containing z only, as desired. Similarly the other two equations do not contain r:

$$H''(z) + 2H(z)H'(z) = -P'(z)$$

$$H'(z) - F(z) = 0.$$

Eliminating F we have the differential equation for H:

$$H'''(z) + 2H(z)H''(z) - H'^2(z) = -k \tag{27}$$

while P and F are given by

$$\begin{aligned} P(z) &= -H'(z) - H^2(z) + \text{const.} \\ F(z) &= H'(z) \end{aligned} \tag{27a}$$

Equation (27) is important in that, as we shall see later, it is the type of differential equation which arises in standard problems of the boundary layer theory.

For the problem at hand the boundary conditions at the plate $z = 0$ are seen to be

B.C._0: $$H(0) = H'(0) = 0$$

since $u = 0$, $w = 0$ at the plate. A third condition is required in order to determine the solution uniquely. This will be the condition at $z = \infty$, which we leave open for the present.

We first consider the case $k = 0$. Here it is possible to guess at the solution and then to show that the result satisfies the problem, rather than to carry out a direct integration of the differential equation. We take

$$H(z) = az^2, \quad a = \text{const.}$$

whence, with $c = 2\nu a$,

$$u = crz, \quad w = cz^2.$$

The equation (27) is seen to be satisfied while (27a) gives

$$P(z) = -2az - a^2 z^4 + \text{const.}, \quad F(z) = 2az.$$

The pressure is then

$$p = -2\rho\nu cz - \frac{1}{2}\rho c^2 z^4 + \text{const.}$$

Finally, we observe that the $B.C._0$ are satisfied.

We may introduce a stream function ψ such that

$$u = r^{-1}\psi_z \quad w = -r^{-1}\psi_r$$

and the continuity equation (26a) is automatically fulfilled. The function ψ gives the flux of the flow through a circle normal to and with center on the z-axis. We find immediately from the values of u and w that $\psi = \frac{1}{2}cr^2z^2$. Hence the stream surfaces are the hyperboloids of revolution $rz = \text{const.}$ Thus in a plane normal to the plate and containing the symmetry axis the fluid flows along hyperbolas giving the same appearance as a two-dimensional non-viscous flow.

The flow can be reversed in its direction; also every kinematically similar flow is possible independent of ν. (This flow is thus an example against the conjecture that either the acceleration or the viscosity terms should vanish for such flows.) It is of no great importance, however, because of its behavior at $z = \infty$. The velocity increases quadratically as $z \to \infty$.

The most important case is that where the flow behaves for $z \to \infty$ like the irrotational flow against a plate. That is, a flow which slides along the plate like the flow of a perfect fluid. This irrotational flow can be derived from the potential $\phi = c(\frac{1}{2}r^2 - z^2)$; it is given by

$$u = cr, \; w = -2cz, \; p = -\frac{1}{2}\rho(c^2r^2 + 4c^2z^2) + \text{const.} \tag{28}$$

With $c > 0$ the flow is towards the plate. The equation of continuity (26a) is fulfilled and also equations (26) in which the viscous terms drop out. The pressure is seen to satisfy Bernoulli's law.

The condition that the flow behaves like (28) for $z \sim \infty$ can be met by assuming

$$H'(\infty) = F(\infty) = b, \quad k = b^2, \quad c = \nu b$$

and further $H''(\infty) = 0$ in such a way that $H(\infty)H''(\infty) = 0$ and $H'''(\infty) = 0$ in order to fulfill (27). Thus we obtain the following boundary value problem for H:

D.E.:

$$H''' + 2HH'' - H'^2 = -b^2 \tag{29}$$

with the boundary conditions at the plate and at infinity

$$\begin{aligned} &\text{B.C.}_0\text{:} \quad && H'(0) = H(0) = 0 \\ &\text{B.C.}_\infty\text{:} \quad && H''(\infty) = 0, \; (H \cdot H''(\infty) = H'''(\infty) = 0) \end{aligned} \tag{29a}$$

Then $B.C._\infty$ together with (27) implies that $[H'(\infty)]^2 = b^2$. The sign of b then determines the direction of flow.

The solutions can be discussed better after we differentiate (29) obtaining

D.E.': $$H'''' + 2HH''' = 0 \tag{29'}$$

with

B.C.'$_0$: $$H(0) = H'(0) = 0, \quad H'''(0) = -b^2,$$

while the $B.C._\infty$ remains the same as in (29a) above. It is seen that a solution of (29') satisfying the B.C.'$_0$ is (provided H were known)

$$H'''(z) = -b^2 e^{-2\int_0^z H(t)dt} .$$

From this solution we can make the following conclusions

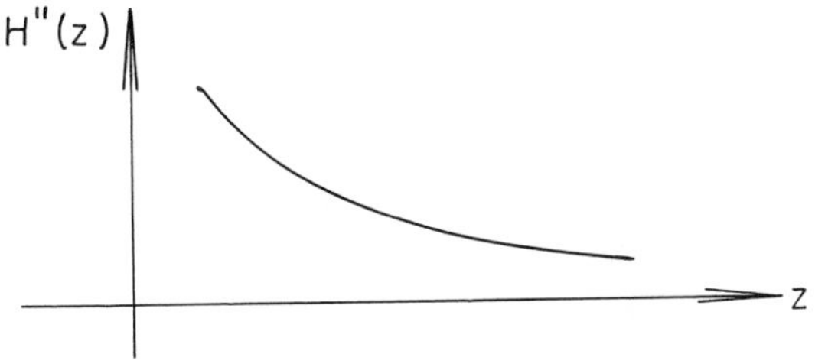

1. $H'''(z) < 0$.

Hence we see from $H''(\infty) = 0$ that

2. $H''(z)$ decreases and

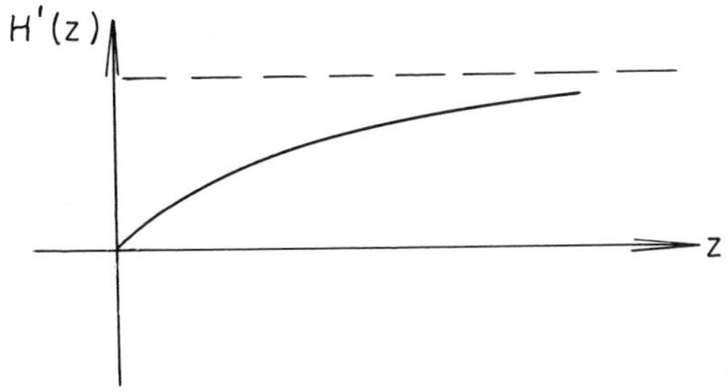

3. $H''(z) > 0$.

From $H'(0) = 0$ we then have

4. $H'(z) > 0$

whence

5. $H'(\infty) > 0$.

Now $H'(\infty) = b$, so we see that the problem has a solution only when $b > 0$, i.e., the flow at infinity is towards the plate. Thus we cannot admit the possibility of $H'(\infty) < 0$, and there is no solution which gives a flow away from the plate coinciding with the irrotational flow at some distance from the plate. Although this has been shown only under the assumption that the flow be of the general form $u = \nu r F(z)$, $w = -2\nu H(z)$, it is likely that the conclusion holds in general.

The solution of the problem can be determined numerically.* Some results are

$$H(z) \sim bz - \eta\sqrt{b}, \quad \eta = .5576, \text{ for } z \sim \infty$$

with

$$w \sim -2cz + 2\eta(c\nu)^{1/2}, \quad u = cr$$

and

$$H''(0) = \zeta b^{3/2}, \quad \zeta = 1.317.$$

Thus, at infinite distance from the plate, the flow represents the irrotational flow towards a plate which moves with the velocity $2\eta(c\nu)^{1/2}$ against the stream. This expresses in a way the retardation due to the viscosity.

A possible explanation of the non-existence of a solution for the problem of flow away from the plate may be seen by comparing it with the flow in two-dimensional channels. The flow towards a plate, after it has come near to the plate, is seen to be very much like a flow in a converging channel with

---

*F. Homann, Zeits. f. angew. Math. u. Mech., 16(1936), p. 153.

decreasing pressure. But the flow away from a plate, as long as it is near to the plate, is more similar to flow in a divergent channel with increasing pressure; it is likely that separation from the plate and back-flow will occur.

The same analogy arises in the investigation of the flow in the neighborhood of the forward and backward stagnation points of a body of revolution in a symmetric uniform flow. No inconsistency occurs for the forward stagnation point; but it is impossible to comply with the condition of finite velocity behind the backward stagnation point. Back-flow will necessarily occur.

Another similar type of problem which has been solved by von Karman* is obtained by allowing the plate to rotate with uniform angular velocity about the z-axis. We can find the solution in the same way as we did with the stationary plate, the only change being in having a new function $G(z)$ which gives the angular velocity of the fluid. Then, with $u = rF(z)$, $v = rG(z)$, $w = H(z)$ and

$$p = \rho[-\frac{1}{2}kr^2 + P(z)]$$

the partial differential equations reduce to ordinary differential equations as before.

Another case that has been considered** arises by keeping the plate fixed but assuming uniform rotation of the fluid for large values of z. This means that $G(\infty)$ is prescribed while $G(0) = 0$. It is found that there can be no flow towards the plate but there is a flow away from the plate with a radial velocity that oscillates in z, the amplitude of the oscillation approaching zero as $z \to \infty$.

Problem 21. Discuss the pressure distribution for divergent flow between plane plates. Find p in terms of $g(\theta)$. Find then the average pressure

*Goldstein, Modern Developments in Fluid Dynamics, Vol. I, pp. 110-113.

**cf. U. T. Bödewadt, Die Drehströmung uber festem Grunde, Zeits. f. angew. Math. u. Mech., 20(1940), p. 241.

$$\overline{p}(r) = \frac{1}{2\Theta}\int_{-\Theta}^{\Theta} p\,d\theta.$$

Will $\overline{p}$ increase or decrease with r? (This depends on a and a_1). Discuss the curve in the Z_0-Θ_0 diagram on which $\overline{p}$ is constant.

Problem 22. Find the expression for the pressure of the flow towards a plate, and compare it with the pressure of the corresponding irrotational flow on the axis $r = 0$.

Answer:

$$p = -\rho[2\nu r^{-1}u + \frac{1}{2}w^2 + \frac{1}{2}c^2r^2]$$

$$p = -\rho[\frac{1}{2}w^2 + \frac{1}{2}c^2r^2]$$

9. The Mathematical Structure of the Boundary Layer Problem

Consider a definite problem and let its Reynolds number approach infinity or what is the same let $\nu \to 0$. The question then arises as to whether the flow converges to the flow of a perfect fluid. We have the Navier-Stokes equation

$$\frac{d\vec{q}}{dt} = -\rho^{-1}\,\text{grad}\,p + \nu\Delta\vec{q}.$$

When $\nu \to 0$ this equation becomes

$$\frac{d\vec{q}}{dt} = -\rho^{-1}\,\text{grad}\,p$$

which is an equation of lower order. This is seen more clearly for the equation of two-dimensional flow which, after elimination of the pressure, is of the fourth order:

$$\nu\Delta\Delta\psi = \psi_y\Delta\psi_x - \psi_x\Delta\psi_y.$$

When $\nu \to 0$, this equation reduces to

$$\psi_y \Delta\psi_x - \psi_x \Delta\psi_y = 0$$

which is of the third order.

Now the solution of a differential equation of the third order cannot satisfy all the boundary conditions that we may impose upon the solution of a differential equation of the fourth order, and accordingly the flow of a perfect fluid cannot satisfy all the boundary conditions that may be imposed upon the flow of a viscous fluid. A perfect fluid cannot be required to adhere to immersed bodies or to walls; it must be allowed to slide.

Consequently, the solution of the equations for $\nu \neq 0$ is expected to converge in the interior of the region but not at the walls when $\nu \to 0$. Hence there must be a thin layer of fluid at the wall where the tangential velocity quickly goes from the value zero at the wall to the value resulting from perfect fluids. This transition region, the boundary layer, will shrink to zero with decreasing ν. The boundary layer theory will be developed for the purpose of investigating any such transition region.

Before developing the theory, however, we consider a more simple analogue which exhibits the mathematical structure of the phenomena. That is we study the simple mathematical problem P_ν of the differential equation

DE_ν $$a - f_y = \nu f_{yy}, \quad a > 0, \quad \nu > 0$$

for a function $f(y)$ defined throughout the interval $0 \leq y \leq 1$ and subjected to the boundary conditions

BC^o $$f(0) = 0$$

BC' $$f(1) = 1.$$

[We note that having a differential equation of the second order we have two boundary conditions, but when $\nu \to 0$ we obtain a differential equation of only the first order and hence its solution can satisfy one less boundary condition. This is the same behavior as occurs when $\nu \to 0$ in the Navier-Stokes equation, but here our equation is ordinary and linear, which is much more simple.]

The solution of P_ν, which we shall later give explicitly, may be denoted by S_ν. The question arises what will the solution S_ν do when the parameter ν approaches zero. One expects that it will converge to a limit function S_0 which satisfies the differential equation

$$\mathrm{DE}_0 \qquad a - f_y = 0, \quad 0 \leq y \leq 1$$

but, in general, S_0 will not satisfy both of the above boundary conditions since the equation is of only the first order. The question is which one of the two boundary conditions will S_0 satisfy. The answer, which could be derived from the explicit solution, is that S_0 satisfies

$$\mathrm{BC}' \qquad f(1) = 1$$

provided ν approaches zero through positive values.

We can say that S_0 is the solution of the problem P_0 consisting of DE_0 and BC', the condition BC^0 being lost. The solution S_0 will approach a certain value F as $y \to 0$, which in general is different from zero. We call P_0 the "interior limit problem".

For small values of ν the solution S_ν vanishes at $y = 0$ but has a value nearly equal to F already for small value of y; i.e., there is a sudden transition over a region whose width shrinks to zero. (See figure following.)

To investigate this transition we magnify the region by contracting the scale; we introduce a new variable $\eta = y/\nu$ and set $f(y) = \phi(\eta)$. The problem P_ν then takes the form

DE_ν $$\nu a - \phi_\eta = \phi_{\eta\eta}, \quad 0 \le \eta \le 1/\nu$$

BC^o $$\phi(0) = 0$$

BC' $$\phi(\tfrac{1}{\nu}) = 1.$$

Now when we let $\nu \to 0$ the function $\phi(\eta)$ will approach a limit function S_b defined for $0 \le \eta \le \infty$ which satisfies the differential equation

DE_b $$-\phi_\eta = \phi_{\eta\eta}, \quad 0 \le \eta \le \infty$$

with the boundary conditions

BC^o $$\phi(0) = 0.$$

Since the differential equation is of the second order the function S_b will not be determined by these two conditions; another boundary condition at $\eta = \infty$ is required. This condition, however, will not be BC': $\phi(\infty) = 1$. When one looks at the diagram showing the transition phenomenon, one sees

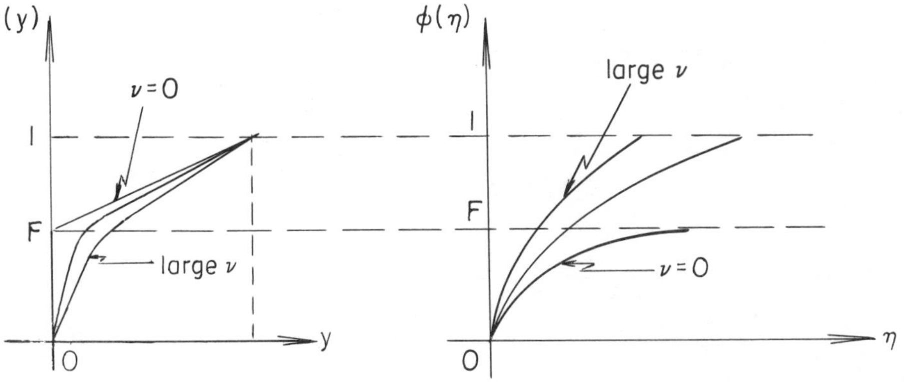

that what is to be expected is that S_b satisfies the boundary condition

$$BC^{\infty} \qquad \phi(\infty) = F$$

where $F = \lim_{y \to 0} S_0$. The conditions DE_b, BC^0 and BC^{∞} constitute the "boundary limit problem" P_b. The boundary condition BC^{∞} for the boundary limit problem at ∞ is derived from the behavior of the solution of the interior limit problem at zero.

It is customary to formulate the boundary limit problem P_b differently. With an arbitrary, but fixed, value of ν different from zero, we introduce the variable $y = \nu\eta$ and set $\phi(\eta) = f(y)$. Then the problem P_b reads

$$DE_b \qquad -f_y = \nu f_{yy}$$

$$BC^0 \qquad f(0) = 0$$

$$BC^{\infty} \qquad f(\infty) = F$$

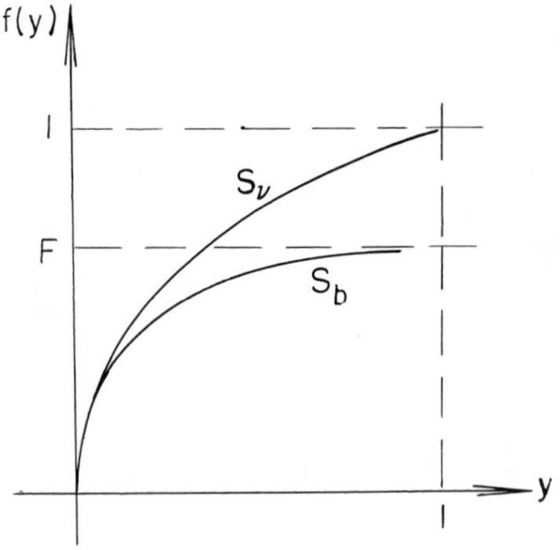

with the interval $0 \leq y \leq \infty$ for y. The solution of the problem P_b in this formulation will then serve as an approximation for the solution of problem P_ν in the neighborhood of $y = 0$. (See figure.)

We can check the preceding statements since the solution S_ν of the present simple problem has the explicit form:

S_ν: $$f(y) = (1-a)\frac{1-e^{-y/\nu}}{1-e^{-1/\nu}} + ay.$$

When $\nu \to 0$ with $y \neq 0$ we obtain

S_0: $$f(y) = 1-a + ay$$

which indeed satisfies DE_0 and BC' as conjectured in the discussion above. It is also apparent that we have $F = 1-a$. When we introduce $\eta = y/\nu$ we have

S_ν: $$\phi(\eta) = (1-a)\frac{1-e^{-\eta}}{1-e^{-1/\nu}} + \nu a\eta.$$

Then when $\nu \to 0$

S_b: $$\phi(\eta) = (1-a)(1-e^{-\eta})$$

which satisfies the conditions DE_b and BC^o and approaches the value $F = 1-a$ as $\eta \to \infty$. On the other hand, when we set $y = \nu\eta$ in S_b we have

S_b: $$f(y) = (1-a)(1-e^{-y/\nu})$$

which approximates S_ν for small values of y.

While in the present case the limits could be obtained from the explicit form of the solution S_ν, this will in general not be the case with the Navier-Stokes equation. It will hardly be possible to solve the problem P_ν for the Navier-Stokes equation. But once we have formulated the proper limit problems P_0 or P_b it will be possible to solve them and take their solutions as approximations for the solution of P_ν.

10. Boundary Layer Equations

Consider a problem P_ν for the flow of a viscous fluid partly bounded by a wall. When we let the Reynolds number approach infinity by letting $\nu \to 0$, we expect that the solution S_ν approaches the flow S_0 of a perfect fluid that slides along the wall. Therefore, at this wall a boundary layer will occur. Indeed, by setting $\nu = 0$ in the Navier-Stokes equation the order of the differential equation reduces from 4 to 3 and the number of boundary conditions that can be satisfied reduces accordingly. The "lost" boundary condition will be that one which states the tangential component of the velocity vanishes, the condition that the normal component vanishes will be retained. From our experience with flows in convergent and divergent channels we know that a boundary layer need not always occur. We may expect a boundary layer when the stream is convergent with decreasing pressure; when the stream is divergent with increasing pressure more and more back-flow regions are to be expected spoiling the convergence.

Boundary layers may occur at streamlines other than walls. Consider for example the flow around a flat plate tangent to the flow. There will be a boundary layer on both sides of the plate. Also in the wake, where the particles that have passed the boundary layers on the plate meet again, such a layer will occur since uniformity of convergence for $\nu \to 0$ cannot be expected there; (this layer should rather be called a wake layer). Further when a jet spurts out of a narrow slit a layer will occur where the jet mixes with the surrounding fluid.

In what follows we consider the two-dimensional case and assume that the boundary layer occurs on a segment "B" of the x-axis. We leave open for the present the boundary condition at such a wall. We assume further that for $\nu \to 0$ the flow converges to that of a perfect fluid outside a neighborhood of the segment B; this interior limit flow is also called the "main flow". We assume that ψ, u, v, p and also u_x are bounded though u_y will become infinite at B. From $\psi_y = u$, $v_y = -u_x$ we conclude that ψ and v will converge uniformly right up to $y = 0$ on B. We also assume (though it may not be necessary) that for the main flow (i.e., in the limit) $\psi = 0$ and $v = 0$ on B; this means B is a streamline for the main flow. The values which u and p assume on B for the main flow may be denoted by $U(x)$ and $P(x)$ respectively. Thus we have for the main flow

$$\psi = 0, \ u = U(x), \ v = 0, \ p = P(x) \quad \text{on} \quad B.$$

The Euler equation (which applies to the main flow) then reduces on B to

$$\rho U U_x = -P_x.$$

The function $U(x)$ is to be determined by solving the corresponding problem P_0 for a perfect fluid permitted to slide along B.

In order to investigate the transition in the boundary layer we introduce a quantity δ which may represent the thickness of the boundary layer and which approaches zero as $\nu \to 0$. We then introduce the new variable $\eta = y/\delta$, consider ψ, u, v, p as functions of x and η and ask for the limits of these functions as $\nu \to 0$. We expect $u(x,\eta)$ to approach a limit different from zero; the functions $\psi(x,\eta)$ and $v(x,\eta)$ are expected to approach zero in accordance with the assumption that $\psi = v = 0$ on B for the main flow. The relations

$$\delta^{-1}\psi_\eta = u, \quad \delta^{-1}v = -\delta^{-1}\psi_x$$

make it likely that

$$\psi^* = \delta^{-1}\psi, \quad v^* = -\psi^*_x = \delta^{-1}v$$

approach finite limits.

The Navier-Stokes equations now become

$$uu_x + v^*u_\eta = -\rho^{-1}p_x + \nu u_{xx} + \nu\delta^{-2}u_{\eta\eta}$$
$$\delta^2 uv^*_x + \delta^2 v^*v^*_\eta = -\rho^{-1}p_\eta + \nu\delta^2 v^*_{xx} + \nu v^*_{\eta\eta}.$$

In order that viscosity and acceleration terms be retained in the limit we stipulate that $\nu\delta^{-2} = \nu^*$ remain fixed; i.e., we set

$$\delta = \left(\frac{\nu}{\nu^*}\right)^{1/2}, \quad \nu^* \text{ constant.}$$

Then in the limit the differential equations are

$$uu_x + v^*u_\eta = -\rho^{-1}p_x + \nu^*u_{\eta\eta}$$
$$0 = -\rho^{-1}p_\eta$$

and we expect them to be satisfied by the limit functions

$$u = \psi^*_\eta \qquad v^* = -\psi^*_x.$$

The behavior of these limit functions for $\eta \to \infty$ will be such that they join on continuously with the main flow. To this end the values of u and p for $\eta = \infty$ will coincide with the values of u and p for the main flow on B. That is we have the boundary conditions

BC^∞ $\qquad u = U(x), \ p = P(x) \quad \text{for} \quad \eta = \infty.$

These conditions are sufficient for $\eta = \infty$; it is not necessary to impose conditions at $\eta = \infty$ on ψ^* and v^*. Indeed, the values of ψ^* and v^* for $\eta = \infty$ are not directly connected with the values of ψ and v for the main flow on B.

It is now convenient to reintroduce variables y, ψ, v by setting

$$y = \delta\eta, \quad \psi = \delta\psi^*, \quad v = \delta v^*$$

with $\delta = \left(\frac{\nu}{\nu^*}\right)^{1/2}$ for any arbitrarily chosen finite value of ν. We then have as our differential equations the original Navier-Stokes equations but without the terms which dropped out in the above limit:

$$u = \psi_y \qquad v = -\psi_x$$

DE
$$uu_x + vu_y = -\rho^{-1}p_x + \nu u_{yy}$$
$$0 = -\rho^{-1}p_y$$

with the boundary conditions for $y = \infty$:

BC^∞
$$u = U(x) \quad p = P(x)$$

The equation $p_y = 0$ together with the condition $p = P(x)$ for $y = \infty$ yields the first important result, i.e., the pressure is constant across the boundary layer or $p = P(x)$ for all y. Then the Euler equation $P_x = -\rho UU_x$ which we have for the main flow enables us to write

DE
$$uu_x + vu_y = UU_x + \nu u_{yy}$$
$$u = \psi_y, \quad v = -\psi_x \tag{30}$$

BC^∞
$$u = U(x) \quad \text{for} \quad y = \infty. \tag{30'}$$

These are the boundary layer equations we have to work with. Although in these equations the range of y is from zero to infinity, the solution will be an approximation to the solution of the original problem P_ν only for y sufficiently small.

11. Flow along a Flat Plate

The problem of determining, by the boundary layer theory, the two-dimensional flow along a flat plate is the simplest such boundary layer problem. It is treated in both of the references given for this chapter.*

*Aerodynamic Theory, Vol. III, p. 84.

Goldstein, Modern Developments in Fluid Dynamics, Vol. I, p. 135.

We represent the plane plate "B" by the half line $y = 0,\ x \geq 0$. The velocity of the main flow is taken to be $(U,0)$ where U is constant. The pressure of the main flow is $P = 0$ and we assume that a boundary layer occurs only at B. For $x < 0$ we assume uniform convergence $u \to U$ up to $y = 0$. Then we have the boundary layer differential equations (30)

DE
$$uu_x + vu_y = \nu u_{yy}$$
$$u = \psi_y, \quad v = -\psi_x$$

with the boundary conditions

BC^0
$$\psi = u = v = 0 \quad \text{for} \quad y = 0,\ x > 0$$

BC^∞
$$u = U \qquad \text{for} \quad x > 0,\ y = \infty.$$

It is necessary to impose "initial conditions" for $x = 0$, i.e., along the y-axis,

IC
$$u = U,\ v = 0,\ \psi = Uy.$$

For symmetry reasons we can restrict ourselves to one side of the plate, $y > 0$.

It was noticed by Prandtl that the solution can be obtained by reducing the differential equation to an ordinary differential equation in the variable

$$\theta = y(\lambda x)^{-\frac{1}{2}}, \quad \lambda = \text{const. length.} \tag{31}$$

We assume

$$u = Uf(\theta), \tag{32}$$

then

$$\psi = (\lambda x)^{1/2} U h(\theta) \tag{32a}$$

where $h'(\theta) = f(\theta)$. Also

$$v = -\psi_x = \frac{1}{2}U\left(\frac{\lambda}{x}\right)^{1/2}[\theta h'(\theta) - h(\theta)]. \tag{32b}$$

With these expressions for u and v we have, since in view of the continuity equation, $u_x = -v_y$:

$$uu_x + vu_y = -u^2\left(\frac{v}{u}\right)_y = -\frac{1}{2}U^2 x^{-1} h(\theta)h''(\theta)$$

$$\nu u_{yy} = \frac{\nu}{\lambda} U x^{-1} h'''(\theta).$$

We now set $\lambda = \frac{2\nu}{U}$ and then the differential equation DE becomes the ordinary equation

$$\text{DE}' \qquad h'''(\theta) + h(\theta)h''(\theta) = 0 \tag{33}$$

while the boundary conditions become

$$\begin{aligned} &\text{BC}^0 \qquad h(0) = h'(0) = 0 \\ &\text{BC}^\infty \qquad h'(\infty) = 1 \end{aligned} \tag{33'}$$

The condition IC is automatically fulfilled since, when $x \to 0$, we have $\theta \to \infty$, $h'(\theta) \to 1$, $\frac{h(\theta)}{\theta} \to 1$ and $u(x,y) \to U$, $\psi \to yU$.

We tacitly assumed above, when we set $\lambda = \frac{2\nu}{U}$, that $U > 0$. If $U < 0$ we can replace U everywhere by $|U|$ and the only change is in BC^∞ which for $U < 0$ becomes $h'(\infty) = -1$. We shall show, however, that $h'(\infty)$ cannot equal -1 if the other conditions are to be satisfied. This is consistent with the fact that with reverse flow a boundary layer is to be expected on the line $y = 0$, $x < 0$ contrary to assumption.

We first discuss the general properties which a solution of (33) must have in order to satisfy the boundary conditions (33'). If $h(\theta)$ were known then (33) gives

$$h''(\theta) = h''(0) \exp\left(-\int_0^\theta h(\bar{\theta})d\bar{\theta}\right). \tag{34}$$

We have two possibilities:

(1) $h''(0) < 0$. In this case $h''(\theta) < 0$ for all θ and we would have $h'(\theta) \leq 0$, $h(\theta) \leq 0$. By (33) $h'''(\theta)$ is also negative; therefore $h''(\theta)$ is a decreasing function of θ. As a result $h'(\theta)$ decreases to $-\infty$ and neither $h'(\infty) = 1$ nor $h'(\infty) = -1$ can be satisfied.

(2) $h''(0) > 0$. In this case $h''(\theta) > 0$ and we have $h'(\theta) \geq 0$, $h(\theta) \geq 0$ while $h'''(\theta) < 0$. Thus again $h''(\theta)$ is a decreasing function of θ; but now, for $\theta > \theta_0 > 0$,

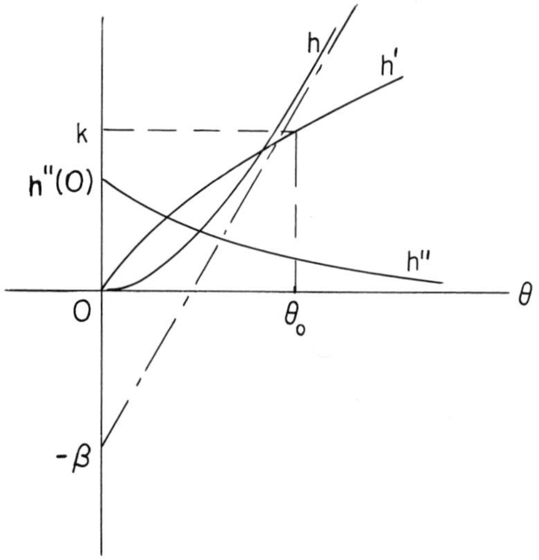

$$h'(\theta) > k = h'(\theta_0)$$

whence

$$h(\theta) > k\theta - k_1;$$

this gives us

$$\int_0^\theta h(\bar{\theta})d\bar{\theta} > \frac{1}{2} k\theta^2 - k_1\theta - k_2$$

and

$$h''(\theta) < h''(0)\ \exp(-\tfrac{1}{2}k\theta^2 + k_1\theta + k_2).$$

As a result, we see that $h''(\theta)$ decreases rapidly like $e^{-\frac{1}{2}k\theta^2}$, so that we certainly have

$$0 < h'(\infty) = \int_0^\infty h''(\theta)d\theta < \infty.$$

An immediate consequence is that $h'(\infty)$ cannot possibly be -1 so that the case $U < 0$ is to be excluded.

We now show that a solution of (33) can exist which fulfills BC^∞. Suppose that $H(\theta)$ is a solution with $H''(0) = 1$. Then $h(\theta) = \kappa H(\kappa\theta)$ is also a solution of (33). It gives $h'(\theta) = \kappa^2 H'(\kappa\theta)$ and as $\theta \to \infty$

$$h'(\infty) = \kappa^2 H'(\infty).$$

Let $\kappa = [H'(\infty)]^{-1/2}$ and $h(\theta)$ will be the solution with $h'(\infty) = 1$. We also have

$$h''(0) = \kappa^3$$

since $H''(0) = 1$, while the equation (34) can be written for $H''(\theta)$ in the form

$$H''(\theta) = \exp[(-\int_0^\theta d\theta)^3 H''(\theta)] \tag{35}$$

which involves the operation $-\int_0^\theta d\theta$ iterated three times.

For the purpose of actually obtaining solutions we have at our disposal two methods. The one usually proposed is the power series development of $H(\theta)$ around $\theta = 0$. The convergence is rapid; when $H'(\theta)$ becomes sufficiently constant this constant is to be identified with $H'(\infty)$. This will occur from about

$\theta = 2$ on. The following values are taken from a more complete table*:

$\theta =$	2.0	2.5	3.0	3.5	4.0	4.5
$\frac{u}{U} = \frac{H'(\theta)}{H'(\infty)} =$	.9269	.9806	.9966	.9996	.99996	1.00000

$H'(\infty)$ is thus found to be 1.6551 whence $\kappa = 0.7773$, $\kappa^3 = 0.4696$ and $\alpha = 2\sqrt{2\kappa^3} = 1.3283$ where α is the constant occuring in the literature**

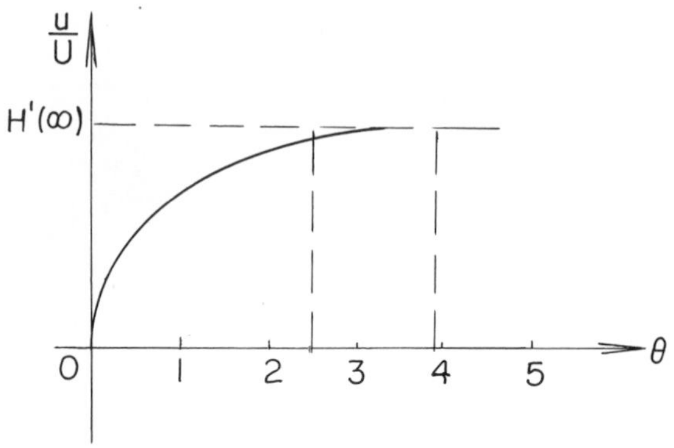

However, as H. Weyl has pointed out, the expansion has a finite radius of convergence which lies between

$(18)^{1/3} = 2.62$ and

$(60)^{1/3} = 3.91$.

Thus it would appear that the results given above have been evaluated beyond the interval of convergence.

Weyl has, therefore, proposed a second method of evaluating $H'(\theta)$. If we set $G(\theta) = H''(\theta)$ we can write (35) as

$$G(\theta) = T\,G(\theta) \tag{36}$$

where T is an operator such that

$$T\,G(\theta) = \exp[(-\int_0^\theta d\theta)^3 G(\theta)] \tag{36'}$$

---

* L. Howarth, Proc. Roy. Soc., 164A (1938), p. 552.

** Goldstein, Modern Developments in Fluid Dynamics, Vol. I, p. 135.

T has the property that, if $G^{(1)}(\theta) < G^{(2)}(\theta)$, then $T\,G^{(1)}(\theta) > T\,G^{(2)}(\theta)$. When we start with $G_0(\theta) = 0$ we obtain $G_1(\theta) = T\,G_0(\theta) = 1$ from

$$G_{n+1}(\theta) = T\,G_n(\theta), \quad n = 0,1,2,3,\ldots .$$

According to the given property of T we then have

$$G_0 < G_1, \; G_0 < G_2 < G_1, \; G_0 < G_2 < G_3 < G_1$$

or in general

$$G_0 < G_2 < G_4 < \cdots < G_5 < G_3 < G_1.$$

Thus this process of iteration will converge. We have $G_2(\theta) = e^{-\theta^3/6}$ and it can be shown that G_3 will behave like $e^{-k\theta^2}$ as $\theta \to \infty$ which, as we showed, is the property of $H''(\theta)$. However, when we use G_2 we obtain

$$H_2'(\infty) = \int_0^\infty e^{-\frac{\theta^3}{6}}\, d\theta = 2\cdot 6^{-2/3} \int_0^\infty e^{-t} t^{-2/3} dt$$

$$= \frac{2}{6^{2/3}}\,\Gamma(\tfrac{1}{3}) \quad \text{with} \quad t = \frac{\theta^3}{6} .$$

Hence $H'(\infty) < 6^{1/3}\Gamma(\frac{4}{3}) = 1.623$ and we have $\kappa = 0.785$ or $\alpha = 1.368$ which are good approximations to the exact values given above.

Since, for a solution, $h'(\infty) = 1$ for $\theta = \infty$, $h(\theta)$ is asymptotically represented by (see figure on p. 256)

$$h(\theta) \sim \theta - \beta, \quad \theta \to \infty, \tag{37}$$

where

$$\beta = \lim_{\theta \to \infty} [\theta h'(\theta) - h(\theta)] \tag{38}$$

In view of $[\theta h'(\theta) - h(\theta)]_\theta = \theta h''(\theta)$ we have

$$\beta = \int_0^\infty \theta h''(\theta)d\theta = 1.215. \tag{38'}$$

Consequently v approaches the limit $\frac{1}{2}\beta(\frac{2\nu U}{x})^{1/2}$ for $y \to \infty$.

On the plate the shearing stress is

$$\tau = \mu u_y\Big|_{y=0} = \frac{\mu}{(\lambda x)^{1/2}} U h''(0).$$

With $h''(0) = \kappa^3$, $\lambda = \frac{2\nu}{U}$ and $\alpha = 2\sqrt{2}\kappa^3$ we obtain

$$\tau = \rho U(\frac{\nu U}{x})^{1/2} \cdot \frac{\kappa^3}{\sqrt{2}} = \frac{1}{4}\alpha\rho U(\frac{\nu U}{x})^{1/2}. \tag{39}$$

The drag per unit area on both sides of a plate of length ℓ is then

$$D = 2 \cdot \frac{1}{\ell}\int_0^\ell \tau dx = \frac{1}{2}\frac{\alpha\rho U}{\ell}\int_0^\ell (\nu U)^{1/2}(x)^{-1/2}dx = \alpha\rho U(\frac{\nu U}{\ell})^{1/2}.$$

The drag coefficient is

$$C_D = \frac{D}{\frac{1}{2}\rho U^2} = 2\alpha(\frac{\nu}{\ell U})^{1/2} = 2\alpha R^{-1/2} \tag{40}$$

with the Reynolds number $R = \frac{\ell U}{\nu}$. Thus the drag is not proportional to R^{-1} as in the case of small Reynolds number but to $R^{-1/2}$.

Problem 23. Find approximately the value of β by means of the iteration method.

Problem 24. Discuss the behavior of the streamlines at $x = 0$ and at $x = \infty$ for the flow over a flat plate. (Hint: Find the slope of the lines at $x = 0$.)

Answer: At $x = 0$, $\psi \sim Uy - \beta(2\nu Ux)^{1/2}$.

$$\text{At} \quad x \sim \infty, \ \psi \sim \frac{1}{4}\alpha\left(\frac{U}{\nu}\right)^{1/2} \cdot y^2/x.$$

12. Displacement Thickness - Momentum Equation

We consider in this section two notions of a general character which refer to the boundary layer theory. The first gives a sort of measure to the thickness of the boundary layer while the second gives us a formula which can be used for finding an approximate solution of the boundary layer equations.

(1) Displacement Thickness

It is rather arbitrary what one should consider to be the "thickness" of the boundary layer. The following definition seems reasonable. The effect of the retardation at the wall causes the streamlines to be displaced through a distance δ^*, the "displacement thickness". It can be defined by

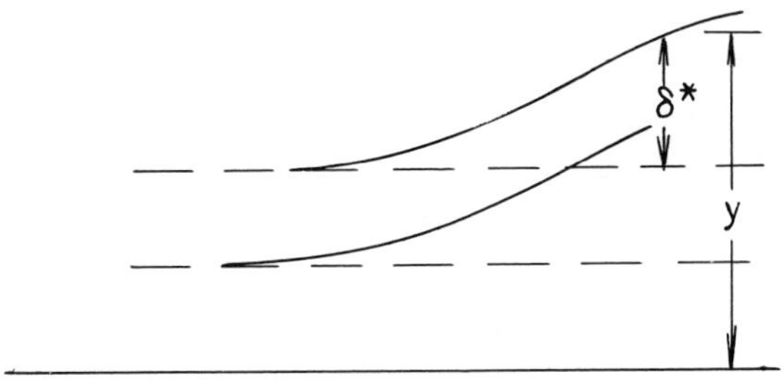

$$\lim_{y \to \infty} [\psi(x,y) - U(x)(y - \delta^*(x))] = 0$$

or

$$\begin{aligned} \delta^*(x) &= U^{-1}(x) \lim_{y \to \infty} [yU(x) - \psi(x,y)] \\ &= U^{-1}(x) \int_0^\infty [U(x) - u(x,y)]dy \end{aligned} \tag{41}$$

Another characteristic of $\delta^*(x)$ is the following. The velocity distribution

$$u = U(x), \quad y \geq \delta^*$$

$$u = 0 \quad , \ y < \delta^*$$

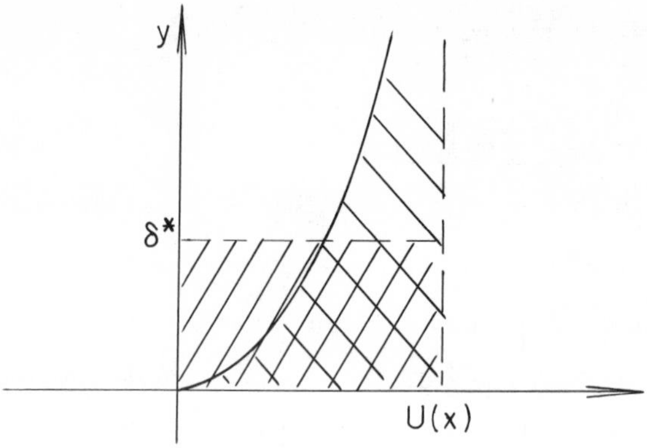

will yield the same flux across the plane $x = \text{const.}$ as the distribution $u(x,y)$; i.e.,

$$\int_0^\infty [U(x) - u(x,y)]dy - U(x)\delta^*(x) = 0.$$

It seems that $3\delta^*$ is the most reasonable definition of the boundary layer thickness.

For the boundary layer along a flat plate $u = Uh'(\theta)$, $U = \text{const.}$, therefore

$$\delta^*(x) = \left(\frac{2\nu x}{U}\right)^{1/2} \int_0^\infty [1 - h'(\theta)]d\theta.$$

The value of the integral is

$$\int_0^\infty [1 - h'(\theta)]d\theta = [\theta - h(\theta)]_0^\infty = [\theta h'(\theta) - h(\theta)]_0^\infty = \beta$$

using (38) since $h'(\infty) = 1$. Thus by (38')

$$\delta^*(x) = \beta\left(\frac{2\nu x}{U}\right)^{1/2} = 1.72\left(\frac{\nu x}{U}\right)^{1/2}. \tag{42}$$

(2) Momentum Equation

Consider a rectangle adjacent to a segment of the wall B in a two-dimensional flow. The shearing force on this segment equals the resultant of the shearing and pressure forces on the rectangle diminished by the momentum transported out of it per unit time. When the rectangle is extended in the y-direction normal to the wall and contracted in the x-direction along the wall to the line $x = x_0$, $0 \leq y \leq \infty$, a simple formula for the shear stress at the wall results. To derive it we proceed as follows instead. We transform the boundary layer differential equation (30), making use of the continuity equation:

$$
\begin{aligned}
-\nu u_{yy} &= UU_x - (uu_x + vu_y) \\
&= UU_x - 2uu_x - (vu)_y \\
&= UU_x - 2uu_x - (vu)_y + Uu_x + (Uv)_y \\
&= UU_x + Uu_x - 2uu_x + [v(U-u)]_y.
\end{aligned}
$$

The integral along the line $x = x_0$ from $y = 0$ to $y = y_0$ yields

$$-[\mu u_y]_0^{y_0} = \rho \int_0^{y_0} (UU_x + Uu_x - 2uu_x)dy + \rho[v(U-u)]_0^{y_0} .$$

We assume that as $y_0 \to \infty$, $\mu u_y \to 0$ and $v(U-u) \to 0$ with the result that for the shearing stress on the wall at $x = x_0$ we obtain

$$\tau(x_0,0) = \mu u_y\Big|_{y=0} = \rho\int_0^{\infty} (UU_x + Uu_x - 2uu_x)dy \tag{43}$$

This is called the momentum equation*.

The principal use of the momentum equation has been in finding approximate solutions of the boundary layer equations (v. Kármán or Pohlhausen). The

*In the literature it is usually given with a finite range of integration for y; for the above form see R. von Mises, Zeits. f. angew. Math. u. Mech., 7(1927), p. 425 and 8(1928), p. 249.

method consists essentially of guessing the velocity as a function of y but it also involves the assumption

$$u = U(x)f(\theta), \quad \theta = y/\delta(x) \qquad (44)$$

$$f(0) = 0, \quad f(\infty) = 1$$

similar to that we made for the problem of the flat plate. This assumption gives

$$UU_x + Uu_x - 2uu_x = UU_x[1 + f - 2f^2] - \delta^{-1}\delta' U^2\theta(f' - 2ff')$$

while

$$\tau = \mu u_y\big|_{y=0} = \mu U\delta^{-1}f'(0). \qquad (45)$$

The momentum equation (43) then reduces to the linear differential equation for δ^2:

$$\tfrac{1}{2}BU(\delta^2)_x + AU_x\delta^2 = C\nu \qquad (46)$$

with

$$A = \int_0^\infty [1 + f(\theta) - 2f^2(\theta)]d\theta$$

$$B = -\int_0^\infty \theta[f'(\theta) - 2f(\theta)f'(\theta)]d\theta = \int_0^\infty [f(\theta) - f^2(\theta)]d\theta$$

$$C = f'(0).$$

This equation is to be solved for δ under the condition that $\delta(0) = 0$; the result inserted into (45) gives τ. The displacement thickness is also obtained:

$$\delta^* = F\delta, \quad F = \int_0^\infty [1 - f(\theta)]d\theta. \qquad (47)$$

Consider the case $U = \text{const.}$ (as for the flat plate). The differential equation for δ^2 reduces to

$$BU(\delta^2)_x = 2C\nu$$

whence

$$\delta^2(x) = \frac{2C\nu}{BU}x \quad \text{or} \quad \delta(x) = \left(\frac{2C}{B}\right)^{1/2}\left(\frac{\nu x}{U}\right)^{1/2}. \tag{48a}$$

This value of δ then gives the results

$$\tau = \rho U\left(\frac{BC}{2}\right)^{1/2}\left(\frac{\nu U}{x}\right)^{1/2}, \quad \frac{\tau}{\frac{1}{2}\rho U^2} = (2BC)^{1/2}\left(\frac{\nu}{Ux}\right)^{1/2},$$

$$\delta^* = (2BC)^{1/2}\frac{F}{B}\left(\frac{\nu x}{U}\right)^{1/2} \tag{48b}$$

We see that the essential parameters are BC and $\frac{F}{B}$ which are invariant under a stretching of the θ scale. Hence, when $f(\theta)$ has been chosen, BC is to be calculated to give the shearing stress. Various assumptions for $f(\theta)$ have been made; they yielded results which check very well with the exact solution*. The method has also proved successful for other boundary layer problems.

Objections can, however, be raised to the use of the momentum equation. In the paper quoted, von Mises has pointed out that if one has to guess $f(\theta)$ why not just as well try to guess BC and $\frac{F}{B}$, i.e., essentially τ and δ^* themselves. Nevertheless, it does appear that BC and $\frac{F}{B}$ are more restricted if $f(\theta)$ is required to have a "reasonably" smooth behavior. The strongest objection is to the assumption that $u = U(x)f(\theta)$. This assumption implies that for different values of x the velocities differ only by a similarity transformation. In many cases such an assumption is far away from the correct situation, for instance, in the neighborhood of a separation point.

*Goldstein, Modern Developments in Fluid Dynamics, Vol. I, p. 156-163.

Problem 25. Solve the boundary layer problem for the flat plate by the momentum method choosing a function $f(\theta)$ that is not constant for large θ. Compare the results with those of the exact solution.

13. Jets and Wakes

(1) Jet in two dimensions.

We consider the following two-dimensional problem. Let the left half of the xy plane be a reservoir of fluid, the y-axis corresponding to the wall of this reservoir. Suppose there is a narrow slit in this wall, its width shrunk to zero and hence represented by a point. We take this point to be $y = 0$ in order that there be symmetry to the x-axis. The fluid in the reservoir is assumed to spurt out through the slit in the form of a jet into fluid of the same kind at rest; there is then a boundary layer where the jet has contact with the stationary fluid.

The main flow is given by

$$u = v = 0, \ p = 0.$$

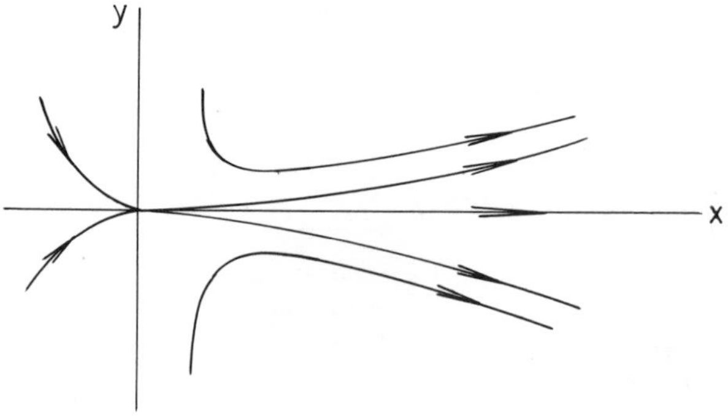

The boundary layer equations are

DE
$$uu_x + vu_y = \nu u_{yy} \tag{49}$$

with

BC^{∞}
$$u = 0 \quad \text{for} \quad y = \infty$$

BC^{0}
$$\psi = 0, \ u_y = 0 \quad \text{for} \quad y = 0 \tag{49'}$$

with the symmetry assumed we consider $y \geq 0$.

One more condition fixing the "strength" of the jet is needed. We cannot expect that the flux of mass transported by the jet is constant since there is no direct connection between ψ^* for $y = 0$ and the stream function of the main flow. Indeed, as we shall see, the mass flux will increase along the jet; i.e., more and more fluid from the outside will be drawn into the jet. We may, however, use the momentum equation (43) which, in view of $U = 0$ and $u_y(x,0) = 0$, reads

$$\int_0^\infty u u_x dy = 0, \text{ or}$$

$$\int_0^\infty u^2 dy = N = \text{const.} \tag{A}$$

Then $2\rho N$ is the total momentum flux per unit breadth across every cross section $x = \text{const.}$ of the jet.

We now show that an explicit solution can be found if we assume that

$$\psi = \omega\lambda\xi^\alpha k(\theta), \quad \theta = \frac{\eta}{\xi^\beta}, \tag{50}$$

where $\eta = y/\lambda$, $\xi = \frac{x}{\lambda}$, λ = constant length and the constants λ, ω, α, β are to be determined. In terms of the unknown function $k(\theta)$ we now find

$$u = \psi_y = \omega\xi^{\alpha-\beta}k'(\theta)$$

$$u_y = \psi_{yy} = \omega\lambda^{-1}\xi^{\alpha-2\beta}k''(\theta)$$

$$u_{yy} = \omega\lambda^{-2}\xi^{\alpha-3\beta}k'''(\theta)$$

$$v = -\psi_x = \omega\xi^{\alpha-1}[\beta\theta k'(\theta) - \alpha k(\theta)]$$

and obtain from (49)

$$uu_x + vu_y = -u^2\left(\frac{v}{u}\right)_y = -\omega^2\lambda^{-1}\xi^{2\alpha-2\beta-1}[(\beta-\alpha)k'^2 + \alpha kk'']$$

$$= \nu u_{yy} = \nu\omega\lambda^{-2}\xi^{\alpha - 3\beta}k'''.$$

This equation for $k(\theta)$ will be free of terms in x if the powers of ξ coincide or

$$\alpha + \beta = 1.$$

Similarly by (A)

$$N = \int_0^\infty u^2 dy = \lambda\omega^2\xi^{2\alpha-\beta}\int_0^\infty k'^2(\theta)d\theta$$

and this expression will also be independent of x if

$$2\alpha - \beta = 0.$$

We then have that $\alpha = \frac{1}{3}$, $\beta = \frac{2}{3}$. Setting $\omega\lambda = 6\nu$ we obtain the differential equation for $k(\theta)$:

DE $$k'''(\theta) + 2k(\theta)k''(\theta) + 2k'^2(\theta) = 0 \tag{51}$$

with

BC^∞ $$k'(\infty) = 0$$

BC^0 $$k(0) = 0,\ k''(0) = 0, \text{ and} \tag{51'}$$

(A) $$\int_0^\infty k'^2(\theta)d\theta = \frac{N}{6\omega\nu}$$

When this boundary value problem for k has been solved we can find ψ and u from:

$$\psi = 6\nu\xi^{1/3}k(\theta), \quad u = \omega\xi^{-1/3}k'(\theta). \tag{52}$$

We replace condition (A) by another condition. Since $k'(0) \neq 0$ we may dispose of ω such that

(B) $$k'(0) = 1.$$

Then the solution is determined by initial values; equation (A) serves to determine ω.

The problem DE, BC^{∞}, BC^{0}, (B) was first treated by Schlichting*. Bickley found a simple explicit form of the solution. The DE:

$$k''' + 2(kk')_{\theta} = 0$$

possesses a first integral

$$k'' + 2kk' = 0,$$

using in the integration the conditions $k(0) = k''(0) = 0$. Integrating again gives, in view of $k'(0) = 1$,

$$k' + k^2 = 1.$$

The well-known solution of this equation with $k(0) = 0$ is then

$$k(\theta) = \tanh \theta \tag{53}$$

which satisfies all the required conditions.

The flux of momentum $2\rho N$ gives ω. We have

---

*Goldstein, Modern Developments in Fluid Dynamics, Vol. I, p. 145.

$$\int_0^\infty [k'(\theta)]^2 d\theta = \int_0^1 (1-k^2)dk = 2/3$$

and then obtain from (A)

$$\omega = \frac{N}{4\nu} \quad \text{whence} \quad \lambda = \frac{24\nu^2}{N}.$$

The most interesting result that can be derived is the total flux of mass per unit breadth $2\rho Q$ transported. Here $Q = \psi(x,\infty)$ so from (52)

$$Q = 6\nu\xi^{1/3} = 6\nu\left(\frac{Nx}{24\nu^2}\right)^{1/3} = (9N\nu x)^{1/3}, \tag{54}$$

which indicates that the amount of fluid drawn into the jet is proportional to $x^{1/3}$.

The velocity on the symmetry axis is given by

$$u(x,0) = \omega\xi^{-1/3} = \left(\frac{3N^2}{8\nu x}\right)^{1/3}$$

One can also solve the problem of the three-dimensional jet symmetric to an axis. It turns out that in this case the flux is independent of N, that is, independent of the pressure forcing the fluid through the opening. This is clearly seen by considering the dimensions of the quantities; they are the same for the flux as for νx and no dimensionless number occurs in the problem. Hence $Q = \text{const.}\ \nu x$.

Problem 26. Discuss the streamlines for the two-dimensional jet.

(2) Wakes.

Consider the boundary layer problem for a flat plate of length ℓ. Then beyond the plate there is a "wake" where the velocity will be different from that of the main flow. We have symmetry with respect to the line representing the plate.

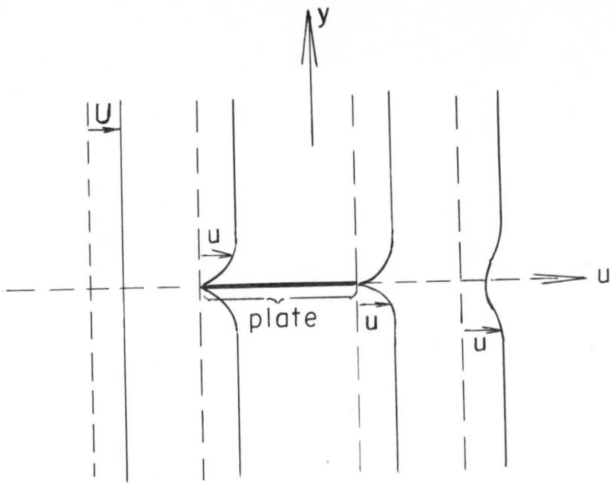

Let us consider the wake behind the plate. We take coordinates such that the plate is represented by the segment $-\ell \leq x \leq 0$ of the x-axis. We have initial conditions at the end of the plate which are obtained from the velocity distribution at $x = \ell$ for the problem of the infinite plate already worked out. This gives

$$u = u_0(y) = Uh'\left(\frac{y}{(\lambda\ell)^{1/2}}\right) \quad \text{at} \quad x = 0. \tag{55}$$

We also make use of the symmetry by considering only the part $y \geq 0$.

In the wake, $x > 0$, we have the same boundary layer equations and conditions at infinity

DE
$$uu_x + vu_y = \nu u_{yy} \tag{56}$$

BC^∞
$$u = U = \text{const.}, \quad y = \infty. \tag{56a}$$

The conditions for $y = 0$ are different, becoming

BC^0
$$\psi = u_y = 0, \quad y = 0. \tag{56b}$$

For $x = 0$ we have the initial condition

IC $$u = u_0(y), \quad x = 0 \tag{56c}$$

The expression (55) for u_0 can be developed* as a series in powers of y ; we take only the first term which is linear giving

$$u_0(y) = \alpha U \frac{y}{(\lambda \ell)^{1/2}} = Cy. \tag{55a}$$

Hence we replace (56c) by

IC $$u = Cy, \ x = 0 \tag{56d}$$

At the same time we must give up the condition BC^∞.

A solution of this simplified problem can be obtained by setting (just as we did in the case of the jet):

$$\psi = \omega\lambda\xi^\alpha k(\theta)$$

where $\theta = \eta\xi^{-\beta}$, $\xi = x/\lambda$, $\eta = y/\lambda$ and the constants α, β, ω, λ are to be determined. Then

$$u = \psi_y = \omega\xi^{\alpha-\beta} k'(\theta)$$

and for (56) to be an equation in k and θ independent of ξ, we find, just as we did for the free jet, that

$$\alpha + \beta = 1. \tag{57}$$

---

*Goldstein, Modern Developments in Fluid Dynamics, Vol. I, p. 572.

To give us another equation in α and β we use the condition (56d). With $y = \lambda\theta\xi^{\beta}$ we write

$$u = \omega\lambda^{-1}y\xi^{\alpha-2\beta}\theta^{-1}k'(\theta).$$

We note that $\theta \to \infty$ when $\xi \to 0$; the condition (56d), therefore, requires firstly

$$\alpha - 2\beta = 0 \tag{57'}$$

and secondly

$$\lim_{\theta \to \infty} \theta^{-1}k'(\theta) = C\,\frac{\lambda}{\omega}.$$

Consequently, $u \to \infty$ as $y \to \infty$ which is correct in view of $u_0(y) \to \infty$ as $y \to \infty$. Relations (57) and (57') yield

$$\alpha = 2/3, \quad \beta = 1/3.$$

We now have the following problem for $k(\theta)$, setting $\omega = \frac{3}{2}U$, $\lambda = \frac{2\nu}{U}$:

DE $$k'''(\theta) + 2k(\theta)k''(\theta) - k'^2(\theta) = 0 \tag{58}$$

BC^0 $$k(0) = 0, \quad k''(0) = 0$$

BC^{∞} $$\theta^{-1}k'(\theta) \to \frac{4\nu C}{3U^2} = \frac{2}{3}\sqrt{2}\,\alpha\left(\frac{\nu}{\ell U}\right)^{1/2} = A, \text{ for } \theta \to \infty$$

The differential equation is the same that we had for flow towards a flat plate in Section 8. For discussion we differentiate (58) and obtain

$$k''''(\theta) + 2k(\theta)k'''(\theta) = 0 \tag{58a}$$

with the additional boundary condition

BC^{o} $$k'''(0) = k'^{2}(0) > 0.$$

We set $k(\theta) = \kappa K(\theta n)$ where $K(\theta)$ is a solution of (58a) such that $K'''(0) = 1$, then the problem for $K(\theta)$ is

DE $$K''''(\theta) + 2K(\theta)K'''(\theta) = 0 \tag{58b}$$

BC^{o} $$K(0) = K''(0) = 0,\ K'(0) = \pm 1,\ K'''(0) = 1 \tag{58c}$$

BC^{∞} $$\theta^{-1}K'(\theta) \to A\kappa^{-3},\ \text{for}\ \theta \to \infty. \tag{58d}$$

From equation (58b) and $K'''(0) = 1$ we derive

$$K'''(\theta) = \exp(-2\int_0^{\theta} K(\theta)d\theta)$$

where $\int_0^{\theta} K(\theta)d\theta \to \infty$ as $\theta \to \infty$ in such a way that $\int_0^{\infty} K'''(\theta)d\theta = K''(\infty)$ is a finite positive quantity.

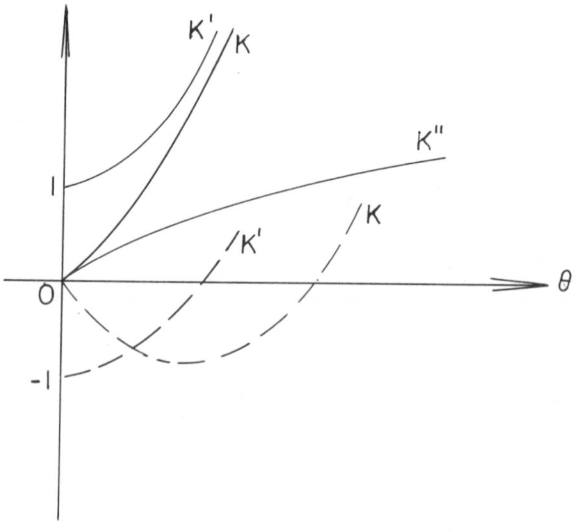

We observe that $\theta^{-1}K'(\theta)$ approaches $K''(\infty)$ as $\theta \to \infty$. This reasoning is independent of the sign of $K'(0)$; hence, both assumptions, $K'(0) = 1$ and $K'(0) = -1$, yield solutions for which $\theta^{-1}K'(\theta)$ approaches the same positive limit. A value for κ can, therefore, be found for which (58d) is satisfied.

Thus it appears that the solution of the problem is not unique, but two wake flows are possible, one involving back-flow. The regular case $K'(0) = 1$

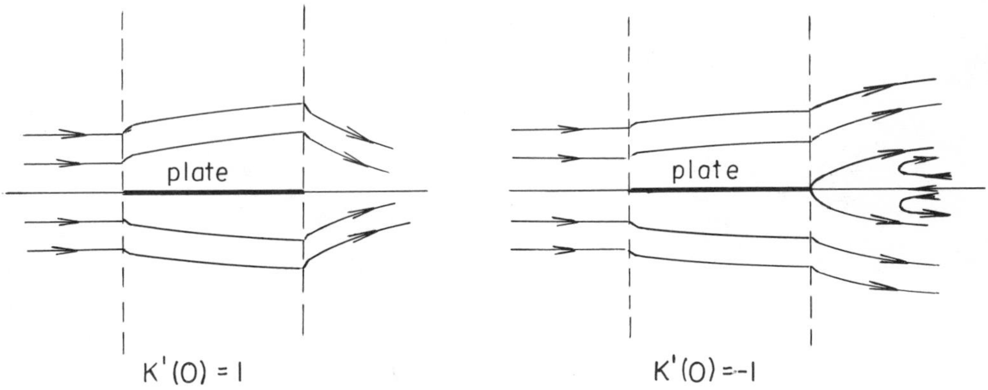

has been treated extensively*. It would be worthwhile to investigate the case $K'(0) = -1$.

An unsymmetrical wake flow occurs when the fluid below the plate is kept at rest. The velocity profile is actually a curve as shown.

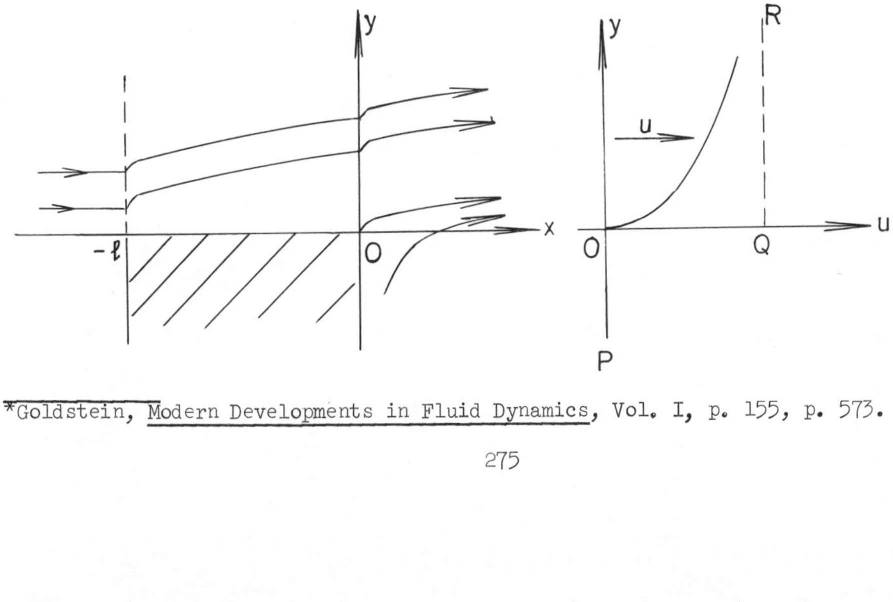

*Goldstein, Modern Developments in Fluid Dynamics, Vol. I, p. 155, p. 573.

To simplify the problem we approximate this velocity distribution by the broken line POQR or set

$$u = 0,\ y < 0 \quad u = U = \text{const.},\ y > 0.$$

Then the differential equation is the same as (56)

DE $$uu_x + vu_y = \nu u_{yy}$$

but the boundary conditions become

BC^{∞} $$u = U = \text{const.},\ y = \infty$$

$BC^{-\infty}$ $$u = 0,\ y = -\infty$$

IC $$u = U \quad \text{for} \quad x = 0,\ y > 0$$

$$u = 0 \quad \text{for} \quad x = 0,\ y < 0.$$

We only consider the right half of the xy plane, as these conditions indicate. The half line $y = 0,\ x > 0$ represents the "wake layer" in which we are interested.

If we set

$$\psi = U(\lambda x)^{1/2}\ h(\theta), \quad \theta = y(\lambda x)^{-1/2}$$

we have

$$u = Uh'(\theta)$$

and the problem becomes

$$\text{DE} \qquad h'''(\theta) + h(\theta)h''(\theta) = 0 \tag{59}$$

$$\text{BC}^{\infty} \qquad h'(\infty) = 1$$

$$\text{BC}^{-\infty} \qquad h'(-\infty) = 0.$$

It turns out that as $\theta \to -\infty$, $h(\theta)$ approaches a finite value $-\kappa$. Then $\kappa U(\lambda x)^{1/2}$ is the area of fluid drawn into the wake per unit time.

On setting $h(\theta) = \kappa H(\kappa\theta)$ where $H(\theta)$ is a solution of (59) such that $H(-\infty) = -1$, we have $\kappa^{-2} = H'(\infty)$ from which to determine κ. The problem for $H(\theta)$ is then

$$\text{DE} \qquad H'''(\theta) + H(\theta)H''(\theta) = 0$$

$$\text{BC}^{-\infty} \qquad H(-\infty) = -1,\ H'(-\infty) = 0,\ H''(-\infty) = 0 \tag{59'}$$

This problem can be solved numerically just as that for the flat plate.

Problem 27. Solve the problem (59') numerically and determine $H'(\infty)$. Use any method such as power series or iteration.

14. von Mises' Equations

In this section we derive a different form of the boundary layer equation which is sometimes used. It is of lower order and similar to the heat equation.

We introduce ψ as a new independent variable, which is possible if $u \neq 0$; we assume $u > 0$. We further introduce $z = U^2 - u^2$ as the new dependent variable, and consider z a function of x, ψ. On differentiating $U^2 - u^2 = z$ with respect to y, x we have

$$-2uu_y = z_\psi \psi_y = uz_\psi \quad \text{or} \quad z_\psi = -2u_y$$

$$2(UU_x - uu_x) = z_x + z_\psi \psi_x = z_x - vz_\psi$$

while

$$u_{yy} = -\frac{1}{2} uz_{\psi\psi} = -\frac{1}{2}(U^2-z)^{1/2} z_{\psi\psi}.$$

Thus

$$z_x = 2(UU_x - uu_x - vu_y)$$

or on using the boundary layer equation (30)

$$z_x = \nu(U^2-z)^{1/2} z_{\psi\psi}. \tag{60}$$

Thus the boundary layer equation is reduced in order and similar to the heat equation. However, it is still non-linear since the coefficient $\nu(U^2-z)^{1/2}$ depends on the function z instead of being constant as in the heat equation. Nevertheless the form (60) makes clear just what the proper boundary and initial conditions should be in analogy to the heat equation. They are

$$\left.\begin{array}{lll} BC^0 & z = U^2(x), & \psi = 0 \\ BC^\infty & z = 0, & \psi = \infty \\ IC & z = z(\psi), & x = 0, \end{array}\right\} \tag{60'}$$

where $z(\psi)$ is a given function.

It is to be noted, however, that for $\psi = 0$ Equation (60) becomes singular.

We now see why it was proper to give initial conditions for the problem of the flat plate. In that problem we took U constant. The boundary conditions are the same as above; i.e., $z = U^2$ at $\psi = 0$ corresponds to $u = 0$ on the wall $(y = 0)$ and $z = 0$ at $\psi = \infty$ corresponds to $u = U$ at $y = \infty$. We had

$$\psi = (\lambda x)^{1/2} U h(\theta), \quad \theta = y(\lambda x)^{-1/2}$$

so

$$u = Uh'(\theta)$$

and hence

$$z = U^2[1-h'^2(\theta)] \tag{61}$$

Thus, we are led to consider h as a new independent variable instead of ψ with the dependent variable

$$z = U^2 Z(h), \quad h = \frac{\psi}{U(\lambda x)^{1/2}} \tag{61'}$$

where $\lambda = \frac{2\nu}{U}$. Then

$$z_x = -U^2 Z_h \frac{h}{2x}, \quad z_{\psi\psi} = \frac{1}{\lambda x} Z_{hh}$$

and the differential equation (59) reduces to the ordinary equation of second order:

DE
$$hZ_h + (1-Z(h))^{1/2} Z_{hh} = 0 \tag{62}$$

with

BC
$$Z(0) = 1, \quad Z(\infty) = 0$$

while IC is automatically satisfied.

This differential equation (62) is of the second order while (33)

$$h'''(\theta) + h(\theta)h''(\theta) = 0 \qquad (33)$$

is of the third order. We could have obtained (62) directly from (33); since θ does not occur in (33) it must be possible to reduce the order by introducing h as independent variable and $f = h'$ as dependent variable. Then

$$h'' = f_h \cdot f = \frac{1}{2}(f^2)_h, \quad h''' = \frac{1}{2} f(f^2)_{hh}$$

and (33) becomes

$$(f^2)^{1/2}(f^2)_{hh} + h(f^2)_h = 0$$

which is the same as (62) with $Z = 1 - f^2$. This checks with (60) as well.

Equation (62) is however highly singular since at $h = 0$ both terms in (62) vanish. Also the independent variable h appears in (62). It thus seems that, although (62) is of lower order than (33), for the purpose of numerical calculations the third order equation (33) is more appropriate. Near a separation point equation (60) cannot be used at all because the condition $u \neq 0$ is violated in the neighborhood.

Problem 28. Determine and solve the von Mises' equation for a spreading two-dimentional jet.

15. Curved Walls and Separation

In the preceding considerations it has always been assumed that the line B where the boundary layer occurs is straight. It is easy to carry out the theory without such an assumption. To this effect it is convenient to introduce parameters s and n in the neighborhood of B, where s is the arc length along B and n is the directed distance along straight normals to B. On the equidistant curves $n =$ const. we have the arc length differential $(1+\kappa n)ds$, where κ is

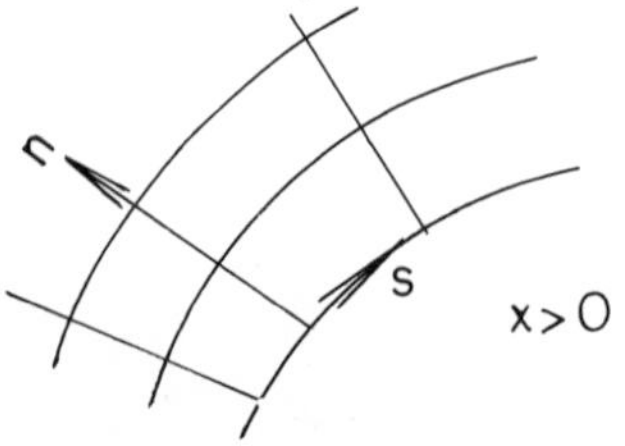

the curvature of B, positive when B is concave towards negative n.

The boundary limit procedure consists in determining the limit of the velocity considered as a function of the variables s and $\eta = \frac{n}{\sqrt{\nu}}$ A detailed analysis then shows that in virtue of the fact that $1+\kappa n = 1+\sqrt{\nu}\kappa\eta \to 1$, the influence of the curvature disappears in the limit; i.e., the resulting boundary layer equation is the same as for a straight boundary. Instead of the variables s, η one customarily introduces $x = s$, $y = \eta\sqrt{\nu}$ with an arbitrary but fixed value of $\nu \neq 0$, the equation is then expressed in the same variables as in the preceding sections.

The curvature has, however, a decisive influence on the main stream. If, for example, the flow around a cylinder is considered, the main flow will be irrotational. The velocity is zero at the forward stagnation point, then it increases over the front part and finally decreases over the rear part; the pressure first decreases and then increases again. Roughly speaking we may say that the main flow is divergent on the rear part; therefore, we may expect separation and back-flow.

The question arises as to what the main flow is; i.e., what the limit of the flow for $\nu \to 0$ is. There are several possibilities:

1. The limit flow is an irrotational flow around the obstacle.

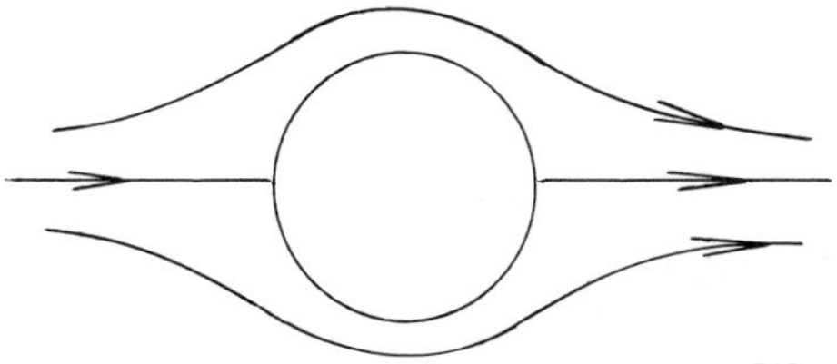

[1]

2. The limit flow is an irrotational flow around the obstacle with free surfaces leaving. A free surface is a streamline with different velocities, but with the same pressure on both sides. There are two types of such flows. Those [2.1] where

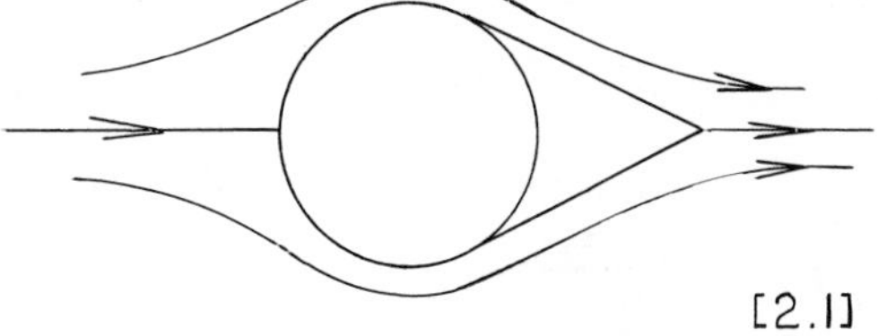

[2.1]

the free surfaces meet and disappear, and those [2.2] where the free surfaces extend to infinity. By varying the distance of the point of detachment from the rear stagnation point of flow [1], we obtain first, a set of flows of type [2.1] and then of type [2.2].

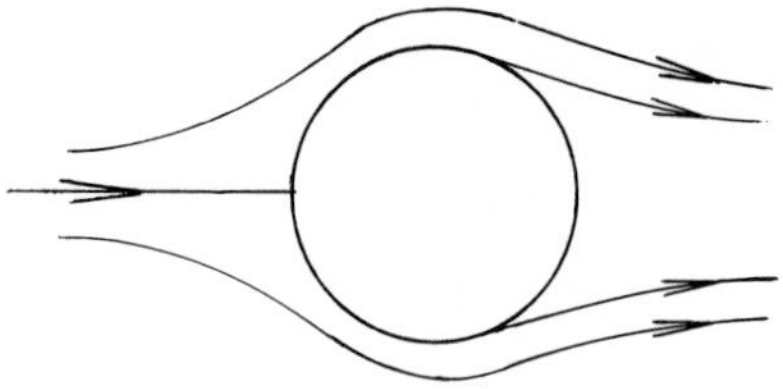

[2.2]

There is a last possible "extreme" situation for the point of detachment. For circular cylinders the extreme situation seems to be at about 60° from the forward stagnation point. For a flat plate, normal to the stream, the extreme situation is detachment at the edge; every point on the rear side and no point on the front side is a possible point of detachment. In the extreme situation the curvature of the free surface vanishes, unless there is a sharp edge as for the plate. It has been conjectured that the flow with this extreme situation for the point of detachment is the interior limit flow, the main stream. Convergence can be expected up to the free surfaces; it is doubtful, though, whether there is convergence in the wake. If there were it should be convergence to "dead water", i.e., fluid at rest. The investigation of divergent flow makes it seem more likely that in the wake there will be an increasing number of regions of forward and backward flow, and that there is no convergence, at least not of the vorticity.

However, even if the question of the interior limit flow were answered, its significance would be restricted. This limit flow would not be attained for $\nu \to 0$ anyway. The reason is that before the limit is reached the flow becomes unstable and passes over into a non-steady flow. The results would then be of significance only up to the "critical" value of the Reynolds number where the breakdown or transition occurs.

In view of all these uncertainties it is preferable to base the boundary layer investigation on a velocity distribution along the wall which is taken from experiment rather than from theory. Then the boundary layer theory gives the right position for the point of separation, i.e., about 81° from the forward stagnation point. The following figure indicates that the velocity distribution for flow of type [2.2] corresponds to experiment except for the infinite slope at the separation point. The points O and P are the front and rear stagnation points.

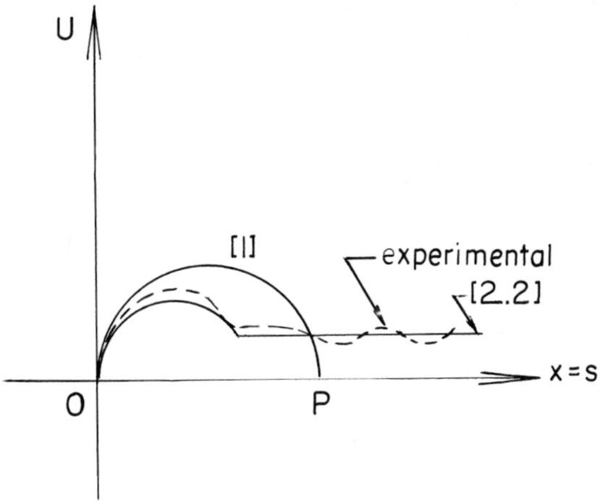

Once one has decided upon a main velocity distribution $U(x)$, one may proceed to solve the boundary layer equation. The question then arises whether the separation effect is contained in the boundary layer equation as such. To study this question take the prospective point of separation to be $x = 0$. Assume for the solution $\psi(x,y)$ the series

$$\psi(x,y) = a(x)y^2 + b(x)y^3 + \cdots .$$

There will be no terms preceding $y^2 a(x)$ since $\psi = 0$ and $\psi_y = u = 0$ at $y = 0$. Insertion of the series into the boundary layer differential equation

$$\psi_y \psi_{xy} - \psi_x \psi_{yy} = UU_x + \nu\psi_{yyy}$$

yields

$$UU_x + 6\nu b(x) + \cdots = 0$$

where only the terms independent of y are written out. The term $a(x)$ has dropped out, since by the Cauchy-Kowalewski theorem $a(x)$ can be given any prescribed value, and we have

$$b(x) = -\frac{1}{6\nu} UU_x.$$

Then

$$\psi(x,y) = a(x)y^2 - \frac{UU_x}{6\nu} y^3 + \cdots$$

$$u(x,y) = 2a(x)y - \frac{UU_x}{2\nu} y^2 + \cdots$$

$$u_y(x,y) = 2a(x) - \nu^{-1}UU_x y + \cdots .$$

The streamlines $\psi = 0$ and $y = 0$ and, for small values of y,

$$y = \frac{6\nu a(x)}{UU_x}$$

unless $U_x \equiv 0$. This representation of the streamline will be valid in the neighborhood of a value of x for which $a(x) = 0$. We assume $a(0) = 0$ and the "separating" streamline $y = \frac{6\nu a(x)}{UU_x}$ will begin at the point $x = 0$, $y = 0$.

We may further assume $u \geq 0$ and $x < 0$ implying that the flow up to the point of separation behaves qualitatively like that for the flat plate. This assumption implies $a(x) > 0$ for $x < 0$ and $U_x < 0$. Then the separating streamline proceeds in the direction of increasing x. This situation of a separating streamline can only thus occur when the velocity of the main flow decreases or the pressure increases.

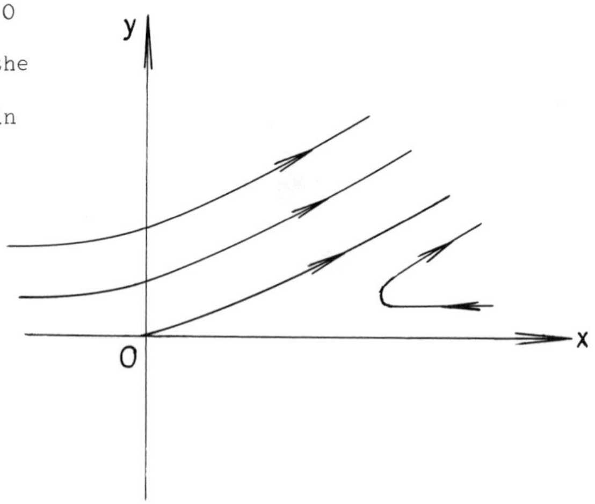

The slope of the separating streamline is given by

$$\left.\frac{6\nu a_x}{UU_x}\right|_{x=0} = \left.\frac{-3\tau_x}{P_x}\right|_{x=0} .$$

Since τ is of the order $\sqrt{\nu}$ in ν, the slope approaches zero as $\nu \to 0$.

A basic assumption for the preceding discussion is that the main flow does not separate at the point where the boundary layer flow separates. The investigation of the boundary layer under the assumption that also the main flow separates is an important problem which has not as yet been attacked.

16. Instability of Vortex Motion

Various flows that are solutions of the Navier-Stokes' equations do not occur since they are unstable. The flow will change into an unsteady flow. For example, the flow around a cylinder gives rise to a vortex pair on the rear side, which, if R exceeds a certain value, will become unstable, break away and form an unsymmetrical Kármán vortex sheet. Under other circumstances the flow will change over into a completely irregular "turbulent" flow.

The first question is when and under what circumstances will the flow be unstable; this means that after a slight disturbance the flow will not return to its previous form. One expects, however, that viscosity will act as a stabilizing factor by damping out the disturbance. One also expects that instability must already be revealed with non-viscous fluid. We shall investigate whether or not this expectation is justified.

Flow of non-viscous fluids need not be irrotational, while irrotational flow is indifferent with respect to irrotational disturbances. Therefore one should admit vorticity from the outset for instability investigations. The simplest classical case of a flow with vorticity is the flow with a "free" surface, i.e., in two dimensions a streamline with the velocity different on both sides.

Let this two-dimensional flow be given by the potential

$$\Phi = U_1 x, \quad u = U_1, \quad v = 0, \quad y < 0$$

$$\Phi = U_2 x, \quad u = U_2, \quad v = 0, \quad y > 0.$$

The line $y = 0$ is then a discontinuity line. The pressure p is the same on both sides; we also assume the density ρ to be the same. The flow is assumed to be irrotational on both sides. The only variation that we consider is that of the position of the discontinuity line.

We may simplify the problem by observing the motion from a system moving with velocity $\frac{1}{2}(U_1 + U_2)$. Then with reference to this system we have

$$u = U, \quad v = 0, \quad y > 0$$

$$u = -U, \quad v = 0, \quad y < 0.$$

Suppose now that the discontinuity line is distorted into the form of a wave. Then one might say that the velocity of the upper fluid is greater at a crest than at a trough and, therefore, the crest will tip forward while the trough falls back. The same argument for the lower fluid yields a tendency in the same direction. This is of course only a rough consideration, the problem can be handled more exactly.

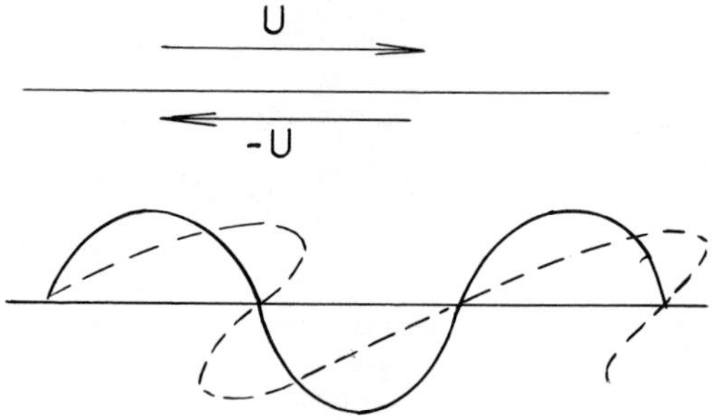

We note that the Bernoulli equation for non-steady irrotational flow is

$$\phi_t + \frac{1}{2}(\phi_x^2 + \phi_y^2) + \rho^{-1}p = C(t).$$

Since we can add to the potential ϕ any arbitrary function of t, we can make $C(t) = 0$. Since the pressure as well as the density is assumed to be the same on both sides, we may write

$$[\phi_t + \frac{1}{2}(\phi_x^2 + \phi_y^2)]_1^2 = 0 \tag{63}$$

as the first condition which must hold on the line, where the bracket refers to the difference of the values at the upper and the lower sides of the discontinuity surface.

This surface may be represented by the function $y = y(t,x)$. The condition that it is a streamline may be expressed as follows: Let the motion of a particle be given by $x = x(t)$, $y = y(t)$; then $y(t) \equiv y(t,x(t))$, whence, by differentiation with respect to t with $x_t = u$, $y_t = v$,

$$v \equiv y_t + uy_x.$$

This relation is now to be considered an identity between the functions $y(t,x)$, $u(t,x,y)$, $v(t,x,y)$. It holds on both sides of the discontinuity line or

$$y_t + u^{(1)}y_x = v^{(1)} \tag{64'}$$

$$y_t + u^{(2)}y_x = v^{(2)}. \tag{64''}$$

From now on we assume that deviations from the undistrubed flow are so small that terms of second order can be neglected; we shall discuss somewhat later what influence this will have. We then replace ϕ, u, v by $\pm Ux + \phi$, $\pm U + u, v$ where the new functions ϕ, u, v refer to the varied potential and velocity; thus, terms of second order in ϕ, u, v can be neglected. Then Equations (63), (64') and (64") become

$$[\phi_t]_1^2 + U(\phi_x^{(1)} + \phi_x^{(2)}) = 0 \tag{65}$$

$$y_t = Uy_x = \phi_y^{(1)} \tag{66'}$$

$$y_t + Uy_x = \phi_y^{(2)}. \tag{66''}$$

These relations must hold on the undisturbed discontinuity line $y = 0$.

The problem is now to obtain potentials $\phi^{(1)}$, $\phi^{(2)}$ which satisfy the transition conditions (65), (66), and the initial value problem of an arbitrarily given function $y = y(0,x)$ prescribing the initial shape of the discontinuity line. If we develop any such function $y(0,x)$ in a Fourier series or represent it by a Fourier integral, it is sufficient to investigate a single harmonic component. For is there is a single harmonic component which does not die out as $t \to \infty$ we have instability; if they all die out there is stability. Therefore we assume $y(0,x) = e^{i\alpha x}$ where α is real; then the solution will be of the form

$$y(t,x) = e^{\beta t} e^{i\alpha x}$$

with β complex. The question is then whether there are solutions for which $\mathcal{R}\beta$ is positive. It follows that the potential function, which fulfills the condition $\Delta\Phi = 0$, has the form

$$\phi^{(1)}(t,x,y) = a^{(1)} e^{\beta t} e^{i\alpha x} e^{|\alpha| y}; \quad y < 0$$

$$\phi^{(2)}(t,x,y) = a^{(2)} e^{\beta t} e^{i\alpha x} e^{-|\alpha| y}, \quad y > 0.$$

Conditions (65), (66) yield

$$\beta(a^{(2)} - a^{(1)}) + iU\alpha(a^{(2)} + a^{(1)}) = 0$$

$$\beta - iU\alpha + |\alpha| a^{(1)} = 0$$

$$\beta + iU\alpha - |\alpha| a^{(2)} = 0.$$

Elimination of $a^{(1)}$ and $a^{(2)}$ gives

$$(\beta + iU\alpha)^2 + (\beta - iU\alpha)^2 = 0$$

or

$$\beta = \pm U\alpha.$$

Thus for any $\alpha \neq 0$ there is a value of β which is real and positive, and therefore the discontinuity surface is unstable.

We are interested in learning why steady flow cannot be maintained when the Reynolds number R is greater than a certain critical value. Such critical values can be calculated for only a few problems, e.g., for rotating Couette flows, where the value obtained checks well with experiment.

We would expect that the presence of viscosity generally damps out any small disturbance of the flow and, therefore, if the flow of a perfect fluid is stable, it is expected to remain stable if the viscosity of the fluid is included. However it has been claimed* that the exact opposite occurs for certain flows, i.e., the viscosity causes the motion to be unstable where it otherwise would be stable if the fluid were considered perfect.

Consider a steady, two-dimensional flow given by $\psi = \Psi(x,y)$ which satisfies the Navier-Stokes' equation

$$U\Delta\Psi_x + V\Delta\Psi_y = \nu\Delta\Delta\Psi. \tag{67}$$

Let $\Psi(x,y) + \psi(t,x,y)$ represent the unsteady flow resulting from a slight disturbance; hence, assume ψ, $\Delta\psi$, $\Delta\Delta\psi$ are relatively small so that it is sufficient to consider only terms of the first order in ψ. Then ψ satisfies the equation

$$\Delta\psi_t + U\Delta\psi_x + V\Delta\psi_y + u\Delta\Psi_x + V\Delta\Psi_y = \nu\Delta\Delta\psi. \tag{67'}$$

As is customary, we restrict ourselves to a simple type of flow given by

$$\Psi = \Psi(y), \quad U = U(y), \; V = 0. \tag{68}$$

This is not consistent with (67) unless Ψ gives a Couette or Poiseuille flow. We may assume though that the variation of Ψ with respect to x is small enough to be neglected. Equation (67') becomes

$$\Delta\psi_t + U\Delta\psi_x - U_{yy}\psi_x = \nu\Delta\Delta\psi. \tag{67''}$$

---

*See sections on instability in Goldstein, Modern Developments in Fluid Dynamics, Vol. I, and Aerodynamic Theory, Durand, ed., Vol. III.

Let the "main flow" given by Ψ be restricted to the channel $y_1 \leq y \leq y_2$. Then we have the boundary conditions

$$\text{B.C.}^{\text{m.f.}} \qquad \Psi = \text{const.}, \quad U = 0 \quad \text{for} \quad y = y_1, y_2, \tag{69}$$

while for the distributing flow ψ we have

$$\text{B.C.}^{\text{d.}} \qquad \psi = u = 0 \quad \text{for} \quad y = y_1, y_2. \tag{69'}$$

As before, we take ψ to be a simple wave motion, i.e., of the form

$$\psi = h(t,y)e^{i\alpha x}. \tag{70}$$

This assumes, as we may do, that every solution can be represented by the superposition of such wave motions. Using (70) the differential equation (67") becomes

$$(h_{yy} - \alpha^2 h)_t + i\alpha U[h_{yy} - \alpha^2 h] - i\alpha U_{yy} h = \nu[h_{yyyy} - 2\alpha^2 h_{yy} + \alpha^4 h] \tag{71}$$

with the boundary conditions

$$h = h_y = 0 \quad \text{at} \quad y = y_1, y_2. \tag{71'}$$

At the time $t = 0$ an initial disturbance

$$h = h_0(y) \tag{71"}$$

may be prescribed. We note that $h(t,y)$ actually has α as a parameter, that is, $h = h(t,y,\alpha)$. It is <u>sufficient</u> to assume $\alpha \geq 0$ since for $\alpha < 0$ we may set $h(t,y,-\alpha) = h(t,y,\alpha)$.

The problem is to prove either that for all initial disturbances the motion dies out or that there is an initial disturbance for which the motion increases. In the first case the main flow is stable, in the second unstable with respect to wave disturbances of wave length $\frac{2\pi}{\alpha}$; we call this α-stability and

α-instability respectively.

It is customary to consider "normal" modes of motion of the form

$$h(t,y) = f(y)e^{\beta t}, \tag{72}$$

and to assume that every motion can be obtained by superposition of such motions. Although we shall see later that this is not true for perfect fluid flow let us assume it temporarily. If, for a particular value of α, we have $\mathscr{R}\beta < 0$ for all possible factors β then we have α-stability; if for at least one β, $\mathscr{R}\beta > 0$ we have α-instability. The border between stability and instability is characterized by $\mathscr{R}\beta = 0$; in this case the main flow is said to be "capable of vibration". It is convenient to write

$$\beta = -i\alpha c, \quad c = c' + ic''.$$

In view of $\alpha \geq 0$ we have α-stability when all $c'' < 0$ and α-instability when one $c'' > 0$. A "neutral" mode, i.e., one where $\mathscr{R}\beta = 0$, occurs when $c'' = 0$ (c real). In this last case the motion $\psi = f(y)e^{i\alpha(x-ct)}$ can be considered as a progressing wave with phase velocity c.

The problem for the normal modes is to find "characteristic" values of c for a given α such that equation (71) with boundary conditions (71') has a solution when $h(t,y)$ is assumed to be of the form $h(t,y) = f(y)e^{\beta t}$. In this case (71) becomes

$$(U-c)(f_{yy} - \alpha^2 f) - U_{yy}f = \frac{\nu}{i\alpha}(f_{yyyy} - 2\alpha^2 f_{yy} + \alpha^4 f) \tag{73}$$

with boundary conditions

$$f = f_y = 0 \quad \text{at} \quad y = y_1, y_2 \tag{73'}$$

We confine ourselves to the case of a perfect fluid. Here equation (73) reduces to

$$(U-c)(f_{yy} - \alpha^2 f) - U_{yy} f = 0 \tag{74}$$

with the boundary conditions

$$f = 0 \quad \text{at} \quad y = y_1, y_2. \tag{74'}$$

In this case certain general statements can be made.*

<u>Theorem 1</u>. If one non-neutral mode exists the flow is unstable. Indeed, if c is a non-real characteristic value corresponding to the function $f(y)$, then $\bar{c}$ is also one corresponding to $\overline{f(y)}$. Obviously either $\operatorname{Im} c$ or $\operatorname{Im} \bar{c}$ is positive and hence there is instability.

<u>Theorem 2</u>: If U_{yy} does not change sign in $y_1 < y < y_2$ then no non-neutral mode exists.

<u>Proof</u>. Assume there exists a non-neutral mode. Then there is a solution f of (74) with c not real. Multiply (74) by $-\bar{f}$ and integrate. We obtain

$$\int_{y_1}^{y_2} \bar{f}[-f_{yy} + \alpha^2 f + \frac{U_{yy}}{U-c} f]dy = 0.$$

Integration by parts yields

$$\int_{y_1}^{y_2} [|f_y|^2 + \alpha^2 |f|^2 + \frac{U_{yy}}{U-c} |f|^2]dy = 0.$$

In particular, considering the imaginary part of this last equation, we get

$$c'' \int_{y_1}^{y_2} \frac{U_{yy}|f|^2}{|U-c|^2} dy = 0.$$

Hence the assumption that c'' does not vanish leads to a contradiction since by

---

*See Goldstein, <u>Modern Developments in Fluid Dynamics</u>, Vol. I, and <u>Aerodynamic Theory</u>, Durand, ed., Vol. III, and their references to the works of Lord Rayleigh and Tollmien.

hypothesis U_{yy} does not change sign.

Theorem 2 shows that for velocity profiles which are convex, e.g., in Poiseuille flow, the only possible normal modes are neutral. This indicates that there is no proper instability in such cases.

Theorem 3. If $f(y)$ is of the form

$$f(y) = [U(y) - c]g(y) \tag{75}$$

then there are no neutral modes, except for $\alpha = 0$ with $c = 0$, and then

$$f(y) = U(y).$$

Proof. Introducing (75) into the differential equation (74) yields

$$\{(U-c)g_y\}_y - \alpha^2(U-c)^2 g = 0.$$

Upon multiplying through by $-\bar{g}$ and integrating we find

$$\int_{y_1}^{y_2}\{[-(U-c)^2 g_y]_y \bar{g} + \alpha^2(U-c)^2|g|^2\}dy = 0.$$

Integrating the first term by parts gives us for this last equation

$$\int_{y_1}^{y_2}(U-c)^2[|g_y|^2 + \alpha^2|g|^2]dy = 0.$$

Clearly for $\alpha \neq 0$ we get $g = 0$ and for $\alpha = 0$, $g = \text{const.}$, which is the desired result.

Taking the imaginary part of this last expression we get the relation

$$c''\int_{y_1}^{y_2}(U-c')[|g_y|^2 + \alpha^2|g|^2]dy = 0.$$

Hence we have

Corollary. $U_{min} \leq c' \leq U_{max}$, if $c'' \neq 0$. This is clear as (75) holds and the last relation above implies that $U-c'$ must change sign.

On the other hand it can be shown that under certain circumstances there are neutral modes.

Theorem 4. Let there be a point $y = y^*$ such that $U_{yy}(y^*) = 0$, $U_y(y^*) \neq 0$. In addition let $U(y) > 0$ for $y_1 < y < y_2$, and assume that $K(y) = -\frac{U_{yy}(y)}{U(y)-U(y^*)}$ is a regular, positive function for $y_1 \leq y \leq y_2$. Then there is a value of α for which a neutral mode exists with $c = U(y^*)$.

Proof. The differential equation (74) takes the form

$$f_{yy} + K(y)f + \lambda f = 0$$

with boundary conditions

$$f = 0 \quad \text{for} \quad y = y_1, y_2.$$

Since $K(y)$ has no singularities this is a usual characteristic value problem. There is a sequence of characteristic values $\lambda_1, \lambda_2, \ldots, \lambda_n, \ldots \to \infty$. Since we wish to identify λ_1 with $-\alpha^2$ we must show that at least one of the λ_i is negative. The lowest of these values λ_1 can be characterized* as:

$$\lambda_1 = \min \frac{\int_{y_1}^{y_2} \{|f_y|^2 - K(y)|f|^2\} dy}{\int_{y_1}^{y_2} |f|^2 dy}$$

To show $\lambda_1 < 0$ it suffices to show that for some function f satisfying the boundary conditions the above quotient is negative; for then the minimum will certainly be below this value. Take $f = U(y)$. Clearly $U(y_1) = U(y_2) = 0$, and

$$\frac{\int_{y_1}^{y_2} \{|U_y|^2 - K(y)|U|^2\} dy}{\int_{y_1}^{y_2} |U|^2 dy} = -\frac{\int_{y_1}^{y_2} \{UU_y + KU^2\} dy}{\int_{y_1}^{y_2} U^2\, dy},$$

*Cf. Courant and Hilbert, Methods of Mathematical Physics, Vol. I, p. 397 ff.

where integration by parts has been used. Since $U_{yy} = -K(y)\{U(y)-c\}$ the above quotient becomes

$$\frac{\int_{y_1}^{y_2}\{-U[K(y)(U(y)-c)] + K(y)U^2\}dy}{\int_{y_1}^{y_2} U^2 dy} = -\frac{c\int_{y_1}^{y_2} K(y)U(y)dy}{\int_{y_1}^{y_2} U^2 dy}\,.$$

Now taking into account that $U(y) > 0$, $K(y) > 0$ and $c = U(y^*) > 0$, we get $\lambda_1 < 0$. Identifying λ_1 with $-\alpha^2$ gives us the desired result.

It has been proved* that for values of α in the neighborhood of the particular value obtained in Theorem 4 there are non-real characteristic values. Hence from Theorem 1 the main flow is unstable with respect to this value of α and hence unstable with respect to arbitrary disturbances. This is an extremely important result in view of the boundary layer theory. If R is large then the flow is like that of a perfect fluid. In the neighborhood of a point of separation the velocity profile has an inflection point ($U_{yy} = 0$). Therefore it is likely that the flow becomes unstable before it reaches the separation point. In fact for larger R the boundary layer past an obstacle becomes turbulent; the point of separation is moved toward the rear of the body and the resistance is lowered.

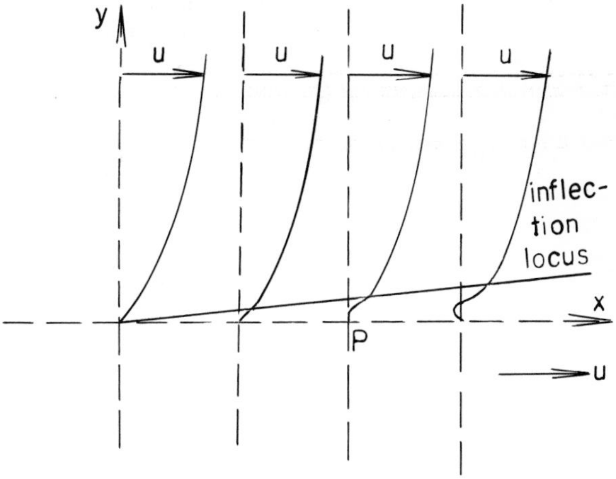

We now wish to investigate those solutions of (74) which give neutral modes. Neutral modes occur when $c'' = 0$ or c is real. We assume in the following discussion that $U(y)$ is analytic. Several cases are to be distinguished.

*Cf. Tollmien: Nachrichten von der Gesellshaft der Wissenschaften Göttingen. Math. u. Phys. Klasse, New Series 1 (1935), 79-114.

I. $U(y) \neq c$ for $y_1 \leq y \leq y_2$. Then equation (74)

$$(U-c)(f_{yy} - \alpha^2 f) - U_{yy} f = 0 \tag{74}$$

is regular in the whole interval. There are two solutions which are regular.

II. There is a value of y' such that $U(y') = c$. A layer $y = \text{const.}$ where $U(y) = c$ is called "critical". For convenience take $y' = 0$, i.e., $U(0) = c$.

II_1. $U_{yy}/U_y = 0$ for $y = 0$. In this case U_{yy}/yU_y is regular at $y = 0$; hence $\frac{U_{yy}}{U-c}$ and the differential equation

$$f_{yy} + (\alpha^2 + \frac{U_{yy}}{U-c})f = 0$$

are regular. In this case also we have two regular solutions.

II_2. $U_{yy}/U_y \neq 0$ for $y = 0$. In this case the differential equation is singular at $y = 0$. To simplify the argument assume $U_y(0) \neq 0$ and set $U_{yy}(0)/U_y(0) = k \neq 0$. Let $C(y) = \alpha^2 y + \frac{yU_{yy}}{U-c}$, and $C(y)$ is regular, $C(0) = k$. Here there is one regular solution and one singular solution. These can be found by power series expansion for f, but without going into details we state that the regular solution $f^*(y)$ is of the form

$$f^*(y) = y + \cdots,$$

and the singular solution $f(y)$ is of the form

$$f(y) = f^0(y) + Kf^*(y) \log (y),$$

where $f^0(y) = 1 + \cdots$ is regular.

The singular solution corresponds to a kind of vortex layer about the point $y = 0$ and should be excluded.

In cases II and II_2 the assumption of Theorem 3 is justified. If U_{yy} never vanishes Theorem 2 states that no non-neutral modes exist; both theorems combined then show that there are no normal modes at all. The fact that no normal modes exist in certain cases indicates that it is not sufficient to consider only normal modes for the investigation of stability. Hence the customary assumption that every motion can be obtained by the superposition of normal modes, which we also started out with, is not justified. A correct treatment of stability, which would be important and interesting to carry out should take into account a continuous spectrum of values for c.

Problem. We consider an example of an explicit expression for unsteady flow with vorticity of a perfect fluid. We ask for a flow which at the time $t = 0$ is given by

$$\psi = \frac{1}{2} y^2 + f(x), \quad u = y, \quad v = -f_x(x) \tag{76}$$

where $f(x)$ is arbitrary. The differential equation to be satisfied is

$$\Delta\psi_t + u\Delta\psi_x + v\Delta\psi_y = 0$$

where $u = \psi_y$, $v = -\psi_x$, the solution of which is

$$\psi = \frac{1}{2} y^2 + \frac{1}{1+t^2} f(x-yt) \tag{77}$$

as can readily be seen by direct substitution. Now (77) reduces to the flow (76) when $t = 0$. We have

$$u = y - \frac{t}{1+t^2} f'(x-yt)$$

$$v = - \frac{t}{1+t^2} f'(x-yt),$$

that is, the flow (77) approaches the plane Couette flow when $t \to \infty$. In fact

$$\psi \to \frac{1}{2} y^2, \quad u \to y, \quad v \to 0.$$

From $\Delta\psi = 1 + f''(x-yt)$ we see that the variations of $\Delta\psi$ in the y-direction concentrate more and more, as $\Delta\psi$ does not approach a limit unless f'' converges.

Problem 29. Consider the same problem for a viscous fluid by taking $f(x) = \sin kx$. Show that $\Delta\psi \to 0$ as $t \to \infty$.

CHAPTER V

COMPRESSIBLE FLUIDS

1. Equations of Motion for Compressible Fluids

Although every fluid is compressible, gases exhibit this property most noticeably, and even here the compressibility property of a gas will have little influence except when the velocity is relatively large. To take compressibility into account we assume that the density ρ is a function of the pressure p, (whereas previously we always considered ρ a constant); it is more convenient, however, to consider the pressure as a function of the density, i.e.,

$$p = p(\rho).$$

First let us consider the effect of this dependence upon the equations of motion for a perfect fluid. Euler's equation is the same as before (Chapter I, Section 1)

$$\rho \frac{d\vec{q}}{dt} = - \text{grad } p; \tag{1}$$

(gravity will be neglected throughout).

This equation can be written

$$\rho \frac{d\vec{q}}{dt} = -a^2 \text{ grad } \rho$$

with

$$a = \left(\frac{dp}{d\rho}\right)^{1/2}; \tag{2}$$

the quantity a which has the dimension of velocity is called the acoustic velocity or the velocity of sound. We note that a, which depends upon the fluid, varies with the density of the fluid. (See Chapter I, Equation (21), p. 27).

To obtain the continuity equation consider a body M whose surface is denoted by S, and let $\vec{n}$ be the outward normal at any point on the surface. The expression

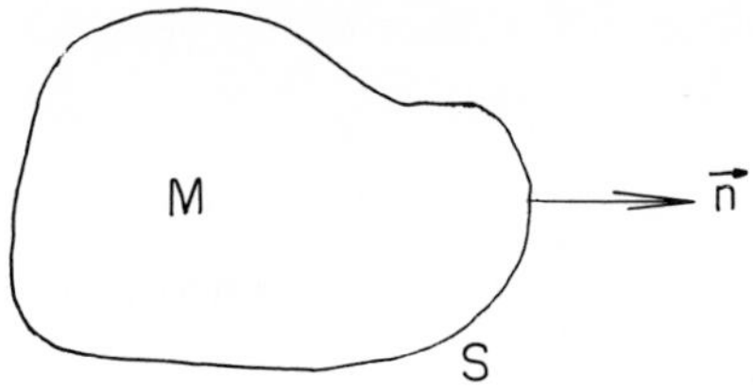

$$-\rho\vec{q}\,\vec{n}\,dS$$

is the mass crossing an element of surface into the body M per unit time. The increase of mass in M per unit time is given by

$$\frac{d}{dt}\int_M \rho\, dM = \rho_t.$$

Stating the law of convervation of mass yields

$$\frac{d}{dt}\int_M \rho\, dM = -\int_S \rho\vec{q}\,\vec{n}\,dS.$$

From this we get the continuity equation (see Chapter I, Equation (5), p. 10):

$$\rho_t + \operatorname{div}(\rho\vec{q}) = 0. \tag{3}$$

Equation (1) allows us to derive Bernoulli's law (for steady flow) along a path $\vec{r} = \vec{r}(s)$ of a particle (s is arc length). Along this path we have

$$\vec{q} = \frac{d\vec{r}}{ds} \cdot \frac{ds}{dt}.$$

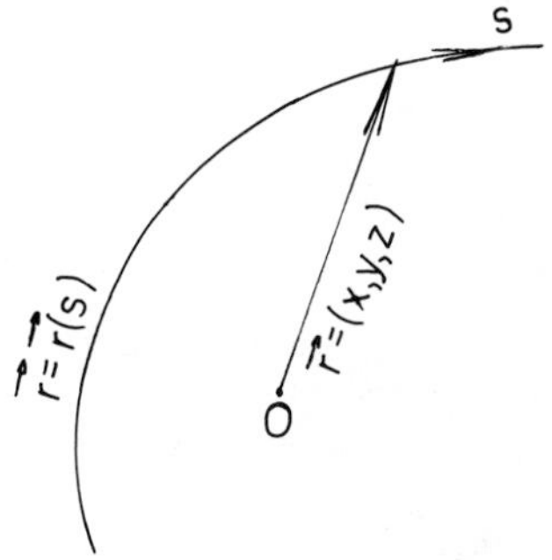

From (1) we get

$$\frac{d\vec{q}}{ds} \cdot \frac{ds}{dt} = -\rho^{-1}\operatorname{grad} p.$$

Multiplying through by $\frac{d\vec{r}}{ds}$ and noticing that $\frac{d\vec{r}}{ds}\operatorname{grad} p = \frac{dp}{ds}$, we get

$$\vec{q}\,\frac{d\vec{q}}{ds} = -\rho^{-1}\frac{dp}{ds}. \tag{1'}$$

We now introduce a function $\mathscr{P}(\rho)$ defined by (compare Chapter I, Equation (2), p. 8)

$$\frac{d\mathscr{P}}{d\rho} = \rho^{-1}\frac{dp}{d\rho}, \quad \mathscr{P}(0) = 0. \tag{4}$$

Inserting this quantity $\mathscr{P}(\rho)$ in (1') we get

$$\vec{q}\,\frac{d\vec{q}}{ds} = -\mathscr{P}(\rho)\,\frac{d\rho}{ds},$$

whence

$$\frac{1}{2}|\vec{q}|^2 + \mathscr{P}(\rho) = \frac{1}{2}q_*^2 = \text{const. along a path of a particle.} \tag{5}$$

This relation is the Bernoulli equation (cf. Chapter I, p. 11, p. 26) and relates the density with the velocity along any path for a particle. The constant q_* is the maximal velocity $|\vec{q}|$ attains; (attained when $\rho = 0$). To every speed $|\vec{q}|$ there is assigned a density ρ and hence an acoustic velocity a. It may occur that on some point on the path the speed $|\vec{q}|$ coincides with the acoustic velocity a assigned to it. Such a speed will be called the <u>critical acoustic velocity</u> a_*. The flow is called <u>subsonic</u> where the speed is less than its acoustic velocity, <u>supersonic</u> where it is greater than its acoustic velocity. These conditions are equivalent to

$$|\vec{q}| < a_* \quad \text{and} \quad |\vec{q}| > a_*$$

when we assume that a is an increasing function of $\mathscr{P}(\rho)$. This may be seen, for example, in an ideal gas where the pressure is an increasing function of the density. When $|\vec{q}| < a_*$, $a_* < a$ and when $|\vec{q}| > a_*$, $a_* > a$.

2. <u>Introductory Concepts of Thermodynamics</u>

Before discussing flows on the basis of the preceding equations we wish to consider certain thermodynamic notions. For every gas there is a relation between the density ρ, the pressure p, and the absolute temperature T. Let us consider a container of gas. Heat can be acquired by conduction from the outside and by violent action from work done inside the container. Processes that are so smooth as to exclude the latter possibility are called quasi-static. If a process, in addition to being quasi-static, is such that there is no heat conduction from or to the outside, it is called adiabatic. In adiabatic changes all work done is stored in potential energy.

Consider a quasi-static process which changes the density from ρ_1 to ρ_2. If we let p be the pressure, T the absolute temperature and h the potential evergy per unit mass, then the work done by the pressure is

$$-\int_{\rho_1}^{\rho_2} p d\rho^{-1}.$$

The increase of heat is then given by

$$h_2 - h_1 + \int_{\rho_1}^{\rho_2} p d\rho^{-1}.$$

It is more convenient to express this relation by the "heat differential":

$$dh + p d\rho^{-1}.$$

Let s denote the "entropy" per unit mass which we define by the relation

$$T ds = dh + p d\rho^{-1}. \qquad (6)$$

For an adiabatic process, since there is no heat conduction and no work transformed into heat, we have

$$ds = 0, \quad s = \text{const.}$$

In the case of non-quasi-static processes without transfer of heat the second law of thermodynamics states $ds > 0$.

Instead of considering the potential energy h it is necessary to introduce another quantity: the heat content per unit mass, i. It is defined by the relation

$$i = h + p\rho^{-1}.$$

Then from (6) we have

$$di = dh + pd\rho^{-1} + \rho^{-1}dp = \rho^{-1}dp + Tds. \tag{7}$$

Since $ds = 0$ for adiabatic processes, we get in this case

$$di = \rho^{-1}dp. \tag{8}$$

The preceding ideas may be exemplified in the case of an ideal gas; here we have

$$p = \sigma\rho^{k} \tag{9}$$

where $\sigma = \sigma(s)$ depends upon the entropy and k is a constant (= 1.4 for air). Since σ = const. for adiabatic processes, we get in this case from (8) and (9)

$$di = k\sigma\rho^{k-2}d\rho,$$

and

$$i = \frac{k}{k-1}\,\sigma\rho^{k-1} = \frac{k}{k-1}\,\sigma^{\frac{1}{k}}\,p^{1-\frac{1}{k}} \tag{10}$$

Indeed, from (10), we get, in general

$$di = \sigma^{\frac{1}{k}}\,p^{-\frac{1}{k}}dp + \frac{1}{k-1}\,p^{1-\frac{1}{k}}\sigma^{\frac{1}{k}-1}d\sigma = \rho^{-1}dp + \frac{1}{k-1}\rho^{k-1}d\sigma,$$

and this agrees with (8), since for adiabatic processes we have $d\sigma = 0$.

The acoustic velocity for adiabatic processes is

$$a = \left(\frac{dp}{d\rho}\right)^{1/2} = (k\sigma\rho^{k-1})^{1/2} = ((k-1)i)^{1/2}. \tag{11}$$

From this we get

$$i = \frac{1}{k-1}\,a^{2} \tag{12}$$

which gives an expression for the heat content in terms of the acoustic velocity.

Consider a high speed flow of a gas. As the gas particles go from one region with a certain pressure to another with different pressure, the density and temperature will change. This change will hardly be affected by heat transfer which is relatively slow. In addition the transformation of work into heat by viscosity can be disregarded except in boundary layers. Consequently we may assume that changes in density due to changes in pressure are adiabatic. The function $p = p(\rho)$ is then determined once the entropy is known; the entropy is assumed to be constant throughout the field of flow. The function $\mathcal{P}(\rho)$ is then to be identified with the heat content per unit mass, i; Bernoulli's equation along a path, therefore, assumes the form

$$\frac{1}{2}|\vec{q}|^2 + i = \frac{1}{2} q_*^2 = i_*. \tag{13}$$

We mention various relations for ideal gases that follow from (13). The critical acoustic velocity defined by $u = a = a_*$ is to be found from $\frac{1}{2} a_*^2 + \frac{1}{k-1} a_*^2 = q_*^2$, where (12) has been used. This can be written

$$a_* = \left(\frac{k-1}{k+1}\right)^{1/2} q_*. \tag{14}$$

The critical acoustic velocity a_* is apparently smaller than the maximum velocity q_* on the streamline. Further, the pressure expressed in terms of the velocity or entropy is

$$p = \left(\frac{k-1}{2k}\right)^{\frac{k}{k-1}} \sigma^{-\frac{1}{k-1}} \cdot (q_*^2 - |\vec{q}|^2)^{\frac{k}{k-1}}. \tag{15}$$

3. Steady Flow in One-Dimensional Treatment

Consider the flow in a rotationally symmetric tube leading out of a container. Let the x-axis be the axis of the tube and let $S = S(x)$ denote the cross-sectional area cut out of the tube by the plane $x = \text{const}$. The tube is assumed to narrow down as it leaves the container and then to widen again; i.e., it is assumed that $S(x)$ has a minimum $S = S_*$ for a certain value $x = x_*$. The container may be characterized by $S = S_0 = \infty$ at $x = x_0$ and the end of the tube

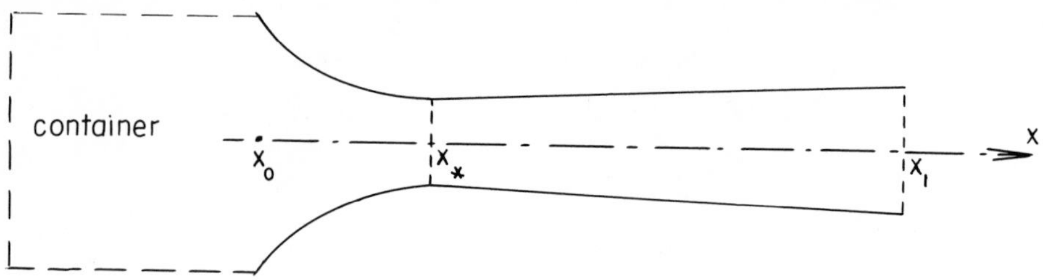

by $S = S_1$ at $x = x_1$, (see figure). Suppose the pressure p_0 and density ρ_0 in the container are given and also the pressure p_1 of the surrounding fluid. The pressure of the outcoming jet is to be p_1. Under these conditions the mass flux Q, i.e., the mass of gas crossing every cross-section $S(x)$ per unit time is to be determined, and the flow as a whole is to be described.

The "one-dimensional" treatment consists in considering only the average values of q, ρ and i over the cross-sections, averaging also the equations, and disregarding deviations from the average values of the quantities involved.

Let $u(x) > 0$ be the average x-component of the velocity; the symmetry makes it clear that the average y- and z-components will be zero. Let $\rho(x)$ be the average density, $p(x)$ the average pressure. The average continuity equation yields

$$\rho u S = Q = \text{const.}, \tag{16}$$

and the average Bernoulli equation gives

$$\frac{1}{2} u^2 + i = \frac{1}{2} q_*^2 = \text{const.} \tag{17}$$

Equations (16) and (17) are to be used to calculate u, ρ, and p in terms of Q, q_* and the entropy; Q, q_*, and the entropy are in turn to be determined by the prescribed values of p_0, ρ_0 and p_1.

However certain results can be obtained from a direct discussion of equations (16) and (17). If we differentiate these equations we obtain

$$\frac{du}{u} + \frac{d\rho}{\rho} + \frac{dS}{S} = 0 \tag{16'}$$

and

$$u\,du + a^2 \frac{d\rho}{\rho} = 0, \tag{17'}$$

where the fact that $di = \rho^{-1}dp = \rho^{-1}a^2 d\rho$ has been used. Elimination of $\frac{d\rho}{\rho}$ from (16') and (17') yields

$$\left(\frac{u^2}{a^2} - 1\right) \frac{du}{u} = \frac{dS}{S} . \tag{18}$$

The right side of this equation vanishes only for $S = S_*$, i.e., $x = x_*$. Then for $x = x_*$ either $du = 0$ or $u = a$; the latter means $u = a = a_*$. Conversely $u = a$ or $du = 0$ implies that $dS = 0$ which means $x = x_*$.

Since $u = u_0 = 0$ in the container the velocity increases as long as S decreases unless the value $u = a_*$ is reached.

Suppose first that the value a_* is not yet reached at x_*. Then $du = 0$ as $u^2/a^2 - 1 \neq 0$ at $x = x_*$. Since $dS > 0$ for $x > x_*$ we have $du < 0$ for $x > x_*$, and we see that the value a_* is never reached. The flow remains subsonic throughout.

Suppose now that the value $u = a_*$ is attained at $x = x_*$. Then there are two possibilities: (i) the quantity $du < 0$ for $x > x_*$. In this case we have subsonic flow throughout except for the one point $x = x_*$ where the critical acoustic velocity is attained. (ii) the flow continues with $du > 0$ in which case $u > a_*$ for $x > x_*$ and the flow changes over to supersonic flow; the density will decrease and the gas expands. Equation (18) shows that the acoustic velocity can never be attained before the narrowest cross-section of the tube is reached. The values of p_1 in cases (i) and (ii) may be denoted by p' and p'' respectively.

We now investigate for ideal gases the distribution of the pressure for these cases. If i_0 is the value of i in the container, equation (16) can be written

$$Q\,S^{-1} = \rho u = \rho[2i_0 - 2i]^{-1/2},$$

where equation (13) has been used.

Since $\frac{p}{p_0} = (\frac{\rho}{\rho_0})^k$ and $\frac{i}{i_0} = (\frac{\rho}{\rho_0})^{k-1}$, we obtain

$$Q\, S^{-1} = \rho_0 (\frac{p}{p_0})^{\frac{1}{k}} (2i_0)^{1/2} (1 - \frac{i}{i_0})^{1/2}$$

$$= \rho_0 (\frac{p}{p_0})^{\frac{1}{k}} (2i_0)^{1/2} (1 - (\frac{p}{p_0})^{\frac{k-1}{k}})^{1/2},$$

or

$$S = Q\, \rho_0^{-1} (2i_0)^{-\frac{1}{2}} (\frac{p}{p_0})^{-\frac{1}{k}} [(1 - \frac{p}{p_0})^{\frac{k-1}{k}}]^{-1/2} . \qquad (18)$$

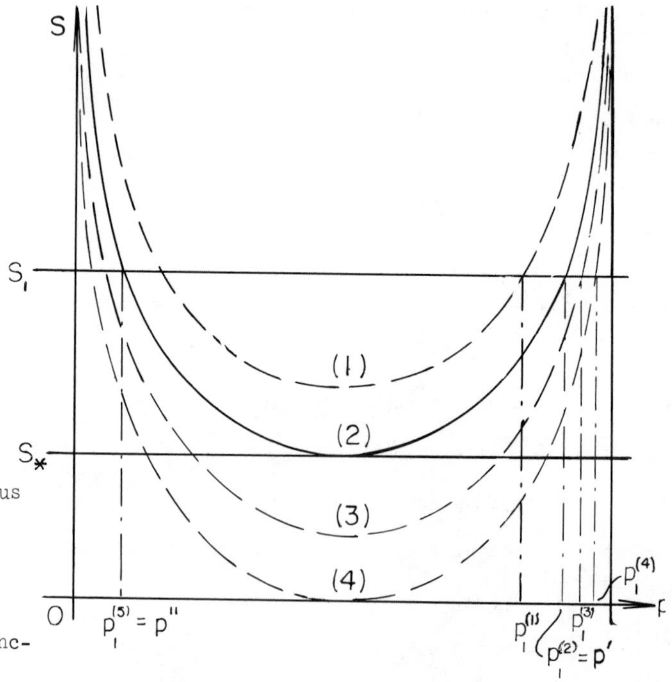

In the container $p = p_0$, $\rho = \rho_0$, and hence $i = i_0$, are known. Equation (18) states that $S = S(p)$ depends on the mass flux Q. The magnitude Q is to be determined from the prescribed value $p = p_1$ at $S = S_1$. The figure shows $S = S(p)$ plotted for various prescribed values of p_1.

If Q is above a certain value Q_* the function $S(p)$ never reaches the value S_*. This is seen in the figure where the prescribed pressure $p_1 = p_1^{(1)}$ corresponding to curve (1) doesn't intersect the line $S = S_*$. For $Q = Q_*$ the value S_* is just attained and curve (2) illustrates this case. Here $S = S(p)$ takes on the value S_1 for $p_1 = p_1^{(2)} = p'$ and $p_1 = p_1^{(5)} = p''$. This means that the same flux Q will be carried if either $p_1 = p'$ or $p_1 = p''$ is the prescribed pressure. For the case $p_1 = p'$ the flow remains subsonic, while for $p_1 = p''$ the flow changes to supersonic flow after passing the cross-section S_*.

If Q is less than Q_* then the value S_* is attained before the curve $S(p)$ attains its minimum value. Such cases correspond to prescribed pressures $p_1 > p'$, and here the flow will remain subsonic throughout. Curves (3) and (4) in the above figure illustrate this. However we notice that these curves also correspond to prescribed pressures $p_1 < p''$. In such cases the fluid will leave the walls of the tube and continue as a free jet with a pressure greater than that of the surrounding fluid.

If the prescribed pressure p_1 is such that $p'' < p_1 < p'$ no continuous flow will be possible. The flow will turn into supersonic flow after passing S_* just as if the prescribed pressure were p'', but somewhere before S_1 the velocity will suddenly drop to a subsonic value and the density will jump to a higher value. From then on it will continue as subsonic flow (unless another jump is necessary). The figure shows the different pressures corresponding to the curves in the preceding diagram. The curve (3') corresponding to (3) is purely subsonic; the curve (2') corresponds to (2) when the prescribed pressure is $p_1^{(2)} = p'$, while (2") corresponds to (2) when the prescribed pressure is $p_1^{(5)} = p''$.

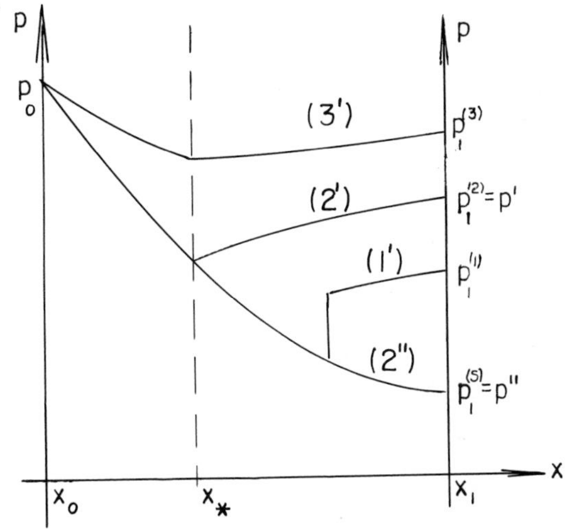

The sudden drop in velocity from a supersonic to subsonic value is illustrated by curve (1') which is obtained from (1) where $p'' < p_1^{(1)} < p'$. This last type of flow with its sudden "compression shocks" will be treated in the next section.

4. Compression Shocks

Consider a one-dimensional adiabatic flow with the x-component of velocity $u = u(x)$, density $\rho(x)$, and pressure $p(x)$. This flow obeys the laws

$$(\rho u)_x = 0 \tag{19$_1$}$$

and

$$\rho u u_x = -p_x; \tag{20$_1$}$$

equation (20_1) can be written

$$(\tfrac{1}{2}u^2)_x = -i_x. \tag{20$_2$}$$

Suppose that at $x = 0$ there is a transition from a state (1) for $x < 0$ to a state (2) for $x > 0$. We wish to investigate whether such a discontinuity is consistent with the equilibrium conditions. The first condition is that the mass flux per unit area, ρu, is the same for the two states, i.e.,

$$[\rho u]_{(1)}^{(2)} = 0 \tag{19}$$

If ρ_1, u_1 are the density and velocity in state (1), similarly ρ_2, u_2 for state (2), condition (19) can be written

$$\rho_1 u_1 = \rho_2 u_2 = m = \text{const.} \tag{19a}$$

The second condition is that the momentum flux per unit area out of the "shock" region, i.e., the transition region, equals the resultant pressure force acting on the shock region:

$$[\rho u^2]_{(1)}^{(2)} = -[p]_{(1)}^{(2)},$$

or

$$[p + \rho u^2]_{(1)}^{(2)} = 0, \tag{20}$$

which in turn can be stated in the form

$$p_1 + \rho_1 u_1^2 = p_2 + \rho_2 u_2^2. \tag{20a}$$

If it is assumed that the transition across the shock region is adiabatic, the following third condition, the energy value equation, would be satisfied:

$$[\tfrac{1}{2}u^2 + i]_{(1)}^{(2)} = 0, \tag{21}$$

or

$$\tfrac{1}{2}u_1^2 + i_1 = \tfrac{1}{2}u_2^2 + i_2 = \tfrac{1}{2}q_*^2. \tag{21a}$$

Now supposing that ρ_1, u_1, p_1 are given we have in trying to determine ρ_2, u_2, p_2, one more equation than we have unknowns. It can be shown that a solution for the system (19), (20) and (21) is possible only if state (2) coincides with state (1). Indeed, "shocks" are non-adiabatic processes; in a shock process heat is generated and the relation $di = \rho^{-1}dp$ does not hold. We replace condition (21) by the condition that the increase of kinetic energy per unit mass is equal to the sum of the work done by the pressure forces per unit mass and the energy per unit mass transformed into heat:

$$[\tfrac{1}{2}u^2]_{(1)}^{(2)} = \int_{(1)}^{(2)} \rho^{-1}dp - \int_{(1)}^{(2)} T ds.$$

This can be written

$$[\tfrac{1}{2}u]_{(1)}^{(2)} = -\int_{(1)}^{(2)} di = -[i]_{(1)}^{(2)}; \quad [\tfrac{1}{2}u^2 + i]_{(1)}^{(2)} = 0. \tag{21'}$$

Thus relation (21) remains valid across a shock process; the values of i_1 and i_2 refer, however, to different entropies. Once the requirement that the process be adiabatic is given up, equations (19), (20), and (21) are three equations for three unknowns, namely for u_2, ρ_2, s_2 when u_1, ρ_1, s_1 are given.

For an investigation of the transition relation we consider a graphical representation due to A. Busemann.* The method consists in drawing adiabatic curves for the pressure, p, as a function of the velocity, u, with a fixed value of

*A. Busemann: Handbuch der Experimental Physik, Bd. 4, 1. Teil, p. 375.

$q_*^2 = u^2 + 2i$. For each value of the entropy there will correspond a curve $p = p(u)$ touching the u-axis at $u = q_*$. For example, for ideal gases we have the relation $p = \sigma\rho^k$; using the equation

$$\frac{1}{2}u^2 + \frac{k}{k-1}\sigma^{\frac{1}{k}} p^{\frac{k-1}{k}} = \frac{1}{2}q_*^2$$

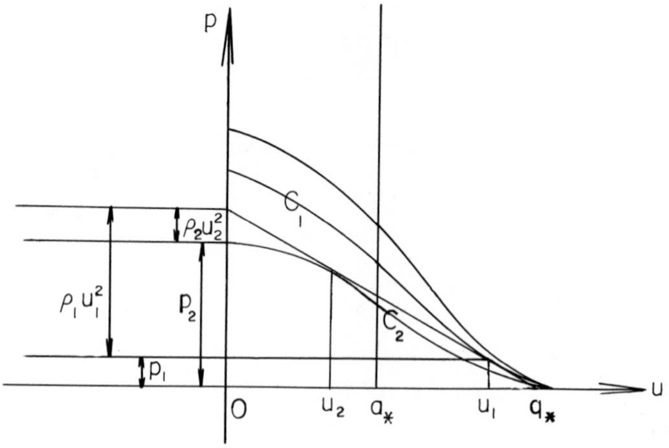

we obtain equation (15):

$$p = (q_*^2 - u^2)^{\frac{k}{k-1}} \left(\frac{k-1}{2k}\right)^{\frac{k}{k-1}} \sigma^{-\frac{1}{k-1}}; \tag{15}$$

this equation, for a fixed value of q_*, gives a set of curves for various values of σ.

Consider a point (u,p) on one of the curves. Since the curves are adiabatic we use relation (20_1) to find the slope of the curve at the point:

$$\frac{dp}{du} = -\rho u.$$

The p-intercept of the tangent line through (u,p) is then

$$p + \rho u^2.$$

Consider now two points (u_1,p_1) and (u_2,p_2) corresponding to both sides of a shock. Condition (21') states that the value of q_* is the same on both sides. Draw the adiabatic curve C_1 through (u_1,p_1) and C_2 through (u_2,p_2) with this value of q_*. (See figure.) By condition (19) the slope of C_1 at (u_1,p_1) must be the same as the slope of C_2 at (u_2,p_2); condition (20) states that the tangent lines at (u_1,p_1) and (u_2,p_2) have the same p-intercept. Hence the tangent lines coincide. According to the second law of thermodynamics the entropy must increase. The point (u_1,p_1) is therefore the one for lower entropy, the point (u_2,p_2) for higher entropy.

By consideration of the curvature of the adiabatic curves it is possible to see whether a point (u_2,p_2) exists when (u_1,p_1) is given. We have

$$d(-\rho u) = -\rho du - ud\rho$$

$$= \frac{dp}{u} - ud\rho$$

$$d(-\rho u) = - ud\rho(1 - \frac{a^2}{u^2}).$$

Since $d\rho < 0$ as u increases we see that the curvature is positive when $u > a$, the acoustic velocity belonging to u, and negative when $u < a$. Hence the curvature of every adiabatic curve for a fixed q_* is positive to the right of the line $u = a_*$, a_* being the critical acoustic velocity, and negative to the left of this line; the curvature vanishes at $u = a_*$ for all the curves, i.e., independently of the entropy.

From the above considerations we see that to every point (u,ρ) with $u > a_*$ one corresponding point (u,ρ) with $u < a_*$ exists and vice versa. From (15) we see that the point with $u < a_*$ lies on the curve with higher entropy. The "shock" is therefore a transition from supersonic velocity to subsonic velocity. Since u decreases across the shock we see from (19) and (20) that the density and pressure increase. The shock is thus a "compression" shock. "Expansion shocks" are excluded by the second law of thermodynamics.

Problem 30. Prove $u_1 u_2 = a_*^2$.

We now turn to a discussion of what occurs inside such a compression shock. This can be done by taking into account heat conduction, and heat generation due to viscosity.

The continuity equation is again

$$(\rho u)_x = 0 \tag{22}$$

or

$$\rho u = m = \text{const.}$$

It was seen (Chapter IV, p. 202) that the viscous stress system is given by

$$(\tau) = 2\mu \text{ Def } \vec{q}.$$

If we consider a motion $u = u(x)$, $v = w = 0$ the stress τ_{xx} reduces to

$$\tau_{xx} = \frac{4}{3}\mu u_x.$$

Hence the Navier-Stokes equation becomes

$$\rho u u_x = -p_x + \frac{4}{3}(\mu u_x)_x. \tag{23}$$

Since $\rho u = \text{const.}$ by (22), integration of (23) yields

$$\rho u^2 + p - \frac{4}{3}\mu u_x = mB = \text{const.} \tag{23'}$$

To obtain the condition of energy balance we observe that the energy flux diminished by the work done by the stresses per unit time per unit area is constant. The energy flux consists of the flux of kinetic energy $\frac{1}{2}mu^2$, the flux of intrinsic energy mh, and the heat transfer which is assumed to be proportional to temperature gradient $-T_x$. This energy flux is to be diminished by $(-p + \frac{4}{3}\mu u_x)u$. Hence we obtain

$$m(\frac{1}{2}u^2 + h) - \lambda T_x - (-p + \frac{4}{3}\mu u_x)u = mA = \text{const.}$$

We now observe that

$$i = h + \rho^{-1}p;$$

further we set

$$i = c_p T, \; c_p \quad \text{constant},$$

which is justified for ideal gases. In addition we assume that

$$\lambda = c_p \mu,$$

a relation that holds approximately for ideal gases. Then we obtain with the help of $i = h + \rho^{-1} p$ and (22')

$$m(\tfrac{1}{2}u^2 + i) - \mu i_x - \tfrac{4}{3}\mu u_x u = mA = \text{const.} \qquad (24')$$

Differentiation of (24') yields

$$m(uu_x + i_x) - (\mu i_x)_x = \tfrac{4}{3} m(\mu u_x u)_x.$$

Taking (22) into account this last equation becomes

$$m(i_x - \rho^{-1} p_x) = (\mu i_x)_x + \tfrac{4}{3} m\mu u_x^2. \qquad (24)$$

In view of $di = \rho^{-1}dp = Tds$, the left member of (24) is the increase of heat per unit volume, per unit time. The right member shows that this consists of the heat obtained by transport and the heat created by viscosity, the latter being positive in accordance with the second law of thermodynamics.

Although equation (24') makes it likely that the heat transfer is of the same order of magnitude as the action of viscosity, we will disregard the heat transfer to simplify the problem. The conditions to be imposed on the problem are

$$u = u_1, \; x = -\infty, \quad u = u_2, \; x = +\infty,$$

and a third condition fixing the origin of the x-axis. From equation (23') we have

$$p = -mu + \tfrac{4}{3}\mu u_x + mB;$$

since

$$i = \alpha\rho^{-1}p = \alpha m^{-1}up, \quad \alpha = \frac{k}{k-1},$$

equation (24') becomes

$$\frac{1}{2}mu^2 + \alpha u\{-mu + \frac{4}{3}\mu u_x + mB\} - \frac{4}{3}\mu u_x u = mA,$$

where we have disregarded the term μi_x referring to heat conduction. A rearrangement of terms yields

$$\frac{4}{3}(\alpha-1)\mu u u_x = (2\alpha-1)(\frac{1}{2}mu^2) - \alpha muB + mA, \tag{25}$$

and we note that

$$\alpha-1 = \frac{1}{k-1}, \quad 2\alpha-1 = \frac{k+1}{k-1}.$$

Since the left member of (25) vanishes for $x = \pm\infty$ the right member can be written

$$\frac{1}{2}(2\alpha-1)m(u-u_1)(u-u_2);$$

the constant A is found to be

$$A = \frac{1}{2}\frac{k+1}{k-1}u_1u_2.$$

Then the equation takes the form

$$\frac{8}{3}\frac{\mu}{m(k+1)}uu_x = (u-u_1)(u-u_2).$$

By integration the solution is found to be

$$\frac{3}{8}\frac{m(k+1)}{\mu}x = \frac{u_1}{u_1-u_2}\log\frac{u_1-u}{u_1-u_2} - \frac{u_2}{u_1-u_2}\log\frac{u-u_2}{u_1-u_2}. \tag{25a}$$

From this solution we see that $u = u_1$, $x = -\infty$ and $u = u_2$, $x = +\infty$

hold, if and only if both u_1, u_2 are of the same sign and $u_1 > u_2$. If u_1, u_2 are positive this is seen immediately from (25a), while for u_1, u_2 negative the fact that $m = \rho u$ is negative is to be taken into account.

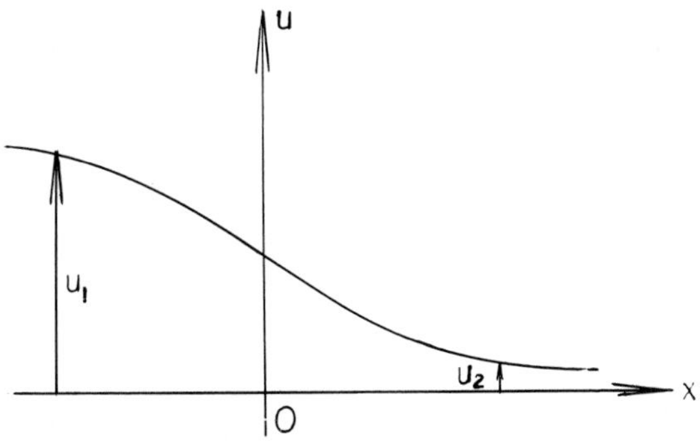

We now see that transitions as discussed in the first part of this section are possible in accordance with the relations (19), (20), and (21). The viscosity does not enter these relations; the influence of viscosity is only to determine the region in which the deviation of the velocity from u_1 and u_2 is appreciable, i.e., the transition region. When $\mu \to 0$ we have $u \to u_1$ for $x < 0$ and $u \to u_2$ for $x > 0$; hence the transition region shrinks to zero.

The latter remark indicates that the occurrence of shocks is nothing but a boundary layer phenomenon. When one considers viscosity throughout the field and then lets it approach zero, the solutions will approach that of a perfect adiabatic flow except in certain regions where the convergence is non-uniform and the limit solutions are discontinuous. For compressible fluids this may occur not only at walls and in wakes but also in the midst of the field of flow. Shocks are just such interior boundary layers.

5. Non-Steady One-Dimensional Flow

For a non-steady, one-dimensional flow we assume $u = u(t,x)$, $v = w = 0$, $\rho = \rho(t,x)$ and an adiabatic relation $p = p(\rho)$ between the pressure and density. Here the equations of motion and continuity equation take the form

D.E.:

$$\begin{aligned} \rho(u_t + uu_x) &= -\frac{dp}{d\rho}\rho_x = -a^2\rho_x \\ \rho_t + u\rho_x + \rho u_x &= 0 \end{aligned} \tag{26}$$

subject to the initial conditions

$$\text{I.C.:} \qquad u = u(x), \quad \rho = \rho(x) \quad \text{for} \quad t = 0 \tag{26'}$$

where $u(x)$ and $\rho(x)$ are given distributions.

First we consider only small disturbances, i.e., replace u, ρ, and a by $U+u$, $\bar{\rho}+\rho$, and $A+a$, where U, $\bar{\rho}$, and A are constants and u, ρ, and a are small. In addition terms of second order will be neglected. Then the differential equations (26) can be written

$$\begin{aligned} \bar{\rho}(u_t + Uu_x) &= -A^2\rho_x \\ \rho_t + U\rho_x + \bar{\rho}u_x &= 0, \end{aligned} \tag{27}$$

or in matrix form

$$\begin{pmatrix} \frac{\partial}{\partial t} + U\frac{\partial}{\partial x} & \bar{\rho}^{-1}A^2\frac{\partial}{\partial x} \\ \bar{\rho}\frac{\partial}{\partial x} & \frac{\partial}{\partial t} + U\frac{\partial}{\partial x} \end{pmatrix} \begin{pmatrix} u \\ \rho \end{pmatrix} = 0.$$

Hence we can eliminate either ρ or u and obtain

$$\left(\left(\frac{\partial}{\partial t} + U\frac{\partial}{\partial x}\right)^2 - \left(A\frac{\partial}{\partial x}\right)^2 \right) \begin{pmatrix} u \\ \rho \end{pmatrix} = 0. \tag{27a}$$

This equation is nothing but the "wave equation"

$$\left(\frac{\partial^2}{\partial t^2} - A^2\frac{\partial^2}{\partial \xi^2} \right) \begin{pmatrix} u \\ \rho \end{pmatrix} = 0$$

after x has been replaced by $x = \xi + Ut$. The solution of this equation is of the form

$$\begin{aligned} \rho &= \bar{\rho}\{f(x - Ut - At) + g(x - Ut + At)\}, \\ u &= A\{f(x - Ut - At) - g(x - Ut + At)\}, \end{aligned}$$

as is easily verified from (27). The initial conditions reduce to the equations

$$\rho(x) = \mathfrak{p}\{f(x) + g(x)\}$$

$$u(x) = A\{f(x) - g(x)\}$$

from which $f(x)$ and $g(x)$ can be determined. The characteristic lines of equation (27a) are

$$x - (U + A)t = \text{const.}$$

$$x - (U - A)t = \text{const.}$$

When $\rho(x)$, $u(x)$ represent "initial disturbances", i.e., differ from zero only on a small segment, the solution represents a propagation of this initial disturbance with velocities $U \pm A$.

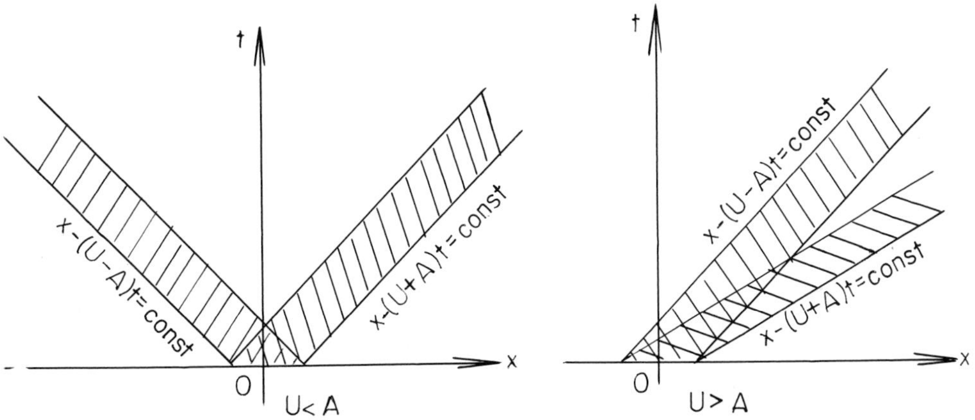

When $U < A$ one set of characteristics has positive slope, the other negative and the disturbance spreads in both the positive and negative x-direction, when $U > A$ both sets of characteristics have positive slope and the disturbance "runs forward". In the latter case no effect will be notices at any point behind the initial disturbance, (see figure). For example, in the supersonic flow out of a container considered in Section 3, a disturbance outside the tube will travel only in the direction of the flow and no effect will be noticed inside the tube; this explains why changes of the outside pressure do not affect the flow inside the tube (up to the shock).

We now pass over to Riemann's treatment of the problem. For simplification we introduce the quantity j,

$$j = \int_0^\rho a\rho^{-1} d\rho, \quad dj = a\rho^{-1} d\rho.$$

The differential equations take the form

$$\text{D.E.:} \quad \begin{aligned} u_t + uu_x + aj_x &= 0 \\ j_t + uj_x + au_x &= 0, \end{aligned} \tag{28}$$

and the initial conditions are

$$\text{I.C.:} \quad u = u(x), \quad j = j(x) \quad \text{for} \quad t = 0. \tag{28'}$$

The differential equations can now be transformed into linear ones by introducing u and j as new independent variables, and x and t as new dependent variables. We have

$$\begin{pmatrix} t_u & t_j \\ x_u & x_j \end{pmatrix} = \begin{vmatrix} u_t & u_x \\ j_t & j_x \end{vmatrix}^{-1} \begin{pmatrix} j_x & -u_x \\ -j_t & u_t \end{pmatrix},$$

and the equations (28) can be written

$$\text{D.E:} \quad \begin{aligned} x_j &= ut_j - at_u \\ x_u &= ut_u - at_j \quad . \end{aligned} \tag{28a}$$

The initial conditions change over to having $x = \xi$ and $t = 0$ on a curve $u = u(\xi)$, $j = j(\xi)$. For the investigation of the characteristics it is convenient to introduce the new variables α, β where

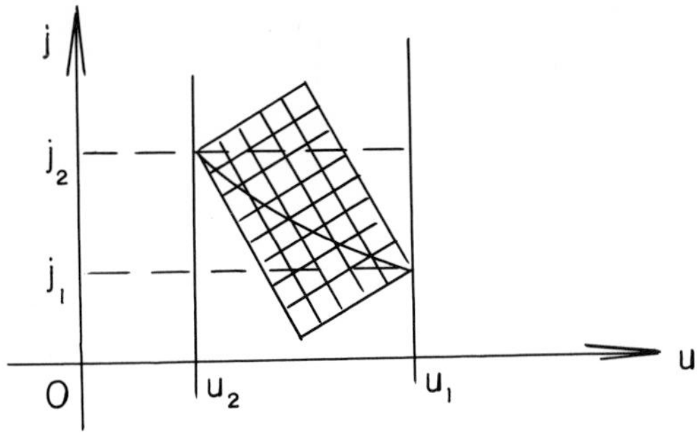

$$\alpha = u + j, \quad \beta = u - j.$$

Then we have

$$x_\alpha = (u - a)t_\alpha$$

$$x_\beta = (u + a)t_\beta.$$

This form of the equations shows that the characteristics for (28a) are given by

$$\alpha = \text{const.}, \quad \beta = \text{const.},$$

or

$$u + j = \text{const.}, \quad u - j = \text{const.}$$

The solution can now be continued into the rectangle obtained by drawing characteristics through the end points of the initial curve.

If we now try to reintroduce t and x as variables. We know that this can be done only in a region where the determinant

$$\begin{vmatrix} x_j & t_j \\ x_u & t_u \end{vmatrix} = a(t_j^{\,2} - t_u^{\,2}) = -4at_\alpha t_\beta$$

does not vanish.

Assume that $t_\alpha = 0$ and hence $x_\alpha = 0$ along some curve. Then the figure shows that the mapping overlaps; the image of $t_\alpha = 0$, $x_\alpha = 0$ is an

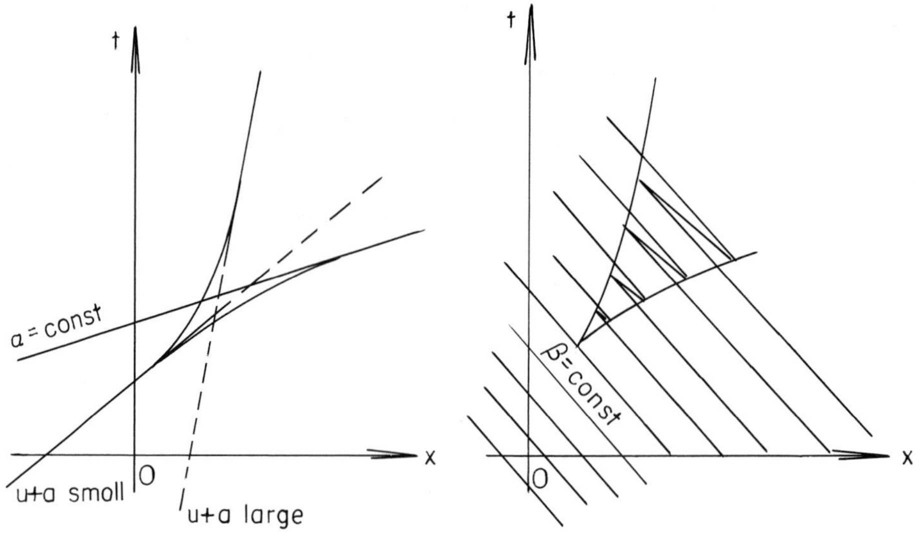

envelope for the characteristics α = const. The characteristics β = const. coil back in the overlapped region. Hence the procedure does not give a solution in the overlapped region beginning at the cusp of the envelope.

Actually the difficulty is overcome by a "shock"; along the shock characteristics of the same kind meet. The propagation of the shock is faster than sound when observed from one side of the discontinuity line and less than that of sound when observed from the other side. Hence for an observer the velocity of the shock will first be supersonic until the shock passes and then it will be subsonic.

6. Two-Dimensional, Steady, Adiabatic Flow

This type of flow is described by the velocity $\vec{q} = \vec{q}(x,y)$ and the density $\rho = \rho(x,y)$; the pressure, p and heat content, i are functions of ρ.

We now consider the equations of motion, continuity equation, Bernoulli law and stream function as was done in two-dimensional incompressible flow except that now we take into account the fact that ρ is a function of x and y instead of constant. (Compare Chapter II, Sections 1 and 2.)

The continuity equation

$$(\rho u)_x + (\rho v)_y = 0 \tag{29}$$

can be solved by the stream function ψ, when

$$\rho u = \psi_y, \quad \rho v = -\psi_x. \tag{30}$$

The difference $\psi(x,y) - \psi(x_0,y_0)$ is the flux of mass per unit breadth across a line joining (x_0,y_0) with (x,y); in fact, we have

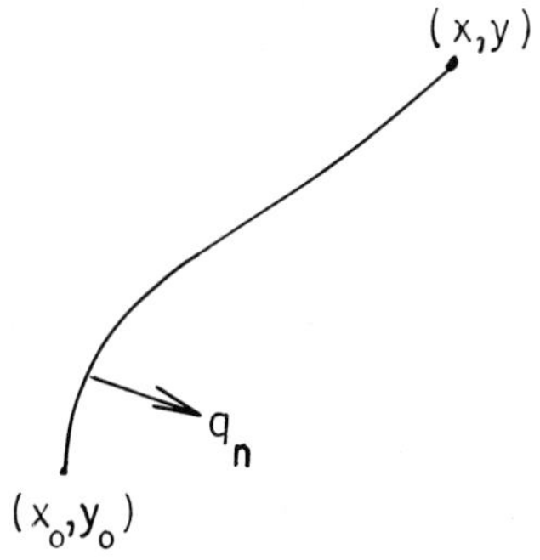

$$\psi(x,y)-\psi(x_0,y_0) = \int_{(x_0,y_0)}^{(x,y)} \rho q_n ds.$$

The Euler equations

$$\begin{aligned} uu_x + vu_y &= \rho^{-1}p_x = -i_x \\ uv_x + vv_y &= -i. \end{aligned} \tag{31}$$

can be written

$$\begin{aligned} [\tfrac{1}{2}(u^2 + v^2) + i]_x &= v\omega \\ [\tfrac{1}{2}(u^2 + v^2) + i]_y &= -u\omega \end{aligned} \quad , \tag{31a}$$

where

$$\omega = v_x - u_y \tag{32}$$

is the vorticity. If the right members of (31a) are written $(\rho v)(\rho^{-1}\omega)$ and $(-\rho u)(\rho^{-1}\omega)$ and the left sides eliminated by differentiation with respect to y and x respectively, we obtain

$$u(\rho^{-1}\omega)_x + v(\rho^{-1}\omega)_y = 0, \tag{33}$$

where equation (29) has been used. From (33) we see that $\rho^{-1}\omega$ is constant along a streamline. If $\omega = 0$ at the beginning of all streamlines, the flow will be irrotational throughout.

For irrotational flow we have $\omega = 0$ and equation (31) yields Bernoulli's equation

$$\frac{1}{2}(u^2 + v^2) + i = i_* = \text{const.} \tag{34}$$

throughout the field of flow. Now we may introduce a potential function ϕ with

$$u = \phi_x, \quad v = \phi_y. \tag{35}$$

Insertion of (35) and (29) yields

$$(\rho\phi_x)_x + (\rho\phi_y)_y = 0. \tag{36}$$

Differentiation of (34) along a streamline yields

$$\phi_x{}^2\phi_{xx} + 2\phi_x\phi_y\phi_{xy} + \phi_y{}^2\phi_{yy} + a^2\rho^{-1}(\rho_x\phi_x + \rho_y\phi_y) = 0.$$

This equation in combination with (36) yields

$$a^2(\phi_{xx} + \phi_{yy}) = \phi_x{}^2\phi_{xx} + 2\phi_x\phi_y\phi_{xy} + \phi_y{}^2\phi_{yy}, \tag{37}$$

which replaces the simple equation

$$\Delta\phi = 0$$

which held for incompressible fluid. From (34) we see that the quantity a^2 is to be considered as a function of $\phi_x{}^2 + \phi_y{}^2$; for ideal gases we have

$$a^2 = (k-1)\{i_* - \frac{1}{2}(\phi_x{}^2 + \phi_y{}^2)\}.$$

If a^2 is large compared with $\phi_x{}^2 + \phi_y{}^2$, i.e., the fluid behaves like an incompressible fluid, then (37) can be approximated by the equation $\Delta\phi = 0$. On the other hand if $|\vec{q}|$ is of the order of magnitude of a then both sides of (37) are of the same order of magnitude.

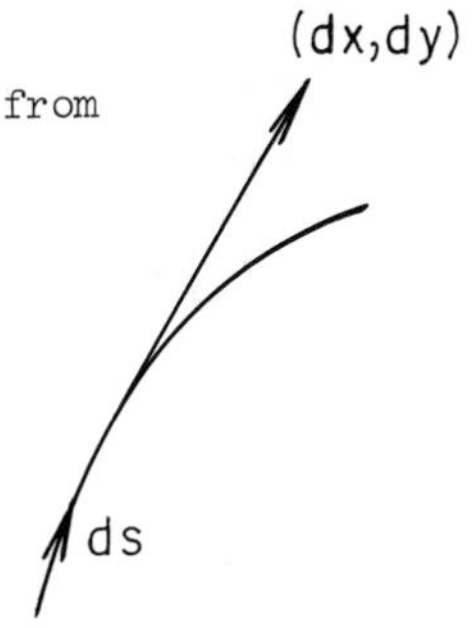

The characteristics of (37) are determined from

$$a^2(dx^2 + dy^2) = (\phi_x dy - \phi_y dx)^2$$

or

$$u\frac{dy}{ds} - v\frac{dx}{ds} = \pm a;$$

that is, the normal component of velocity in

$$q_n = \pm a. \tag{38}$$

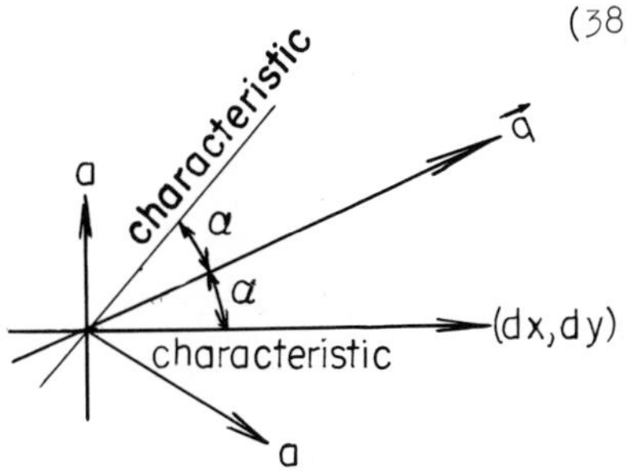

If α is the angle made by the characteristic with the direction $\vec{q}$, (called "Mach's angle"), we have

$$\sin\alpha = \frac{a}{|\vec{q}|};$$

hence the characteristic directions are real only where

$$|\vec{q}| > a.$$

Upon introducing "Mach's number"

$$M = \frac{|\vec{q}|}{a} = \frac{1}{\sin\alpha} \tag{39}$$

the condition for real characteristics becomes

$$M > 1. \tag{40}$$

The flow is called supersonic when $M > 1$, subsonic when $M < 1$; the differential equation is hyperbolic in the former case, elliptic in the latter.

7. Nearly Constant Parallel Flow

We consider a flow which differs only slightly from the "main flow" $u = U$, $v = w = 0$, $\rho = \mathfrak{p}$, $p = P$, $i = I$, where $U, \mathfrak{p}, P$ and I are constants. We require that $u = U_1$, $v = 0$, $\rho = \mathfrak{p}$, for $x = -\infty$, and impose possibly other conditions at infinity.

The flow may further be subjected to the boundary conditions:

Type I:

$$\psi = 0 \quad \text{on} \quad y = h(x),\ h(x) = 0 \quad \text{for} \quad x < 0$$

$$\text{and} \quad h(x) \quad \text{is small for} \quad x \geq 0.$$

Instead of these boundary conditions, the following ones may be prescribed.

Type II:

$$\psi = 0 \quad \text{on} \quad y = h_1(x) \quad \text{and} \quad y = h_2(x) \quad \text{for} \quad |x| \leq c/2,$$

$$\text{where} \quad h_1(x) \leq h_2(x),\ |x| \leq c/2 \quad \text{and}$$

$$h_1(\tfrac{c}{2}) = h_2(\tfrac{c}{2}),\ h_1(-\tfrac{c}{2}) = h_2(-\tfrac{c}{2}).$$

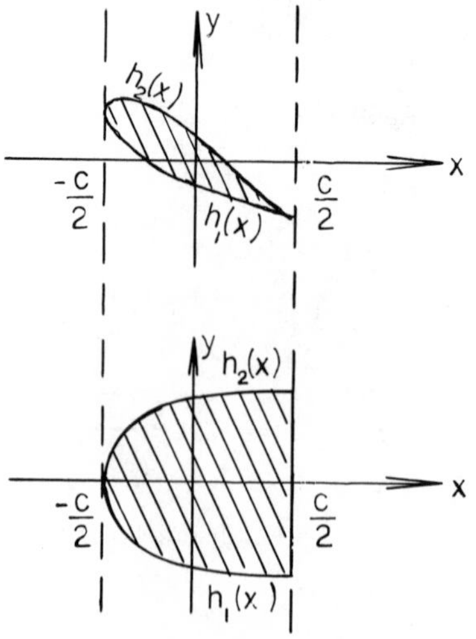

The flow is then defined outside the region enclosed by the two curves $y = h_1(x)$, $y = h_2(x)$. In the conditions of type I the function $h(x)$ may represent a roughness of a wall, while in type II the functions $y = h_1(x)$, $y = h_2(x)$ may be the boundary of a wing profile or a projectile.

We replace the quantities

$$u,\ v,\ \phi,\ \psi,\ \rho,\ p,\ i$$

by

$$U + u,\ v,\ Ux + \phi,\ \rho Uy + \psi,\ \mathfrak{p}+\rho,\ P + A^2\rho,\ I + \mathfrak{p}^{-1}A^2\rho$$

respectively, where the last two quantities are obtained from the expansions

$$p = p(\rho) = P + (\frac{dp}{d\rho})_\rho \rho + \cdots = P + A^2\rho + \cdots$$

$$i = i(\rho) = I + (\frac{di}{d\rho})_\rho \rho + \cdots = I + \mathfrak{p}^{-1}A^2\rho + \cdots$$

and where $A = a(\mathfrak{p})$ is the acoustic velocity of the main flow.

Terms of higher order than the first are neglected. Then the equations derived in Section 6 simplify considerably. Equations (30), (34), (35) and (37) become

$$\psi_y = \mathfrak{p}u + U\rho,\ \psi_x = -\mathfrak{p}v; \tag{41}$$

$$Uu + \mathfrak{p}^{-1}A^2\rho = 0 \tag{42}$$

$$u = \phi_x,\ v = \phi_y \tag{43}$$

and

$$A^2(\phi_{xx} + \phi_{yy}) = U^2\phi_{xx} \tag{44}$$

respectively. If $M = U/A$ is the Mach number for the main flow then (44) is elliptic when $M < 1$ and hyperbolic when $M > 1$.

Equation (42) can be written

$$\rho = -Uu\mathfrak{p}A^{-2}; \tag{45}$$

insertion of this value of ρ in (41) yields

$$\psi_y = \mathfrak{p}u(1-M^2),\ \psi_x = -\mathfrak{p}v. \tag{46}$$

To be consistent terms of order higher than the first should be neglected in the boundary conditions also. After the above replacements are made the boundary conditions of the first type take the form

$$\mathfrak{p}Uy + \psi = 0 \quad \text{for} \quad y = h(x),$$

that is,

$$\mathfrak{p}Uh(x) + \psi(x,h(x)) = 0.$$

The first order terms of this last expression yield

$$\mathfrak{p}Uh(x) + \psi(x,0) = 0$$

or

$$\psi(x,0) = -\mathfrak{p}Uh(x). \tag{47}$$

The flow is therefore defined for $y \geq 0$ throughout. Similarly, the boundary conditions of the second type reduce to, when first order terms are taken

$$\begin{aligned} \psi(x,0) &= -\mathfrak{p}Uh_2(x), \quad |x| \leq \tfrac{c}{2}, \quad y \downarrow 0 \\ &= -\mathfrak{p}Uh_1(x), \quad |x| \leq \tfrac{c}{2}, \quad y \uparrow 0. \end{aligned} \tag{48}$$

The flow is now defined in the whole plane but has a discontinuity along $y = 0$, $|x| \leq \frac{c}{2}$.

We now take up separately the cases where the Mach number is less than one and greater than one.

Case I. $M < 1$. In this case, where the differential equation (44) is elliptic we may complete the boundary conditions by requiring that Φ, ψ, u, v, ρ become zero as $x^2 + y^2 \to \infty$. Equation (44) can be reduced easily to the potential equation by the transformation

$$x' = x, \quad y' = y(1-M^2)^{1/2}$$

$$\phi' = \phi, \quad \psi' = \rho^{-1}\psi/(1-M^2)^{1/2} \tag{49}$$

$$u' = u, \quad v' = v/(1-M^2)^{1/2}.$$

Then equations (46), (43) and (44) become

$$u' = \psi'_{y'}, \quad v' = -\psi'_{x'} \tag{46'}$$

$$u' = \phi'_{x'}, \quad v' = \phi'_{y'} \tag{43'}$$

$$\phi'_{x'x'} + \phi'_{y'y'} = 0 \tag{44'}$$

respectively. Hence the problem is transformed to that of a potential flow.

We apply this transformation to investigate the influence of compressibility on the lift of an airfoil. The boundary conditions (48) correspond to the "thin wing" method (cf. Chap. II, page 110), replacing the flow around the airfoil with chord c and angle of attack β by the flow which on the segment

$$y = 0, \quad -\frac{c}{2} < x < c/2$$

has a vortex distribution such that the vertical component of the velocity is continuous and given by

$$\frac{v}{U} = -\beta(\approx - \tan\beta).$$

When we set

$$h_1(x) = h_2(x) = -\beta x, \ |x| \le \frac{c}{2},$$

the boundary conditions (48) become

$$\psi = -\rho U\beta x, \ |x| \le c/2, \ y = 0,$$

which is equivalent to

$$\frac{v}{U} = -\frac{\psi_x}{\rho U} = \beta, \ |x| \le \frac{c}{2}, \ y = 0.$$

Under the transformation (49) the boundary conditions (48) go over into

$$\psi'(x,0) = U\beta' x, \quad |x| \leq \frac{c}{2}$$

with

$$\beta' = \beta/(1 - M^2)^{1/2}.$$

The irrotational flow with

$$Ux + \phi', \quad Uy' + \psi'$$

as potential and stream functions then corresponds to the flow around an airfoil with the angle of attack β' and the chord $c' = c$.

The circulation of such a potential flow by the Kutta condition is

$$\Gamma' = 2\pi c\beta'.$$

Since this circulation Γ' is the jump of the potential on encircling the segment (now taken clockwise) and since $\phi' = \phi$, we have $\Gamma' = \Gamma$; hence

$$\Gamma = 2\pi c\beta/(1-M^2)^{1/2}. \tag{50}$$

We see that the influence of the compressibility is to increase the circulation.

We now wish to investigate how the lift depends upon the circulation when the compressibility is taken into account. We take a large circle C about the profile and determine the lift per unit breadth, $\mathcal{L}$, as

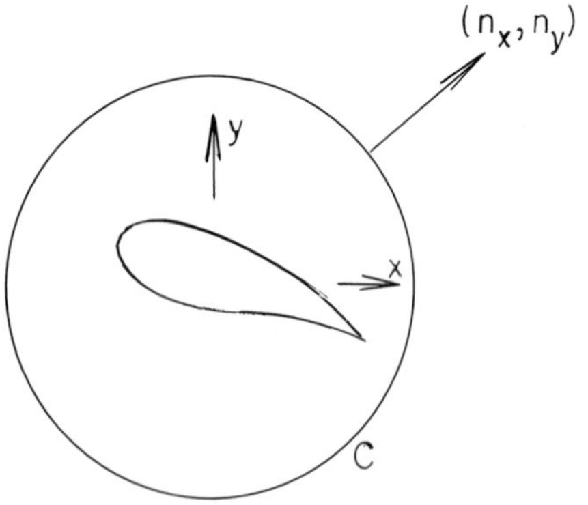

$$\mathcal{L} = -\oint_C (P+p)y_n ds - \oint_C (p+\rho)q_n v ds.$$

If the first order approximation is taken we find

$$\mathcal{L} = -\oint_C A^2 \rho n_y ds - \oint \mathfrak{p} U v n_x ds.$$

Application of (45) yields

$$\mathcal{L} = \mathfrak{p} U \oint_C [u n_y - v n_x] ds = -\mathfrak{p} U \oint_C (udx + vdy).$$

However the expression under the integral sign is precisely $(-d\Gamma)$, and we find

$$\mathcal{L} = \mathfrak{p} U \Gamma.$$

This means that the Kutta-Joukowski condition holds for compressible fluids as well as for incompressible fluids when the flow is subsonic and a parallel flow at infinity. The formula for the lift now becomes

$$\mathcal{L} = 2\pi c U \, \mathfrak{p} \beta / (1-M^2)^{1/2}. \tag{51}$$

This "Prandtl-Glauert" formula shows that the influence of compressibility is to increase the lift. However, it is valid only when the Mach number is not too large, say below 0.5 or 0.6. For higher values of M the above approximation fails.

Case II. $M > 1$. In this case $U > A$ and we have supersonic flow; the differential equation is hyperbolic and boundary conditions of a different kind must be prescribed.

To the differential equations

$$\begin{aligned} u &= \phi_x, \quad v = \phi_y \\ \mathfrak{p}(1-M^2)u &= \psi_y, \quad -\mathfrak{p} v = \psi_x \end{aligned} \tag{52}$$

we impose the boundary conditions

$$\phi, \psi, u, v, \rho \to 0 \quad \text{as} \quad x^2 + y^2 \to \infty \text{ provided that } x \le x_0 \text{ for every } x_0. \tag{52'}$$

We cannot prescribe what happens for $x \to +\infty$.

We introduce "Mach's angle", α, for the main flow; it is defined by the relation

$$\sin \alpha = \frac{1}{M} \quad \text{where} \quad M = \frac{U}{A} .$$

The differential equations (52) can be solved explicitly

$$\begin{aligned} \mathfrak{p}^{-1}\psi &= F(x-y \cot \alpha) + G(x+y \cot \alpha) \\ \phi &= \tan \alpha\{F(x-y \cot \alpha) - G(x+y \cot \alpha)\}. \end{aligned} \tag{53}$$

The characteristic lines of (52) are

$$y = -x \tan \alpha$$

$$y = x \tan \alpha.$$

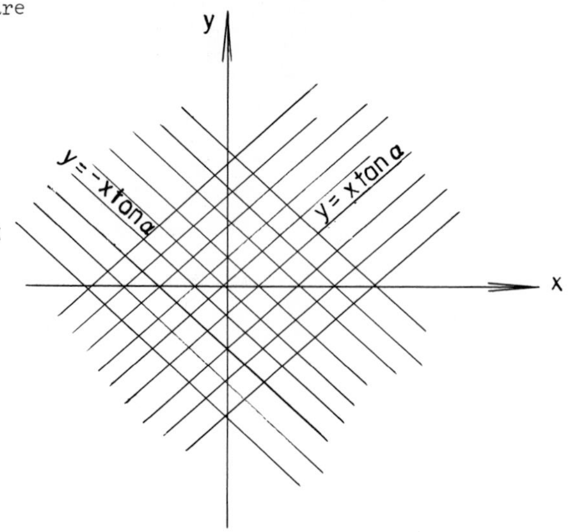

The function F is constant along the characteristics $y = x \tan \alpha$, while G is constant along $y = -x \tan \alpha$.

Now assume that we have prescribed boundary conditions of the first type, i.e.,

$$\psi(x,0) = -\mathfrak{p}Uh(x), \; y = 0, \; h(x) = 0, \; x < 0,$$

and also that the flow for $y > 0$ only is desired. Then, since $G = \text{const.}$ along $y = -x \tan \alpha$ and since G approaches zero as we go to infinity along any of these characteristics, we see that $G \equiv 0$. We then have

$$F(x) = -Uh(x),$$

and the solution of the problem is

$$\psi = -\rho U h(x-y\cot\alpha) \qquad u = -U\tan\alpha\, h'(x-y\cot\alpha)$$
$$\phi = -U\tan\alpha\, h(x-y\cot\alpha) \qquad v = Uh'(x-y\cot\alpha). \tag{54}$$

The flow is constant along the characteristic lines

$$y = x\tan\alpha$$

and consequently the deviations ϕ, ψ, u, v from the main flow do not die out everywhere at infinity. We see that the flow will be unsymmetrical even though the "roughness" function $h(x)$ is symmetrical. Hence there will be a resistance in addition to the ordinary resistences due to skin friction, etc., due to the fact that the flow is supersonic. This is the so-called "wave-resistance".

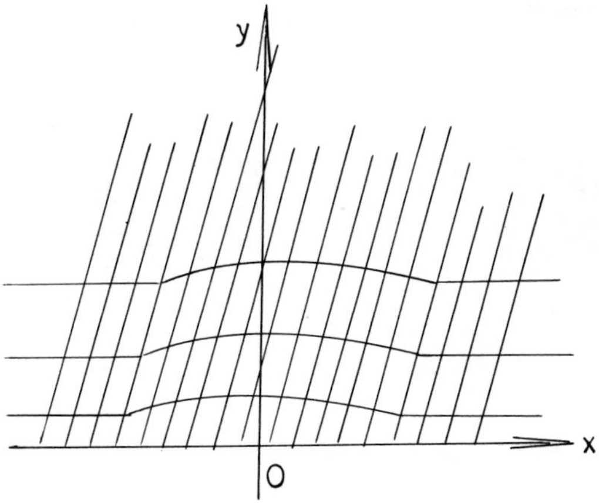

We now calculate the "wave drag" per unit breadth, D. This drag is given by

$$D = -\int_{y=h(x)} (P+p)\, n_x\, ds = \int_{-\infty}^{\infty} (P+p(x,h(x)))\, h'(x)\, dx.$$

In our first approximation, with $p = P + A^2\rho$, we have

$$D = A^2 \int_{-\infty}^{\infty} \rho h'(x)\, dx = -\rho U \int_{-\infty}^{\infty} u(x,0)\, h'(x)\, dx$$
$$= \rho U^2 \tan\alpha \int_{-\infty}^{\infty} [h'(x)]^2 dx \tag{55}$$

where $\rho = -\rho U u A^{-2}$ and $u = -U\tan\alpha h'(x-y\cot\alpha)$ have been used. Equation (55) shows that the "wave drag" is proportional to the integral of the square of the slope of our "roughness function" $h(x)$. This indicates that this drag can be kept small by keeping the slope of $h(x)$ small.

As a special case we now consider briefly the solution for the flow past a wedge. Here the function $h(x)$ is given by

$$h(x) = \gamma x, \quad x > 0$$
$$= 0, \quad x < 0.$$

The quantity γ, the angle made by the wedge with the x-axis is considered small so that $\gamma \approx \tan \gamma$.

Then the solution is

$$u = \begin{cases} -U\gamma \tan \alpha & \text{for } x > y \cot \alpha \\ 0 & \text{for } x < y \cot \alpha \end{cases}$$

$$v = \begin{cases} U\gamma & \text{for } x > y \cot \alpha \\ 0 & \text{for } x < y \cot \alpha. \end{cases}$$

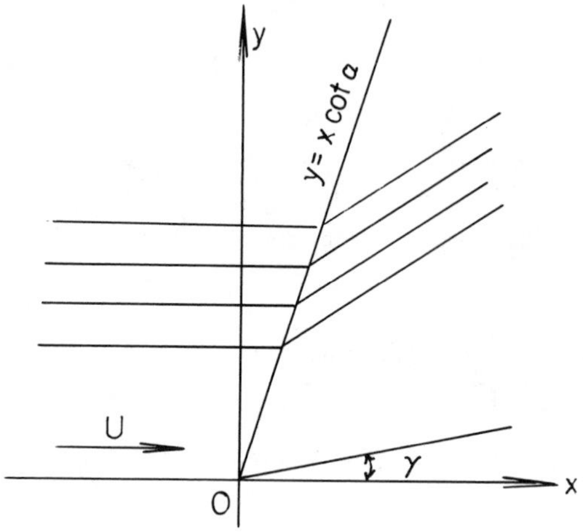

From the form of solution it is obvious that we must have $\gamma < \alpha$ in order for the solution to have any meaning. The flow possesses a discontinuity along the line $y = x \cot \alpha$.

The preceding treatment, which consists in linearizing the problem by considering only terms of the first order, can be extended to the case of three-dimensional flow around a body of revolution. The differential equation then reduces to the wave equation where the axis of the body corresponds to the time axis; then a wave resistance can be calculated. This treatment with particular reference to projectiles has been carried out by v. Kármán and Moore.*

We now consider boundary conditions of type II which refer to a flow around a contour, e.g., a wing profile. We then have different representations of the solution for $y > 0$ and for $y < 0$. While for $y > 0$ we have $G \equiv 0$ and

*Trans. of the Amer. Soc. of Mech. Eng., APM-54-27, p. 303-310.

$$F(x) = \begin{cases} -Uh_2(x), & |x| \leq c/2 \\ 0 & , |x| > c/2, \end{cases}$$

for $y < 0$ we have $F \equiv 0$ and

$$G(x) = \begin{cases} -Uh_1(x), & |x| \leq c/2 \\ 0 & , |x| > c/2. \end{cases}$$

In formulas (54) the signs of ϕ and u are to be reversed for $y < 0$.

The drag, (i.e., the x-component of the force), per unit breadth can be calculated as before and is

$$D = \rho U^2 \tan\alpha \int_{-c/2}^{c/2} [(h_1'(x))^2 + (h_2'(x))^2]\, x. \qquad (56)$$

To obtain the lift, (i.e., the y-component of the force), per unit breadth, $\mathcal{L}$, we note that

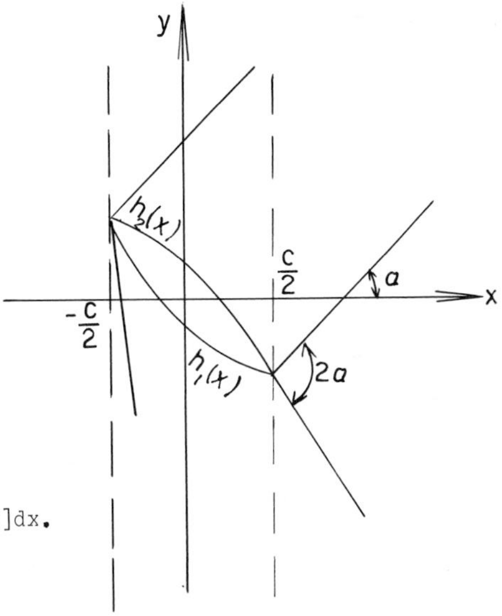

$$\mathcal{L} = -\int_{y=h_2(x)} (P+p)n_y\, ds + \int_{y=h_1(x)} (P+p)n_y\, ds$$

$$= -\int_{-c/2}^{c/2} [(P+p(x,h_2(x)) - (P+p(x,h_1(x))]dx.$$

The first approximation yields

$$\mathcal{L} = -A^2 \int_{-c/2}^{c/2} [\rho_2(x,0) - \rho_1(x,0)]dx = \rho U \int_{-c/2}^{c/2} [u_2(x,0) - u_1(x,0)]dx$$

where u_1, u_2 denote the values of u for $y < 0$ and $y > 0$ respectively. If the value of u given in (54) is inserted in this last formula we find

$$\mathcal{L} = -\rho U^2 \tan\alpha \int_{-c/2}^{c/2} [h_2'(x) + h_1'(x)]dx$$

$$= 2\rho U^2 \tan\alpha\beta c \tag{57}$$

where

$$\beta = \frac{1}{c}[h_1(-\frac{c}{2}) - h_1(\frac{c}{2})] = \frac{1}{c}[h_2(-\frac{c}{2}) - h_1(\frac{c}{2})]$$

is the average angle of attack.

Some interesting remarks can be made concerning formulas (56) and (57). In designing an airwing for supersonic flow we see that the wave drag is proportional to h'^2. Hence it would be preferable to have a sharp leading edge with a small slope than a rounded one with a large slope. The lift is proportional to the average angle of attack and has been determined independently of the circulation. We note the fact that formula (57) states that as Mach's angle α approaches $\frac{\pi}{2}$; i.e., as the velocity $U \to A$ we get an infinite lift. This indicates that the formula for $\mathcal{L}$ is not valid for values of U only slightly above the critical acoustic velocity. Hence we see that the formulas for lift we derived in both the supersonic and subsonic cases fail to hold for velocities near the critical acoustic velocity.

8. Flow In and Around Corners. Oblique Shocks

We now take up Prandtl's solution for the flow in and around a corner. Suppose that a wall is given by

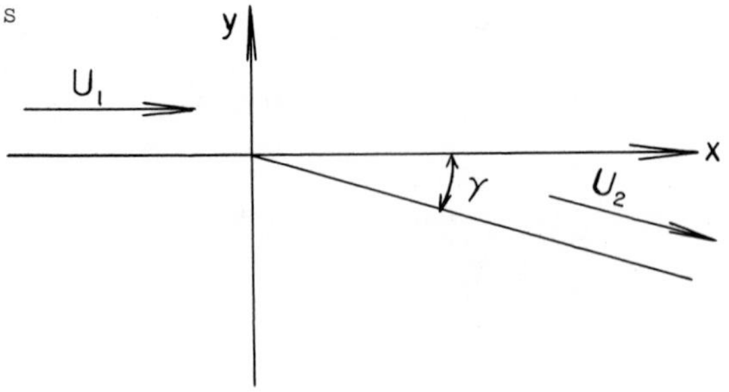

$$y = 0 \text{ for } x < 0$$

$$y = -(\tan\gamma)x \text{ for } x > 0,$$

and in addition the flow at $-\infty$ is given by U_1, while the flow at $+\infty$ is of constant magnitude U_2 and has the direction $-\gamma$. For the solution of this problem it is convenient to use polar coordinates r, θ. Let θ = const. be the family of rays through the origin, and if P is any point let q_t, q_n denote the tangential and normal components of the velocity $\vec{q}$ with respect to the ray containing P. Then the differential equations for the flow can be written

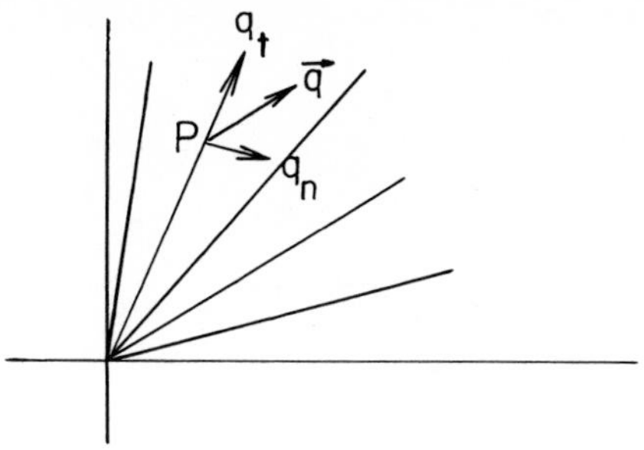

$$q_t = \phi_r, \quad q_n = -r^{-1}\phi_\theta,$$
$$\rho q_t = r^{-1}\psi_\theta, \quad \rho q_n = \psi_r. \tag{58}$$

Bernoulli's condition becomes

$$\tfrac{1}{2}(q_t^2 + q_n^2) + i = \tfrac{1}{2}q_*^2 = \text{const.} \tag{59}$$

where

$$\rho di = a^2 d\rho. \tag{60}$$

The assumption about the flow is that q_t, q_n and ρ depend on θ only. To obtain the solution we set

$$\phi = rq_t, \quad \psi = r\rho q_n; \tag{61}$$

then two of the equations (58) are satisfied. The other two equations give

$$q_n = -q_{t/\theta} \tag{62}$$

and

$$q_t = \rho^{-1}(\rho q_n)_\theta = q_{n/\theta} + q_n \rho^{-1} \rho_\theta$$

$$= q_{n/\theta} + q_n a^{-2} i_\theta$$

$$= q_{n/\theta} - q_n a^{-2}(q_n q_{n/\theta} + q_t q_{t/\theta})$$

or

$$(1 - q_n^2 a^{-2}) q_t = (1 - q_n^2 a^{-2}) q_{n/\theta},$$

where (59) and (60) have been used.

In case $q_t = q_{n/\theta}$ we would have

$$q_t = U \cos(\theta - \theta_0),\ q_n = U \sin(\theta - \theta_0),$$

$$u = U \cos\theta_0,\quad v = U \sin\theta_0.$$

The flow would be a parallel flow with $i = \text{const}$. Therefore we choose the other possibility

$$q_n = a. \tag{63}$$

The function $q_t(\theta)$ is then to be determined from

$$q_t^2 + q_{t/n}^2 + 2I(q_{t/n}) = q_*^2$$

where I is such a function that $i = I(a)$. For ideal gases we have

$$i = \frac{1}{k-1} a^2.$$

When we set

$$m = \left(\frac{k-1}{k+1}\right)^{1/2},$$

(for air, $k = 1.4$ and $m = 0.4$), the differential equation for q_t is

$$q_t^2 + m^{-2} q_{t/n}^2 = q_*^2$$

with the solution

$$\left.\begin{aligned} q_t &= q_* \cos m(\theta+\theta_*), \\ &\text{whence} \\ q_n &= mq_* \sin m(\theta+\theta_*). \end{aligned}\right\} \quad (64)$$

Since

$$a^2 = k\sigma\rho^{k-1}$$

we see from (61) that the streamlines are the curves

$$ra^{\frac{1}{m^2}} = \text{const.}$$

Hence when θ decreases ρ decreases and q increases; the solution represents an "expansion".

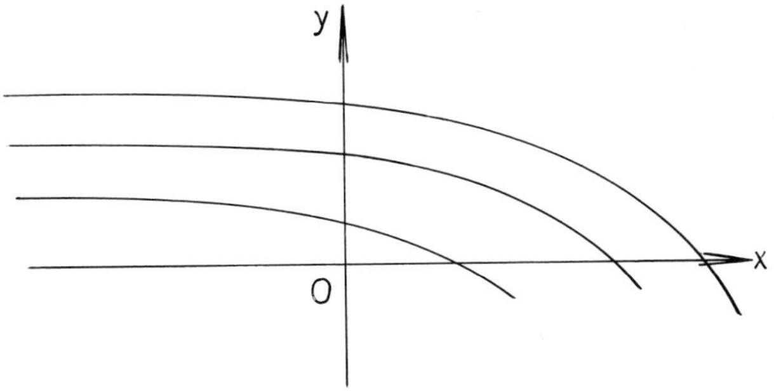

Since, from (63), the normal component of velocity equals the acoustic velocity, the angle the ray makes with the velocity is always just Mach's angle α. Hence the rays are characteristics. The other set of characteristic curves could be determined.

<u>Problem 31</u>: Prove that the other set of characteristics is given by

$$r^2 = \text{const.}\ q_t^{-1}\, q_n^{-\frac{1}{m^2}}.$$

To use the preceding solution for the flow around a corner one has to determine two angles θ_1 and θ_2 such that the flow is given by

$$u = U_1, \; v = 0$$

for $\theta_1 \leq \theta \leq \pi$, and by

$$u = -U_2 \sin \gamma, v = U_2 \cos \gamma$$

for $-\gamma \leq \theta \leq \theta_2$. The angle θ_1 is determined by the condition that it must be Mach's angle for the flow with velocity $U_1 \sin \theta_1 = \frac{A_1}{U_1} = M_1^{-1}$. The conditions for entering the sector are

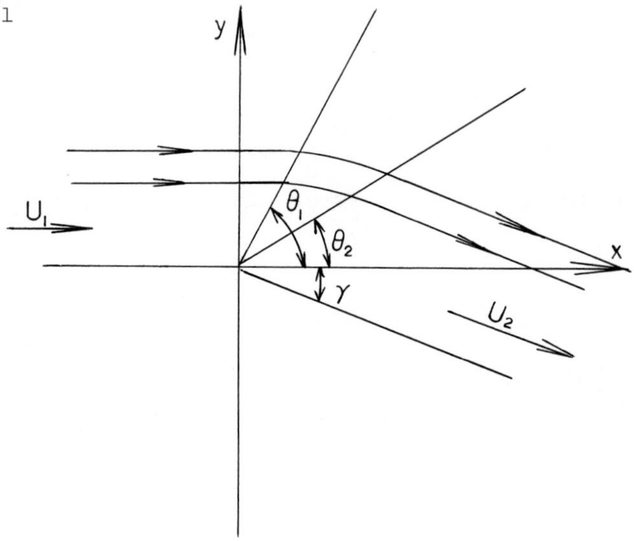

$$U_1 \cos \theta_1 = q_* \cos m(\theta_1+\theta_*)$$

$$U_1 \sin \theta_1 = mq_* \sin m(\theta_1+\theta_*),$$

whence

$$\tan \theta_1 = m \tan m(\theta_1+\theta_*).$$

From this the angle θ_* is to be determined in such a way that

$$0 < m(\theta_1+\theta_*) < \frac{\pi}{2} .$$

The conditions for passing into parallel flow again are

$$U_2 \cos(\theta_2+\gamma) = q_* \cos m(\theta_2+\theta_*)$$

$$U_2 \sin(\theta_2+\gamma) = mq_* \sin m(\theta_2+\theta_*).$$

Division of these equations yields

$$\tan(\theta_2+\gamma) = m \tan m(\theta_2+\theta_*).$$

Since θ_* is known from the conditions for entering, this equation serves to determine θ_2. A solution will exist if $\theta_* > \gamma$; if γ is too large no flow around the corner exists. We can see this by letting $\theta = -\theta_*$; then $q_n = 0 = a$ and the density is zero. Hence as γ approaches θ_* the fluid thins out near the wall. Once θ_2 is found the new velocity U_2 can be determined from the above conditions for leaving the sector.

The flow in a corner is quite different from the flow around a corner. The solution obtained above for flow in a sector cannot be applied to flow in a corner. The change of angle, as we shall see, is effected by an oblique shock. Just as we studied flow in a sector to find the flow around a corner we will investigate oblique shocks to obtain the flow in a corner.

An oblique shock consists in the transition from a state (1) to a state (2) upon crossing a certain line. Let q_t and q_n be the tangential and normal components of velocity with respect to this line. As in the case of straight shocks the conditions on the flow are (compare equations (19), (20) and (21)):

$$[\rho q_n]_{(1)}^{(2)} = 0 \tag{65}$$

$$[p+\rho q_n^2]_{(1)}^{(2)} = 0 \tag{66}$$

$$[\rho q_n q_t]_{(1)}^{(2)} = 0 \tag{67}$$

$$[\tfrac{1}{2}(q_t^2 + q_n^2) + i]_{(1)}^{(2)} = 0 . \tag{68}$$

Equation (65) refers to the flux of mass, (66) to the difference of pressure resultant and flux of normal velocity, (67) to the flux of tangential velocity, (68) to the flux of energy. In view of (65) equation (67) can be simplified to

$$[q_t]_{(1)}^{(2)} = 0 \tag{67'}$$

i.e., the tangential velocity is continuous across the shock. Equation (68) can now be written

$$\frac{1}{2}q_n^2 + i = \frac{1}{2}(q_*^2 - q_t^2). \qquad (68')$$

This equation appears just like the one for a straight shock except that the constant on the right side is different. The results for an oblique shock are similar to those for a straight shock: the entropy and density increase while the velocity decreases. However an important difference is that the velocity need not drop to a subsonic value.

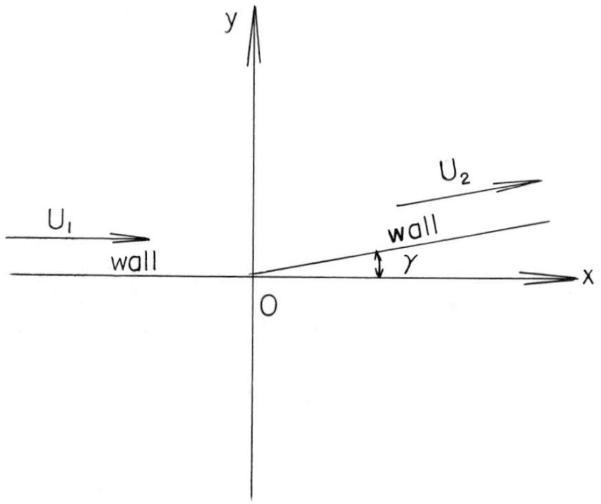

To investigate the flow in a corner* let one wall be represented by the negative x-axis with a constant flow U_1 at $-\infty$, and let the other wall be represented by $y = x \tan \gamma$, $\gamma > 0$. We wish to see what states (2) are possible when the state (1) is prescribed. That is, we are given $u = U_1$, $v = 0$, $\rho = \rho_1$ for state (1) and wish to find u, v, ρ for state (2). For this purpose we find the shock curve in the (u,v) plane. Using equations (65), (66), (67) and (68') we can calculate u, v, ρ. If we let β denote the angle the shock line makes with the u-axis, we have

$$q_t = u \cos \beta + v \sin \beta$$

$$q_n = -v \cos \beta + u \sin \beta$$

From equation (67') we obtain

$$(u - U_1)\cos \beta + v \sin \beta = 0,$$

*For a more detailed discussion, cf. A. Busemann, Gas Dynamik, Handbuch der Experimentalphysik, p. 431 ff.

and from (65) we get

$$-\rho v \cos \beta + (\rho u - \rho_1 U_1) \sin \beta = 0.$$

Elimination of β from these last two equations yields

$$(u-U_1)(\rho u - \rho_1 U_1) + \rho v^2 = 0,$$

from which we obtain

$$\rho = \frac{\rho_1 U_1 (u-U_1)}{(u-U_1)u+v^2} \tag{69}$$

Equation (66) states

$$p-p_1 = \rho_1 q_n^{(1)}(q_n - q_n^{(1)}) = \rho_1 U_1 \sin \beta[(u-U_1) \sin \beta - v \cos \beta]$$

$$= \rho_1 U_1 (u-U_1)$$

or

$$p = p_1 + \rho_1 U_1 (u-U_1). \tag{70}$$

Rewriting equation (68') we have

$$i = \frac{1}{2}[q_*^2 - (u^2 + v^2)]. \tag{71}$$

Since, for ideal gases

$$\rho i = \frac{k}{k-1} p,$$

we can solve equations (69), (70) and (71) for v. Without going into detail we state that the result is

$$v^2 = (U_1-u)^2 \frac{(u - \frac{a_*^2}{U_1})}{\frac{a_*^2}{U_1} + \frac{2}{k+1} U_1 - u} : \tag{72}$$

Equation (72) shows that v vanishes only when $u = U_1$ and when $uU_1 = a_*^2$. Hence if the circle $u^2 + v^2 = a_*^2$ is constructed, (separating subsonic flow, $q < a_*$ from supersonic flow $q > a_*$), the value u such that $uU_1 = u_*^2$ must lie within the circle. The curve represented by equation (72), a strophoid, is shown in the figure. Two types of shocks are possible: one from supersonic to subsonic, and the other remaining supersonic, (points A and B in figure). The angle β, the angle of the oblique shock line, can be determined. The normal through the origin to the line through U_1 and B represents the direction of q_t. Since this is the same as the direction of the oblique shock line, the angle it makes with the u-axis is just β. The flow in a corner can be considered a parallel flow only if a transition from supersonic flow to supersonic flow is used.

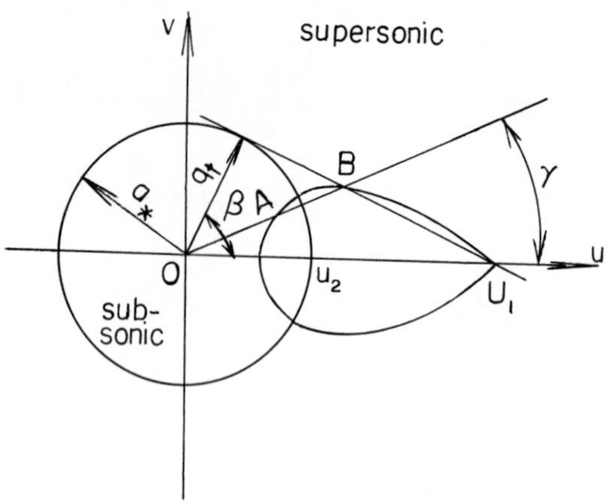

From the above figure we can see what happens as γ changes. If γ is too large no shock is possible. Consequently, if γ is too large no supersonic flow in a corner of the type prescribed is possible. Indeed, if γ approaches a right angle we cannot expect parallel supersonic flow before the corner and parallel flow beyond the corner. As $\gamma \to 0$, A approaches U_2, i.e., the shock area approaches the straight shock, the point B will approach the point U_1. The angle β will then approach a limit angle α_1, this angle is the Mach angle for

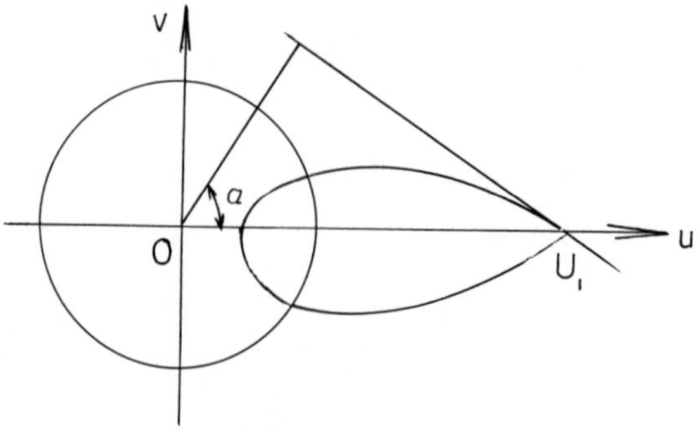

the flow U_1. This result is easily obtained from equation (72) by calculating $\frac{dv}{du}$ at $u = U_1$:

$$\left|\frac{dv}{du}\right|_{u=U_1} = \left(\frac{U_1^2 - a_*^2}{a_*^2 - \frac{k-1}{k+1}U_1^2}\right)^{1/2} = \left(\frac{U_1^2 - a_1^2}{a_1^2}\right)^{1/2} = \cot \alpha_1,$$

where $U_1^2 + \frac{2}{k-1}a_1^2 = \frac{k+1}{k-1}a_*^2$ has been used. Hence we see that as $\gamma \to 0$ the "jumps" get smaller and approach in direction the characteristic direction of the flow U_1.

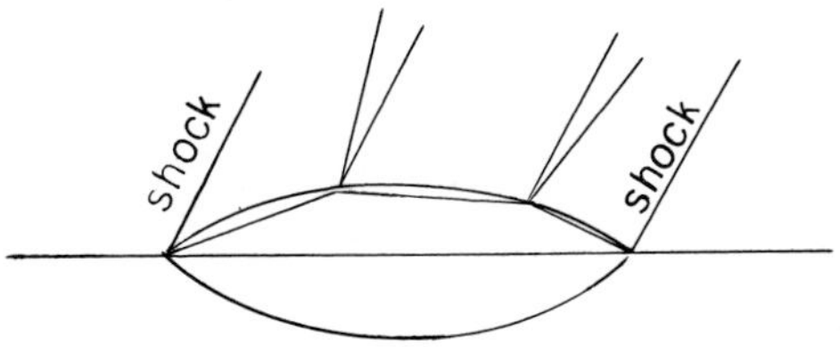

Some remarks may be made about the supersonic flow about wing profiles. Consider a biconvex profile with both a sharp leading and trailing edge. If the boundary of the profile is approximated by a polygon then some facts can be ascertained about the flow. The flow about this polygon can be considered as a succession of flows in and around corners with parallel flow between. For example at the leading and trailing edges we will have flow in a corner and consequently a shock, while at the other corners we will have sector flow. The fact that the profile is curved would introduce certain modifications. It is interesting to note that this approximate method yields a flow pattern quite different from that obtained in the first order approximation method, Section 7, case II, page 329.

There are other exact solutions for the flow of a compressible fluid; they are obtained by a transformation of the differential equations into linear equations with u and v as independent variables. For a discussion of this method see the report of T. v. Kármán presented to the Institute of Aeronautical Sciences in January, 1941. See also the paper by Tsien: Two-Dimensional Subsonic Flow of Compressible Fluids; Journal of the Aeronautical Sciences, 6, 399 ff. (1939).

9. Non-Adiabatic Flow and Boundary Layer

In this section we consider a steady, two-dimensional flow with the heat conduction and viscosity taken into account.*

The continuity equation becomes

$$(\rho u)_x + (\rho v)_y = 0. \tag{73}$$

Newton's law takes the form

$$\begin{aligned} \rho(uu_x + vu_y) &= -p_x + \tau_{xx/x} + \tau_{xy/y} \\ \rho(uv_x + vv_y) &= -p_y + \tau_{yx/x} + \tau_{yy/y} \end{aligned} \tag{74}$$

where

$$(\tau) = \mu\begin{pmatrix} 2u_x & u_y + v_x \\ v_x + u_y & 2v_y \end{pmatrix} - \frac{2}{3}\mu\begin{pmatrix} u_x + v_y & 0 \\ 0 & u_x + v_y \end{pmatrix}.$$

(Cf. Chapter IV, Section 2).

In a manner similar to the way we derived the energy relation with heat conduction and viscosity for the one-dimensional case, we now have the relation

$$\oint_S [\rho q_n(\tfrac{1}{2}q^2 + h) - \lambda T_n + pq_n - \vec{q}(\tau)\vec{n}]ds = 0,$$

where $\vec{q}(\tau)\vec{n} = u\tau_{xx}n_x + v\tau_{yx}n_x + u\tau_{xy}n_y + v\tau_{yy}n_y$. The first term refers to the transport of kinetic and intrinsic energy, the second the heat flux, and the last two terms to the work done by the stresses. The integrand being the normal component of a vector, the above statement is equivalent to the statement that the divergence of the vector vanishes, or

$$\{\rho u(\tfrac{1}{2}q^2 + h)\}_x + \{\rho v(\tfrac{1}{2}q^2 + h)\}_y - (\lambda T_x)_x - (\lambda T_y)_y$$

---

*Cf. H. Bateman: Report of the Committee on Hydrodynamics, Part IV, Compressible Fluids, Bull. of the National Research Council (1931).

$$+ (pu)_x + (pv)_y - (u\tau_{xx/x} + u\tau_{xy/y} + v\tau_{yx/x} + v\tau_{yy/y})$$

$$- \mu\{2(u_x^2 + v_y^2) + (u_y + v_x)^2 - \tfrac{2}{3}(u_x + v_y)^2\} = 0.$$

By taking into account equation (73) and equations (74) (after multiplying through by $u + v$ respectively and adding), we have

$$\rho u h_x + \rho v h_y - (\lambda T_x)_x - (\lambda T_y)_y$$

$$+ \rho u \rho_x^{-1} + \rho v \rho_y^{-1} - \mu\Phi = 0$$

where

$$\Phi = (u_x - v_y)^2 + (u_y + v_x)^2 + \tfrac{1}{3}(u_x + v_y)^2. \qquad (75)$$

Making use of the relation

$$dh = -p d\rho^{-1} + T ds$$

where s is the entropy and T the absolute temperature, we obtain

$$\rho u T s_x + \rho v T s_y = (\lambda T_x)_x + (\lambda T_y)_y + \mu\Phi. \qquad (76)$$

Equation (76) shows that the increase of heat per unit volume, per unit time is partly due to heat conduction and partly due to viscosity; the "dissipation" $\mu\Phi$ is positive in accordance with the second law of thermodynamics.

We now introduce the assumptions, justified for ideal gases:

$$i = c_p T \qquad (77)$$

$$\lambda = c_p \mu \qquad (78)$$

where i is the heat content per unit mass and c_p a constant. Then (76) reduces to

$$\rho u T s_x + \rho v T s_y = (\mu i_x)_x + (\mu i_y)_y + \mu\Phi. \qquad (79)$$

The heat content per unit mass, i, may be considered a given function of p and s such that

$$di = \rho^{-1}dp + Tds. \tag{80}$$

Equation (73) may be solved by the introduction of a stream function ψ with

$$\rho u = \psi_y, \quad \rho v = \psi_x. \tag{81}$$

Then equations (74) and (79) represent three equations in the three unknowns ψ, p, and s. These three equations represent the general equations for the two-dimensional steady flow of fluids; all other cases so far considered were special cases of this set of equations.

We now wish to investigate what happens when the viscosity approaches zero, i.e., $\mu \to 0$. Equations (74) reduce to

$$\rho(uu_x + vu_y) = -p_x$$

$$\rho(uv_x + vv_y) = -p_y$$

and equation (79) becomes

$$us_x + vs_y = 0.$$

This last equation states that the entropy s is constant along each streamline. Equation (80) then becomes

$$di = \rho^{-1}dp,$$

i.e., the flow is adiabatic along each streamline. Consequently Bernoulli's relation

$$\frac{1}{2}q^2 + i = i_* = \text{const.}$$

is valid along each streamline.

If s = const. at the beginning of the streamlines the flow is adiabatic throughout. If i_* is constant at the beginning of the streamlines the flow is irrotational throughout. Therefore the assumption of adiabatic, irrotational flow is reasonable. This assumption however, cannot be retained after the flow has passed a shock. Beyond the shock the entropy s and Bernoulli's quantity i_* will, in general, not be constant across the streamlines and the flow will, in general, have acquired vorticity. The flow beyond a shock can no longer be considered adiabatic and irrotational although viscosity and heat conduction may again be disregarded.

When the limit process $\mu \to 0$ is performed with the solution there may be regions of non-uniform convergence. If the convergence is non-uniform along a line which runs across the streamlines then we call this line a shock line. This can occur only when it is not stipulated that the density be constant. If, on the other hand, the convergence is non-uniform along a streamline, the region around the streamline is called a boundary layer.* This may occur for both variable and constant ρ.

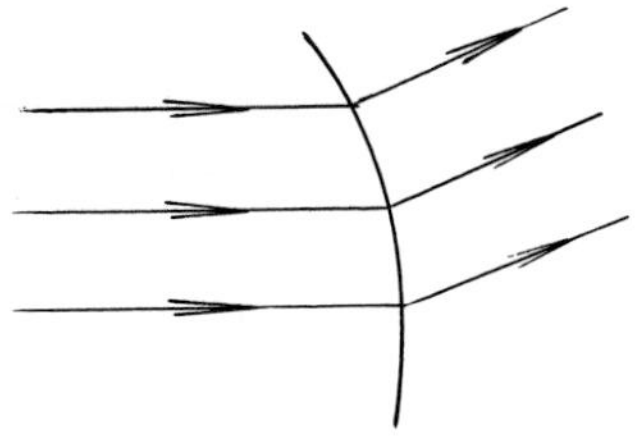

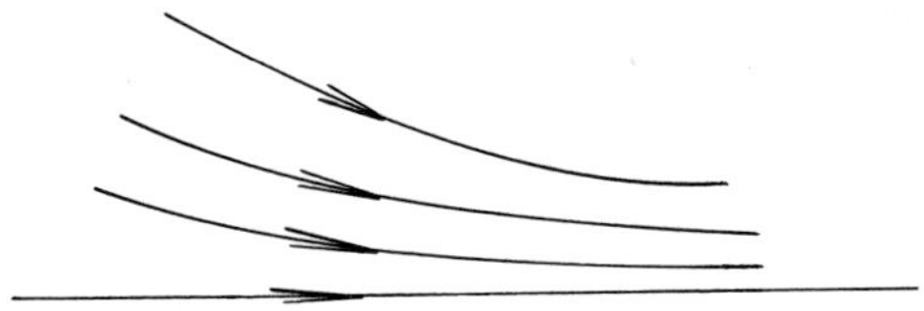

For the boundary layer theory the full set of equations (73), (74) and (79) must be considered. We take up the case of the flat plate given by $y = 0$, $x \geq 0$, assuming that the main flow is given by

$$u = U = \text{const.}, \quad v = 0, \quad p = P = \text{const.}, \quad \rho = \mathfrak{p} = \text{const.}$$

In this case equations (74) become

*Cf. V. Karman and Tsien: Boundary Layer in Compressible Fluids, Journal of the Aeronautical Sciences, 5, 227-232 (1938).

$$\rho u u_x + \rho v u_y = -p_x + (\mu u_y)_y$$

$$0 = -p_y,$$

from which we deduce

$$p = P = \text{const.}$$

throughout. Hence equation (80) reduces to

$$di = Tds,$$

and equation (79) takes the form

$$\rho u i_x + \rho v i_y - (\mu i_y)_y - \mu u_y^2 = 0. \tag{83}$$

In view of $p = \text{const.}$, (82') can be written

$$\rho u u_x + \rho v u_y - (\mu u_y)_y = 0. \tag{82}$$

The boundary conditions are

B.C.$_\infty$: $u = U, \quad \rho = \mathfrak{p}, \quad i = I \quad \text{for} \quad y = \infty$

B.C.$_0$: $u = \psi, \quad v = 0, \quad i = I_0 \quad \text{for} \quad y = 0.$

We see that prescribing the heat content is equivalent to prescribing the temperature T since $i = c_p T$. (Cf. (77).)

Examination of equations (82) and (83) shows that they would be exactly the same except for the factor μu_y^2. It turns out that i is a quadratic function of u, without going into details we state the result:

$$i = \frac{1}{2}u(U-u) + I_0 \frac{U-u}{U} + I \frac{u}{U}\,. \tag{84}$$

Since $\rho i = \frac{k}{k-1} p$ and $p = \text{const.}$, we see that ρ can be expressed in terms of i:

$$\rho = \frac{\text{const.}}{i} = \frac{kP}{k-1}\, i^{-1},$$

and, in view of (84), ρ can be expressed in terms of u. This being done it suffices for the solution to consider only Equations (74) and (82).

Some interesting remarks can be made from an examination of (84). The quantity i being a quadratic function of u will have a maximum:

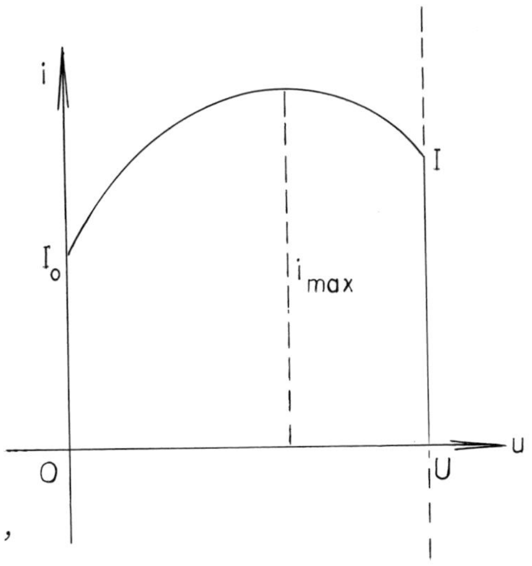

$$i_{max} = \frac{1}{8}U^2 + \frac{1}{2}(I+I_0) + \frac{1}{2}\frac{(I-I_0)^2}{U^2}$$

corresponding to

$$u = \frac{1}{2}U + \frac{I-I_0}{U},$$

provided that this last value of u is assumed inside the boundary layer; this is the case if

$$|I-I_0| < \frac{1}{2}U^2.$$

For high values of U^2/I, therefore, considerable heat will be created in the middle of the boundary layer.

The heat flux into the wall is given by

$$\mu i_y\big|_{y=0} = \{\tfrac{1}{2}U^2 + (I-I_0)\}U^{-1}\tau_0$$

where $\tau_0 = \mu u_y\big|_{y=0}$ is the shearing stress at the wall. There is no heat absorption if $\frac{1}{2}U^2 = I_0-I$. If $\frac{1}{2}U^2 > I_0-I$ the effect of increasing the velocity is to increase the heat flux into the wall. In case one wishes to have the fluid cool the wall, i.e., $I_0 > I$, the cooling effect diminishes as the velocity increases, vanishes when $\frac{1}{2}U^2 = I_0-I$, and actually heats the wall when $\frac{1}{2}U^2 > I_0-I$. If the wall is to cool the fluid, i.e., $I_0 < I$, the heat absorption by the wall increases with the velocity; however, considerable heat is created in the boundary layer so that the cooling effect is diminished.

INDEX